D1000638

Molecular Electronics, Circuits, and Processing Platforms

Nano- and Microscience, Engineering, Technology, and Medicine Series

Series Editor
Sergey Edward Lyshevski

Titles in the Series

Logic Design of NanoICS
Svetlana Yanushkevich

MEMS and NEMS: Systems, Devices, and Structures
Sergey Edward Lyshevski

Microelectrofluidic Systems: Modeling and Simulation
Tianhao Zhang, Krishnendu Chakrabarty, and Richard B. Fair

Micro Mechatronics: Modeling, Analysis, and Design with MATLAB®
Victor Giurgiutiu and Sergey Edward Lyshevski

Microdrop Generation
Eric R. Lee

Molecular Electronics, Circuits, and Processing Platforms
Sergey Edward Lyshevski

Nano- and Micro-Electromechanical Systems: Fundamentals of Nano- and Microengineering
Sergey Edward Lyshevski

Nano and Molecular Electronics Handbook
Sergey Edward Lyshevski

Nanoelectromechanics in Engineering and Biology
Michael Pycraft Hughes

Molecular Electronics, Circuits, and Processing Platforms

SERGEY EDWARD LYSHEVSKI

CRC Press is an imprint of the
Taylor & Francis Group, an informa business

CRC Press
Taylor & Francis Group
6000 Broken Sound Parkway NW, Suite 300
Boca Raton, FL 33487-2742

© 2008 by Taylor & Francis Group, LLC
CRC Press is an imprint of Taylor & Francis Group, an Informa business

No claim to original U.S. Government works
Printed in the United States of America on acid-free paper
10 9 8 7 6 5 4 3 2 1

International Standard Book Number-10: 1-4200-5529-1 (Hardcover)
International Standard Book Number-13: 978-1-4200-5529-0 (Hardcover)

This book contains information obtained from authentic and highly regarded sources. Reprinted material is quoted with permission, and sources are indicated. A wide variety of references are listed. Reasonable efforts have been made to publish reliable data and information, but the author and the publisher cannot assume responsibility for the validity of all materials or for the consequences of their use.

No part of this book may be reprinted, reproduced, transmitted, or utilized in any form by any electronic, mechanical, or other means, now known or hereafter invented, including photocopying, microfilming, and recording, or in any information storage or retrieval system, without written permission from the publishers.

For permission to photocopy or use material electronically from this work, please access www.copyright.com (http://www.copyright.com/) or contact the Copyright Clearance Center, Inc. (CCC) 222 Rosewood Drive, Danvers, MA 01923, 978-750-8400. CCC is a not-for-profit organization that provides licenses and registration for a variety of users. For organizations that have been granted a photocopy license by the CCC, a separate system of payment has been arranged.

Trademark Notice: Product or corporate names may be trademarks or registered trademarks, and are used only for identification and explanation without intent to infringe.

Library of Congress Cataloging-in-Publication Data

Lyshevski, Sergey Edward.
Molecular electronics, circuits, and processing platforms / Sergey Edward Lyshevski.
p. cm. -- (Nano- and microscience, engineering, technology, and medicine ; 11)
Includes bibliographical references and index.
ISBN-13: 978-1-4200-5529-0 (alk. paper)
ISBN-10: 1-4200-5529-1 (alk. paper)
1. Molecular electronics. 2. Nanoelectronics. 3. Molecular integrated circuits. I. Title. II. Series.

TK7874.8.L97 2007
621.381--dc22 2007010374

Visit the Taylor & Francis Web site at
http://www.taylorandfrancis.com

and the CRC Press Web site at
http://www.crcpress.com

Contents

Preface

This book introduces a wide range of topics on molecular electronics, molecular integrated circuits (^{M}ICs), and molecular processing platforms (^{M}PPs). The basic fundamentals are coherently documented, reporting a spectrum of solved and open problems. Recent advances in science and engineering promise to lead to fundamental breakthroughs in the way molecular devices and systems are understood, synthesized, designed, and utilized. Those trends and opportunities promise to change the fundamental principles of many human-made devices and systems by achieving performance limits, novel functionality, and superb capabilities. It is important to devise novel molecular devices (^{M}devices) that are based on a new device physics and uniquely utilize specific phenomena and effects. Those ^{M}devices provide a novel device-level solution for various platforms designed within novel organizations and enabling architectures. In general, molecular electronics is a revolutionary paradigm based on sound science, engineering, and technology. The development and deployment of this paradigm will redefine and enable electronics, neuroscience, informatics, cybernetics, and so forth. The envisioned advances toward molecular electronics will have a significant positive impact in aerospace, biotechnology, electronics, health, informatics, medicine, and other areas of critical importance. Molecular electronics is viewed among the most significant frontiers to be developed in this century. Those developments are well supported by sound theories and practice. In particular, superb biomolecular processing platforms (^{BM}PPs) in living organisms are the undeniable evidence of the soundness and viability of molecular processing hardware as well as molecular electronics paradigm. These ^{BM}PPs far surpass any envisioned microelectronics-centered solutions and advances.

The fundamentals of quantum physics, molecular dynamics, chemistry, and other disciplines have been largely developed. However, many open problems pertaining to molecular electronics remain to be addressed and solved. Therefore, focused fundamental, applied, and experimental research is needed to support far-reaching developments. The purpose of this book is to coherently cover the various concepts, methods, and technologies to approach and solve a wide spectrum of problems including devising, synthesis, analysis, design, and optimization of ^{M}devices, ^{M}ICs, and ^{M}PPs. For ^{M}devices, device-level physics is examined elaborating novel concepts. System-level analysis and design are performed for ^{M}ICs and ^{M}PPs. Fabrication aspects and some synthesis technologies are covered. The emphases of this book are on the fundamental multidisciplinary principles of molecular electronics as well as on practical applications of the basic theory in engineering practice and technology developments.

There is a wide spectrum of problems, different possible solutions, and distinct concepts to be applied. Therefore, one may have valuable suggestions and reservations. Please do not hesitate to provide me with your feedback. I am committed to integrate the suggested topics and examples in the future. At the same time, it appears that it is impossible to cover all topics because there is a wide spectrum of multidisciplinary themes, variety of unsolved problems, debates on the emerging technologies and their feasibility, as well as other uncertainties. This book is written in textbook style, with the goal to reach the widest possible range of readers who have an interest in the subject. Specifically, the objective is to satisfy the existing growing interest of undergraduate and graduate students, engineers, professionals, researchers, and instructors in the fields of molecular electronics, ^{M}ICs, and ^{M}PPs. Efforts were made to coherently deliver fundamental theory and technologies important to study, understand, and research advanced devices and circuits in a unified and consistent manner. The author believes that a coherent coverage is achieved.

The synthesis and design at the device and system levels, supported by the fabrication technologies, must be supported by sound basic, analytic, and numerical methods. Novel concepts should be developed and applied in design examining complex phenomena, devising novel topologies/organizations/architectures, evaluating performance, and so forth. Advanced interdisciplinary research is being carried out. Our overall goal is to expand and research frontiers through pioneering fundamental and applied multidisciplinary studies advancing the envisioned developments. This book develops and delivers the basic theoretical foundations. Synthesis and design of ^{M}devices, ^{M}ICs, and ^{M}PPs are illustrated with analysis of their performance and capabilities. It is the author's goal to substantially contribute to these basic issues, efficiently deliver the rigorous theory, and integrate the challenging problems in the context of well-defined applications. The emphasis is also on the analysis of possible directions and emerging technologies, development of basic theory for attaining fundamental understanding of molecular electronics, as well as applying the theory toward hardware implementation. It should be emphasized that, no matter how many times the material is reviewed and efforts are spent to guarantee the highest quality, the author cannot guarantee that the manuscript is free from minor errors, typos, and other shortcomings. If you find something that you feel needs correction, adjustment, clarification, and/or modification, please notify me. Your help and assistance are greatly appreciated and deeply acknowledged.

Acknowledgments

Many people have contributed to this book. First, my thanks go to my beloved family. I would like to express my sincere acknowledgments and gratitude to many of my colleagues and peers. It gives me great pleasure to acknowledge the help I received from many people in the preparation of this book. The outstanding Taylor & Francis team, especially Nora Konopka (acquisitions editor, electrical engineering), Jessica Vakili (project coordinator), and Glenon Butler Jr. (project editor), who helped me tremendously and assisted me by providing valuable and deeply treasured feedback.

The opportunities to perform the funded research for different agencies and the US Department of Defense laboratories under numerous grants and contracts have had a significant positive impact. I sincerely acknowledge the partial support from *Microsystems and Nanotechnologies* under the US Department of Defense, Department of the Air Force (Air Force Research Laboratory) contracts 8750024 and 8750058.

Disclaimer: Any opinion, findings, and conclusions or recommendations expressed in this book are those of the author and do not necessarily reflect those of the US Department of Defense or Department of the Air Force.

I express my sincere gratitude to MathWorks, Inc. for supplying the MATLAB® environment (MathWorks, Inc., 24 Prime Park Way, Natick, MA 01760-15000; http://www.mathworks.com).

Many thanks to all of you.

Sergey Edward Lyshevski
Department of Electrical Engineering
Rochester Institute of Technology

Author

Sergey Edward Lyshevski was born in Kiev, Ukraine. He received his MS (1980) and PhD (1987) degrees from Kiev Polytechnic Institute, both in Electrical Engineering. From 1980 to 1993, Dr Lyshevski held faculty positions at the Department of Electrical Engineering at Kiev Polytechnic Institute and the Academy of Sciences of Ukraine. From 1989 to 1993, he was the Microelectronic and Electromechanical Systems Division head at the Academy of Sciences of Ukraine. From 1993 to 2002, he was with Purdue School of Engineering as an associate professor of electrical and computer engineering. In 2002, Dr. Lyshevski joined Rochester Institute of Technology as a professor of electrical engineering.

Dr. Lyshevski serves as a full professor faculty fellow at the U.S. Air Force Research Laboratories and Naval Warfare Centers. He is the author of more than ten books (including *Logic Design of NanoICs* co-authored with S. Yanushkevich and V. Shmerko, CRC Press, 2005; *Nano- and Microelectromechanical Systems: Fundamentals of Micro- and Nanoengineering*, CRC Press, 2004; *MEMS and NEMS: Systems, Devices, and Structures*, CRC Press, 2002) and is the author or coauthor of more than 300 journal articles, handbook chapters, and regular conference papers.

His current research activities are focused on molecular electronics, molecular processing platforms, nanoengineering, cognitive systems, novel organizations/architectures, new nanoelectronic devices, reconfigurable super-high-performance computing, and systems informatics. Dr Lyshevski has made significant contributions in the synthesis, design, application, verification, and implementation of advanced aerospace, electronic, electromechanical and naval systems.

He has given more than 30 invited presentations (nationally and internationally) and serves as an editor of the CRC Press book series on *Nano- and Microscience, Engineering, Technology, and Medicine.*

1

Electronics and Emerging Paradigms

1.1 Introduction

This section introduce the basic core definitions and premises. Molecular (nano) electronics focuses on fundamental, applied, and experimental research and technology developments in devising and implementing novel high-performance enhanced-functionality atomic or molecular devices, modules and platforms (systems), and high-yield *bottom-up* fabrication. Molecular electronics centers on the following:

1. Invention of novel devices based on a new device physics
2. Utilization of the exhibited unique phenomena, effects, and capabilities
3. Devising of enabling topologies, organizations, and architectures
4. *Bottom-up* high-yield fabrication

At the device level, the key differences between molecular and microelectronic devices are the following:

1. Device physics and phenomena exhibited
2. Effects, capabilities, and functionality utilized
3. Topologies and organizations attained
4. Fabrication processes and technologies used

In microelectronic devices, individual molecules and atoms do not depict the overall device physics and do not define the device performance, functionality, and capabilities. In contrast, in molecular devices, individual molecules and atoms define the overall device physics and depict the device performance, functionality, capabilities, and topologies.

There are fundamental differences at the system level. In particular, molecular electronics leads to novel organizations, advanced architectures, technology-centric super-large-scale integration (SLSI), and so forth. Everybody agrees that solid-state integrated circuits (ICs) and processors are not biomolecular processing platforms (^{BM}PPs) found in living organisms. Existing ^{BM}PPs provide not only an evidence of their soundness and supremacy but also a substantiation of achievable super-high-performance by envisioned molecular ICs and processing platforms. Additional introductory details are reported in Section 1.5.

In theory, the fundamentals, operation, functionality, and organization or architecture of molecular processing platforms (^{M}PPs) can be devised by making use of ^{BM}PPs through *biomimetics* and *bioprototyping*. Considering a neuron as a module (system) that consists of processing-and-memory primitives, we cover basic fundamentals, study *biomolecular processing hardware* and *software*, analyze electrochemomechanically induced transitions, introduce *information* and *routing* carriers, and so forth. Owing to a great number of unsolved fundamental, applied, and technological problems, it is impossible to accomplish coherent *biomimetics* or *bioprototyping* and devise (discover, design, and implement) synthetic bio-identical or biocentered processing and memory platforms. To some extent, one may attempt to resemble ^{BM}PPs by designing innovative molecular ICs and platforms. The book gradually and coherently introduces various topics. I hope that most readers will find this book appealing and exciting, while some, unfortunately, may find it to be subjective.

Molecular electronics and processing platforms, as a cutting-edge revolutionary endeavor, are researched. The emergence of molecular electronics and processing platforms is pervasive, persuasive, and irreversible. These revolutionary high-risk high-payoff areas require immense research and technological development efforts that largely depend on readiness, commitment, acceptance, investment, infrastructure, innovations, and market needs. Multidisciplinary science, engineering, and technology require a high-degree interaction and association between disciplines. Molecular electronics has spurred enthusiasm and numerous challenging problems. The author would like to be engaged in constructive debates with the ultimate objective to discover, research, verify, apply, and implement sound solutions.

1.2 Historical Overview: From Atoms to Theory of Microscopic Systems and Fabrication

This book focuses on molecular electronics and molecular ICs (^{M}ICs), which implies utilization of molecular devices (^{M}devices) and molecular gates (^{M}gates) engineered from molecules. Molecular electronics is related to nanotechnology if it is to be soundly defined. Unfortunately, *nano* has become the "buzz-prefix" and favorable magnification word to many fantasists who have brought an incredible number of futuristic and speculative viewpoints.

FIGURE 1.1
Democritus (460–371 BC), Epicurus (341–270 BC), and Aristotle (384–322 BC).

The *nano*-pretended futurists and analysts have been overoptimistically painting rosy illusionary pictures and are chasing stratospheric ideas, which unlikely may be materialized in the observable future. To avoid discussions on the terminology and to prevent confusions, the word nano is used as it directly related.

All matter is composed of atoms. When did the philosophy of atom and atomic composition of matter originate? Around 440 BC, Leucippus of Miletus envisioned the atom. In Greek, the prefix "*a*" means not, and the word "*tomos*" means cut. The word atom therefore comes from the Greek word *atomos*, meaning uncut. Leucippus and his student Democritus (460–371 BC) of Abdera further refined and extended this far-reaching prediction. The ideas of Leucippus and Democritus were further elaborated by Epicurus (341–270 BC) of Samos. Aristotle (384–322 BC) questioned and opposed this concept. These Greek philosophers are shown in Figure 1.1. Though the original writings of Leucippus and Democritus are lost, their concept is known from a poem entitled *De Rerum Natura* (*On the Nature of Things*) written by Lucretius (95–55 BC). Attempting to be consistent with the translated text *On the Nature of Things*, the major postulates by those philosophers with a minimum level of refinements can be summarized as follows:

1. All matter is composed of atoms, which are "units of matter too small to be seen." These atoms cannot be further "split into smaller portions." Democritus quotes Leucippus: "the atoms hold so that splitting stops when it reaches indivisible particles and does not go on infinitely." Democritus reasoned that if matter could be infinitely divided, it was also subject to "complete disintegration from which it can never be put back together."
2. There is a "void," which is "empty space," between atoms.
3. Atoms are "solid and homogeneous" with "no internal structure." (In 1897 Thomson discovered an electron, thus departing from this hypothesis, and it is well known now that atoms consist of neutrons, protons, and electrons.)

4. Atoms differ in their sizes, shapes, and weights. According to Aristotle: "Democritus and Leucippus say that there are indivisible bodies, infinite both in number and in the varieties of their shapes . . . ," and "Democritus recognized only two basic properties of the atom: size and shape. But Epicurus added weight as a third. For, according to him, the bodies move by necessity through the force of weight."

The genius predictions of Leucippus, Democritus, and Epicurus are truly amazing. They envisioned the developments of science many centuries ahead.

Although a progress in various applications of nanotechnology has been recently announced, many of those declarations have been largely acquired from well-known theories and accomplished technologies of material science, biology, chemistry, and other matured disciplines established in olden times and utilized for centuries. Atoms and atomic structures were envisioned by Leucippus of Miletus around 440 BC, and the basic atomic theory was developed by John Dalton in 1803. The periodic table of elements was established by Dmitri Mendeleev in 1869, and the electron was discovered by Joseph Thomson in 1897. The composition of atoms was discovered by Ernest Rutherford in 1910 using the experiments conducted under his direction by Ernest Marsden in the scattering of α-particles. The quantum theory was largely developed by Niels Bohr, Louis de Broglie, Werner Heisenberg, Max Planck, and other scientists in the beginning of the twentieth century. Those developments were advanced by Erwin Schrödinger in 1926. For many decades, comprehensive handbooks on chemistry and physics coherently reported thousands of organic and inorganic compounds, molecules, ring systems, nitrogenous bases, nucleotides, oligonucleotides, organic magnets, organic polymers, atomic complexes, and molecules with the dimensionality on the order of 1 nm. In last 50 years, meaningful methods have been developed and commercially deployed to synthesize a great variety of nucleotides and oligonucleotides with various linkers and spacers, bioconjugated molecular aggregates, modified nucleosides, as well as other inorganic, organic and biomolecules. The aforementioned fundamental, applied, experimental, and technological accomplishments have provided essential foundations for many areas including modern biochemistry, biophysics, chemistry, physics, and electronics.

Microelectronics has achieved phenomenal accomplishments within 50 years. The discovered microelectronic devices, ICs, and high-yield technologies have matured and progressed ensuring various high-performance electronics products. Many electronics-preceding processes and materials were advanced and fully utilized. For example, crystal growth, etching, thin-film deposition, coating, and photolithography have been known and used for centuries. Etching was developed and performed by Daniel Hopfer from 1493 to 1536. Modern electroplating (electrodeposition) was invented by Luigi Brugnatelli in 1805. Photolithography was invented by Joseph Nicéphore Niépce in 1822, and he made the first photograph in 1826. In 1837

Boris von Jacobi (Moritz Hermann von Jacobi) introduced and demonstrated silver, copper, nickel, and chrome electroplating. In 1839 John Wright, George Elkington, and Henry Elkington discovered that potassium cyanide could be used as an electrolyte for gold and silver electroplating. They patented this process, receiving the British Patent 8447 in 1840. In the fabrication of various art and jewelry products, as well as Christmas ornaments, those inventions and technologies have been used for many centuries.

By advancing microfabrication technology, feature size has been significantly reduced. The structural features of solid-state semiconductor devices have been scaled down to tens of nanometers, and the thickness of deposited thin films can be less than 1 nm. The epitaxy fabrication process, invented in 1960 by J. J. Kleimack, H. H. Loar, I. M. Ross, and H. C. Theuerer, led to the growing of silicon films layer after layer identical in structure with the silicon wafer itself. Technological developments in epitaxy continued to result in the possibility of depositing uniform multilayered semiconductors and insulators with precise thicknesses so as to improve ICs performance. Molecular beam epitaxy is the deposition of one or more pure materials on a single crystal wafer, one layer of atoms at a time, under high vacuum, forming a single-crystal epitaxial layer. Molecular beam epitaxy was originally developed in 1969 by J. R. Arthur and A. Y. Cho. The thickness of the insulator layer (formed by silicon dioxide, silicon nitride, aluminum oxide, zirconium oxide, or other high-k dielectrics) in field-effect transistors (FETs) was gradually reduced from tens of nanometers to less than 1 nm.

The aforementioned, as well as other meaningful fundamental and technological developments, were not referred to nanoscience, nanoengineering, and nanotechnology until recent years. Recently, the use of the prefix *nano* in many cases has become an excessive attempt to associate products, areas, technologies, and theories with *nano*. Primarily focusing on atomic structures, examining atoms, researching subatomic particles and studying molecules, biology, chemistry, physics, and other disciplines have been using the term *microscopic* even though they have dealt with the atomic theory of matter using pico- and femtometer atomic or subatomic dimensions, employing quantum physics.

De Broglie's postulate provides the foundation of the Schrödinger equation, which describes the behavior of *microscopic* particles within the *microscopic* structure of matter made of atoms. Atoms are composed of nuclei and electrons, and a nucleus consists of neutrons and protons. The proton is a positively charged particle, the neutron is neutral, and the electron has a negative charge. In an uncharged atom, there is an equal number of protons and electrons. Atoms may gain or lose electrons and become ions. Atoms that lose electrons become cations (positively charged) or become anions when they gain electrons (negatively charged). Considering protons, neutrons, and electrons, it should be emphasized that there exist smaller *microscopic* particles (quarks, antiquarks, and others) within protons and neutrons.

The basic units of length that were utilized for *microscopic* particles are: $1\ \text{nm} = 1 \times 10^{-9}$ m, $1\ \text{Å} = 1 \times 10^{-10}$ m, $1\ \text{pm} = 1 \times 10^{-12}$ m, and

1 fm = 1×10^{-15} m. The reader may recall that the following prefixes are used: yocto for 1×10^{-24}, zepto for 1×10^{-21}, atto for 1×10^{-18}, femto for 1×10^{-15}, pico for 1×10^{-12}, nano for 1×10^{-9}, micro for 1×10^{-6}, milli for 1×10^{-3}, and centi for 1×10^{-2}. The atomic radius of a hydrogen atom is 0.0529 nm, that is, 0.0529×10^{-9} m. The atomic mass unit (amu) is commonly used. This atomic mass unit is 1/12 the mass of the carbon atom, which has 6 protons and 6 neutrons in its nucleus, that is, 1 amu = $1/12\ m(^{12}C)$, where ^{12}C denotes the isotope of carbon, and 1 amu = $1.66053873 \times 10^{-27}$ kg. The mass of the positively charged proton is 1.00727646688 amu = $1.67262158 \times 10^{-27}$ kg, while the mass of the electrically neutral neutron is 1.00866491578 amu = $1.67492716 \times 10^{-27}$ kg. The electron (classical electron radius is $2.817940285 \times 10^{-15}$ m) has a mass 0.000548579911 amu = $9.10938188 \times 10^{-31}$ kg and possesses a charge $1.602176462 \times 10^{-19}$ C. The weight of the hydrogen atom is 1.00794 amu or $1.66053873 \times 10^{-27}$ kg.

The *microscopic* theory has been used to examine *microscopic* systems (atoms and elementary particles) such as baryons, leptons, muons, mesons, partons, photons, quarks, and so forth. The electron and π-meson (pion) have masses $\sim 9.1 \times 10^{-31}$ and $\sim 2 \times 10^{-28}$ kg, while their radii are $\sim 2.8 \times 10^{-15}$ and $\sim 2 \times 10^{-15}$ m. For these subatomic particles, the *microscopic* terminology has been used for more than 100 years. The femtoscale dimensionality of subatomic particles has never been a justification to define them to be *"femtoscopic"* particles or to classify these *microscopic* systems to be *"femtoscopic."*

In electronic devices, the motion of charged *microscopic* particles (electrons, ions, etc.) results in current, charge, potential, and other variations. In general, one may utilize various electrochemomechanical transitions and interactions to ensure overall functionality by making use of specific phenomena and effects. The ability to control the motion and behavior of *microscopic* particles within device physics is very important in defining device performance and capabilities. One distinguishes electronic, electrochemical, electromechanical, mechanical, photonic, and other classes of devices, which are fabricated using different technologies. Furthermore, different subclasses, families, and subfamilies exist within the aforementioned classes.

1.3 Devices, Circuits, and Processing Platforms

With our focus on electronics, one may be interested in analyzing the major trends [1–4] and defining microelectronics and nanoelectronics. Within 60 years, microelectronics was well established and matured with more than a $150 billion market per year. With the definition of microelectronics being clear [1], nanoelectronics should be defined emphasizing the underlined premises. The focus, objective, and major themes of nanoelectronics, which is inherently molecule-centered, are defined by means of the following [5,6]: Nanoelectronics focuses on fundamental, applied, and experimental research and technology developments in devising and implementing novel high-performance enhanced-functionality atomic or molecular devices, modules and platforms (systems), as well as high-yield *bottom-up* fabrication. As was emphasized in Section 1.1, molecular (nano) electronics centers on the following:

1. Invention of novel devices based on a new device physics
2. Utilization of exhibited unique phenomena, effects, and capabilities
3. Devising of enabling topologies, organizations, and architectures
4. *Bottom-up* fabrication

Other features at the device, module, and system levels are emerging as subproducts of these four major themes. Compared with the solid-state semiconductor (microelectronic) devices, ^{M}devices exhibit new phenomena and offer unique capabilities that should be utilized at the device and system levels. In order to avoid discussions on terminology and definitions, the term molecular, and not the prefix *nano*, is generally used in this book.

At the device level, IBM, Intel, Hewlett-Packard, and other leading companies have been successfully conducting pioneering research and pursuing technological developments focusing on *solid* molecular electronics devices (^{ME}devices), molecular wires, molecular interconnect, and so forth. Basic, applied, and experimental developments in *solid* molecular electronics are reported in [5–11]. Unfortunately, it seems that a limited progress has been accomplished in molecular electronics, *bottom-up* fabrication, and technology developments. These revolutionary high-risk high-payoff areas have emerged recently, and require time, readiness, commitment, acceptance, investment, infrastructure, innovations, and market needs. Among the most promising applications, which will lead to revolutionary advances, we emphasize the devising and designing of

- Molecular signal or data processing platforms and molecular memory platforms
- Molecular information processing platforms

Our ultimate objective is to contribute to the developments of molecular electronics in order to radically increase the performance of processing (computing) platforms. Molecular electronics guarantees information processing preeminence, computing superiority, and memory supremacy.

In general, molecular electronics spans from new device physics to synthesis technologies, and from unique phenomena/effects/capabilities/functionality to novel topologies, organizations, and architectures. This book proposes innovative *solid* and *fluidic* ^{M}devices coherently examining their device physics. We report a unified synthesis taxonomy in the design of three-dimensional (3D) ^{M}ICs, which are envisioned to be utilized in processing and memory platforms for a new generation of processors, computers, and so forth. The design of ^{M}ICs is accomplished by using a novel technology-centric concept based on the use of neuronal hypercells ($^{\aleph}$hypercells) consisting of ^{M}gates. These ^{M}gates are comprised of interconnected multiterminal ^{M}devices. Some promising ^{M}devices are examined in sufficient details. Innovative approaches in design of ^{M}PPs, implemented by ^{M}ICs, are documented. The performance estimates, fundamentals, and *design rules* are reported.

There are many possible directions in the development of molecular electronics, circuits, and processing platforms. At the device level, one needs to investigate various molecular primitives (^{M}primitives), such as ^{M}devices, ^{M}gates, and $^{\aleph}$hypercells. With these efforts to derive possible solutions and feasible technologies, one may use various cyclic molecules as 3D-topology multiterminal ^{ME}devices, chiropticene molecular switch, protein associative processing primitives, bacteriorhodopsin holographic associative volumetric memories, and so forth. For example, a chiropticene molecular switch is based on a conformational transition. The switching of a chiropticene molecule is triggered by light (photon) excitation, whereas the state is controlled by an electric field. Using the different device physics of ^{M}devices, chemical, electric, mechanical, optical, and other phenomena, as well as quantum and other effects, can be utilized. For example, even a chemical reaction–diffusion paradigm, which is not covered in this book, may be employed. In fact, the distributed character and nonlinear dynamics of chemical reactions can map logical operations. This book emphasizes and describes electrochemomechanically induced transitions in ^{M}devices and describes ^{BM}PPs.

A fundamental theory, coherently supported by enabling solutions, and technologies are further developed and applied. The basic and applied research is expanded toward technology-centric CAD-supported ^{M}ICs design theory and practice. Advancement and progress are ensured by using new sound solutions, and a need for a SLSI design is emphasized. The fabrication aspects are covered. The results reported further expand the horizon of the molecular electronics theory and practice, biomolecular processing, information technology, design of processing/memory platforms, and molecular technologies.

Device physics and system design center on the modern science and engineering, while the progress in chemistry and biochemistry can be utilized to accomplish *bottom-up* fabrication. There are well-established molecular, polymeric, supramolecular, and other motifs. A great number of unsolved fundamental, applied, and technological problems must be overcome to synthesize complex functional molecular aggregates that perform specific operations and tasks. Recent developments and discoveries of organic chemistry and biochemistry can be utilized in molecular electronics, but sound practical innovations and further progress are urgently needed.

1.4 Electronics and Processing Platforms: Retrospect and Prospect

We have emphasized a wide spectrum of challenges and problems. It seems that the devising of ^{M}devices, *bottom-up* fabrication, design, SLSI, and technology-centric CAD developments are among the most complex issues. Before turning to molecular electronics, let us turn our attention to the retrospect, and then, focus on the prospect and opportunities.

The history of data retrieval and processing tools is traced back to thousands years. To enter the data, retain it, and perform calculations, people used a mechanical "tool," called an *abacus*. The early *abacus*, known as a counting board, was a piece of wood, stone, or metal with carved grooves or painted lines between which movable beads, pebbles, or wood/bone/stone/metal disks were arranged. When these beads are moved around according to the "programming rules" memorized by the user, some recording and arithmetic problems are solved and documented. The *abacus* was used for counting, tracking data, and recording facts even before the concept of numbers was invented. The oldest counting board, found in 1899 on the island of Salamis, was used by the Babylonians around 300 BC. As shown in Figure 1.2a, the Salamis *abacus* is a slab of marble marked with 2 sets of 11 vertical lines (10 columns), a blank space between them, a horizontal line crossing each

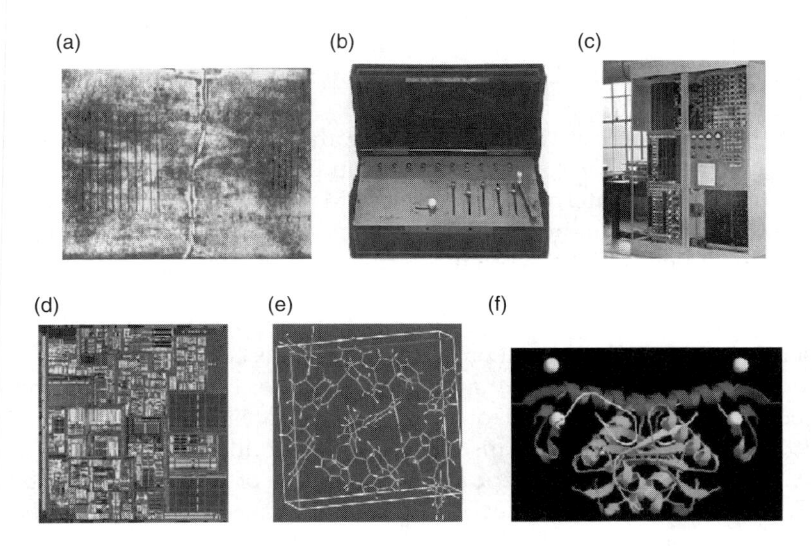

FIGURE 1.2
(See color insert following page 146.) From the *abacus* (300 BC) to Thomas' "Arithmometer" (1820), from the electronic numerical integrator and computer (1946) to the 1.5 × 1.5 cm 478-pin Intel® Pentium® 4 processor with 42 million transistors (2002; http://www.intel.com/), and toward 3D *solid* and *fluidic* molecular electronics and processing.

set of lines, and Greek symbols along the top and bottom. Another important invention around the same time was the astrolabe for navigation.

In 1623, Wilhelm Schickard built his "calculating clock," which is a six-digit machine that can add, subtract, and indicate overflow by ringing a bell. Blaise Pascal is usually credited with building the first digital calculating machine. He made it in 1642 to assist his father, who was a tax collector. This machine was able to add numbers entered with dials. Pascal also designed and built a "Pascaline" machine in 1644. These five- and eight-digit machines used a concept different from the Schickard's "calculating clock." In particular, rising and falling weights instead of a gear drive were used. The "Pascalian" machine can be extended to more digits, but it cannot subtract. Pascal sold more than 10 machines, and several of them still exist. In 1674, Gottfried Wilhelm von Leibniz introduced a "stepped reckoner" using a movable carriage to perform multiplications. Charles Xavier Thomas applied Leibniz's ideas and in 1820 made a mechanical calculator; see Figure 1.2b. In 1822, Charles Babbage built a six-digit calculator that performed mathematical operations using gears. For many years, from 1834 to 1871, Babbage carried out the "analytical engine" project. His design integrated the stored-program (memory) concept envisioning that the memory may hold more than 100 numbers. The proposed machine had a read-only memory in the form of punch cards. These cards were chained, and the motion of each chain could be reversed. Thus, the machine was able to perform the conditional manipulations and integrated coding features. The instructions depended on the positioning of metal studs in a slotted barrel, called the "control barrel." Babbage only partially implemented his ideas in designing a proof-of-concept programmable calculator because his innovative initiatives were far ahead of the technological capabilities and theoretical foundations. But the ideas and goals were set.

In 1926, Vannevar Bush proposed the "product integraph," which is a semi-automatic machine for solving problems in determining the characteristics of electric circuits. International Business Machines introduced in 1935 the "IBM 601" and made more than 1500 of them. This was a punch-card machine with an arithmetic unit based on relays that performed multiplication in 1 sec. In 1937, George Stibitz constructed a 1-bit binary adder using relays. Alan Turing published a paper reporting "computable numbers" in 1937. In this paper he solved mathematical problems and proposed a mathematical model of computing known as the *Turing machine*. The idea of electronic computer is traced back to the late 1920s. However, the major breakthroughs appear later. In 1937, Claude Shannon in his master's thesis outlined the application of relays. He proposed an "electric adder to the base of two." George Stibitz, in 1937, developed a binary circuit based on Boolean algebra. He built and tested the proposed adding device in 1940. John Atanasoff completed a prototype of a 16-bit adder using vacuum tube diodes in 1939. The same year, Zuse and Schreyer examined the application of relay logic. Schreyer completed a prototype of the 10-bit adder using vacuum tubes in 1940, and he built memory using neon lamps. Zuse demonstrated the first operational programmable calculator in 1940. The calculator had floating-point numbers with a 7-bit exponent, 14-bit mantissa, sign bit, 64-word memory with 1400 relays, and arithmetic and control units comprising 1200 relays. Howard Aiken proposed a calculating machine that solved some problems of relativistic physics. He built the "Automatic Sequence Controlled Calculator Mark I." This project was finished in 1944, and "Mark I" was used for calculations in ballistics problems. This electromechanical machine was 15 m long, weighed 5 t, and had 750,000 parts (72 accumulators with arithmetic units and mechanical registers with a capacity of 23 digits and signs). The arithmetic was fixed point, with a plug board determining the number of decimal places. The input–output unit included card reader, card puncher, paper tape reader, and typewriter. There were 60 sets of rotary switches, each of which could be used as a constant register, e.g., as a mechanical read-only memory. The program was read from a paper tape, and data could be read from the other tapes, card readers, or constant registers. In 1943, the U.S. government contracted John Mauchly and Presper Eckert to design the Electronic Numerical Integrator and Computer, which likely was the first electronic digital computer built. The Electronic Numerical Integrator and Computer was completed in 1946; see Figure 1.2c. This machine performed 5000 additions or 400 multiplications per second, showing enormous capabilities for this time. The Electronic Numerical Integrator and Computer weighed 30 t, consumed 150 kW, and had 18,000 vacuum tube diodes. John von Neumann with colleagues built the Electronic Discrete Variable Automatic Computer in 1945 using the so-called "von Neumann computer architecture."

Combinational and memory circuits are comprised of microelectronic devices, logic gates, and modules. Textbooks on microelectronics coherently document the developments starting from the discoveries of semiconductor devices to the design of ICs. The major developments are reported below. Ferdinand Braun invented the solid-state rectifier in 1874. The silicon diode was created and demonstrated by Pickard in 1906. The field-effect devices were patented by von Julius Lilienfeld and Oskar Heil in 1926 and 1935, respectively. The functional solid-state bipolar junction transistor (BJT) was built and tested on December 23, 1947, by John Bardeen and Walter Brattain. Gordon Teal made the first silicon transistor in 1948, and William Shockley invented the unipolar FFT in 1952. The first ICs were designed by Kilby and Moore in 1958.

Microelectronics has been utilized in signal processing and computing platforms. First, second, third, and fourth generations of computers have emerged, and tremendous progress has been achieved. The Intel® Pentium® 4 processor, illustrated in Figure 1.2d, and Core™ Duo processor families were built using advanced Intel® microarchitectures. These high-performance processors are fabricated using 90 and 65 nm complementary metol-oxide-semiconductor (CMOS) technology nodes. The CMOS technology matured to fabricate high-yield high-performance ICs with trillions of transistors on a single die. The fifth generation of computers will utilize further-scaled-down microelectronic devices and enhanced architectures. However, further progress and developments are needed. New solutions and novel enabling technologies are emerging.

The suggestion to utilize molecules as a molecular diode, which can be considered as the simplest two-terminal *solid* ^{ME}device, was introduced by M. Ratner and A. Aviram in 1974 [12]. This visionary idea has been further expanded through meaningful theoretical, applied, and experimental developments [5–11]. Three-dimensional molecular electronics and ^{M}ICs were proposed in [5,6]. These ^{M}ICs are designed as aggregated ${}^{\aleph}$hypercells, which are comprised of ^{M}gates engineered utilizing 3D-topology multiterminal solid ^{ME}devices; see Figure 1.2e. Figure 1.2f schematically illustrates the ion–biomolecule–protein complex as *fluidic* biomolecular or ^{M}devices [5,6], which are considered in Chapter 3.

The US Patent 6,430,511 "Molecular Computer" was issued in 2002 to J. M. Tour, M. A. Reed, J. M. Seminario, D. L. Allara, and P. S. Weiss. The inventors envisioned a molecular computer as formed by establishing arrays of input and output pins, "injecting moleware," and "allowing the moleware to bridge the input and output pins." The proposed moleware includes molecular alligator clip-bearing 2-, 3-, and molecular 4-, or multiterminal wires, carbon nanotube wires, molecular resonant tunneling diodes, molecular switches, molecular controllers that can be modulated via external electrical or magnetic fields, massive interconnect stations based on single nanometer-sized particles, and dynamic and static random access memory (DRAM and SRAM) components composed of molecular controller/nanoparticle or fullerene hybrids." Overall, one may find a great deal of motivating conceptual ideas expecting the fundamental soundness and technological feasibility.

1.5 Molecular Processing Platforms: Evidence of Feasibility, Soundness, and Practicality Substantiation

Questions regarding the feasibility of molecular electronics, ^{M}ICs, and ^{M}PPs arise. There does not exist conclusive evidence on the overall soundness of *solid* ^{M}ICs, as there were no analogs for solid-state microelectronics and ICs in the past. In contrast, different ^{BM}PPs exist in the nature. We will briefly focus our attention on the most primitive biosystems. Prokaryotic cells (bacteria) lack extensive intracellular organization and do not have cytoplasmic organelles, while eukaryotic cells have well-defined nuclear membrane as well as a variety of intracellular structures and organelles. However, even a couple of microns long single-cell *Escherichia coli* (*E. coli*), *Salmonella typhimurium*, *Helicobacter pylori*, and other bacteria possess ^{BM}PPs exhibiting signal or data processing and memory capabilities, as well as information processing features. These bacteria also have molecular sensors, $\sim 50 \times 50 \times 50$ nm motors, as well as other numerous biomolecular devices and systems made from proteins. The biophysics of distinct devices and systems (sensors, actuators, etc.) is based on *specific* phenomena effects and mechanisms. Although bacterial motors (largest devices) have been studied for decades, baseline operating phenomena and mechanisms are still unknown [13]. The biophysics (phenomena, effects, and mechanisms) ultimately behind this processing and memory have not been understood at the device and system levels. Furthermore, the fundamentals of biomolecular processing, memories, and device physics are not well understood even for single-cell bacteria possessing ^{BM}PPs. The processing, memory storage, and memory retrieval in these systems are likely performed utilizing biophysical mechanisms involving electrochemomechanically induced transitions and interactions in ion (~0.2 nm)-biomolecule (~1 nm)-protein (~10 nm) complexes in response to stimuli. Fluidic processing and ^{M}PPs, which resemble ^{BM}PPs, were first proposed in [5,6]. Figure 1.2f schematically illustrates the ion–biomolecule–protein complex. Various existing ^{BM}PPs establish evidence of biomolecular device physics ensuring the overall feasibility and soundness of *synthetic* and *fluidic* ^{M}PPs. This soundness is also extended to the *solid* molecular electronics.

Assume that, in prokaryotic cells and neurons, processing and memory storage are performed by electrochemomechanical transitions in biomolecules such as folding, induced potential, charge variation, bonding change, and so forth. These changes are accomplished by the binding/unbinding of ions and/or biomolecules, enzymatic activities, photon absorption, and so forth. Experimental and analytic results show that protein folding is accomplished within nanoseconds and requires $\sim 1 \times 10^{-19}$ to 1×10^{-18} J of energy. Real-time 3D image processing is ordinarily accomplished even by primitive insects and vertebrates that have less than 1 million of neurons. To perform these and other immense processing tasks, less than 1 μW is consumed. However, real-time 3D image processing cannot be performed by even envisioned processors with trillions of transistors, device switching speed ~1 THz, circuit speed ~10 GHz, device switching energy $\sim 1 \times 10^{-16}$ J, writing energy $\sim 1 \times 10^{-16}$ J/bit, read time ~10 nsec, and so forth. This is an undisputable evidence of superb biomolecular processing that cannot be surpassed by any envisioned microelectronics enhancements and innovations. The existence of ^{BM}PPs provides the foundation for the overall soundness of envisioned ^{M}PPs.

Different concepts are emerging examining solid and fluidic molecular electronics, biomolecular processing, and so forth. All bio-organisms are composed of micron-sized cells that contain thousands of different molecules. Those biomolecules are utilized to perform reproduction, energy conversion, actuation, sensing, processing, memory storage, and so forth. In the living organisms, the ^{BM}PPs are formed by neuronal aggregates. The attempts to prototype ^{BM}PPs face significant challenges. In addition to unsolved fundamental problems (phenomena and effects utilized by neurons and biomolecules, mechanisms exhibited, topologies/organizations/architectures embedded, means of reconfigurability, etc.), synthesis is a critical problem. Controlled by the designer self-replication, though performed in biosystems through complex and not fully comprehended mechanisms, is a far-reaching target for many, at least, decades to come. This can be potentially accomplished utilizing biochemistry and biotechnology.

In Chapter 3, we cover neuronal processing-and-memory primitives, which constitute *biomolecular processing hardware*, electrochemomechanically induced transitions/interactions/events, introduce *information* and *routing* carriers, and focus on other issues directly and indirectly related to processing. Owing to the extraordinary complexity of the neuronal biophysics, there is a great degree of uncertainty that affects overall analysis of devices and systems. For example, the device physics is based on the biophysical phenomena and effects exhibited and utilized by biomolecules in neurons. The biophysics remains to be researched examining transitions, interactions and other effects caused or affected by the *microscopic* particles (ion, electron, etc.). Even basic observed phenomena, known for centuries, exhibited by leaving systems have not been fully understood. For example, bioluminescence is the production and emission of photons (light) by living organisms as the result of chemical reaction, energy conversion and electrochemomechanical transitions. An enzyme-catalyzed chemoluminescence reaction has been widely examined in the literature, and the pigment luciferin is known to be oxidized with presence of the enzyme luciferase. Bacterial luciferin is a reduced riboflavin phosphate oxidized in association with a long-chain aldehyde, oxygen, and a luciferase (enzyme). The structure of this biomolecule is:

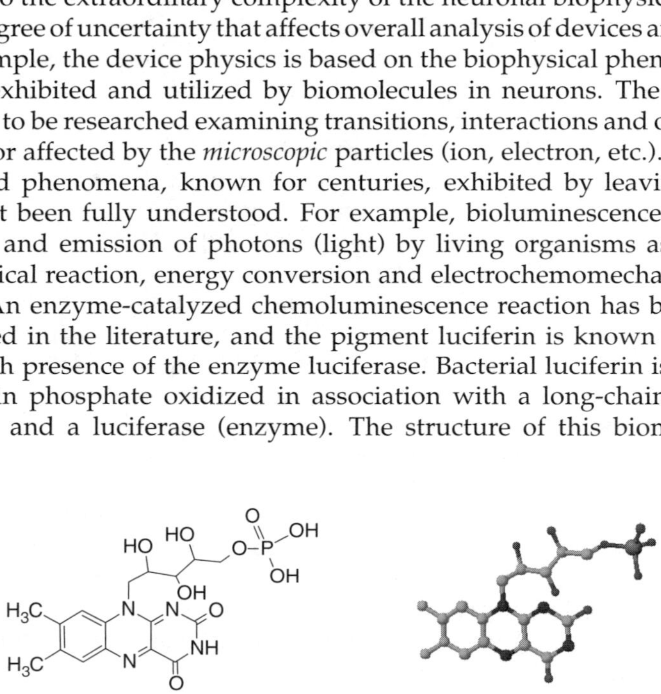

Bioluminescence is not fluorescence, which occurs when the energy from a source of light is absorbed and reemitted as another photon. Bioluminescence is performed by an incredible range of organisms, from bacteria and single-cell protists to fish and squid, from mushrooms to beetles, and so forth. The example illustrated in Figure 1.3 is siphonophore, which uses red light to lure fish to its tentacles, *Melanocetus johnsoni* (anglerfish), and firefly. Beetles in the *Coleoptera* family *Lampyridae* are called "fireflies" or "lightning bugs" because of their crepuscular use of bioluminescence to attract mates and prey. The firefly emits phonons of wavelengths from ~500 to ~670 nm (yellow, red, and green colors). There are more than 2000 species of firefly, and the reader has likely seen these amazing beetles. There are many hypotheses (oxygen control, neural activation, etc.) on how fireflies control the *on* and *off* bioluminescence *states*. The debate continues on basic bioluminescence reaction reported by William McElroy and Bernard Strehler in their papers published

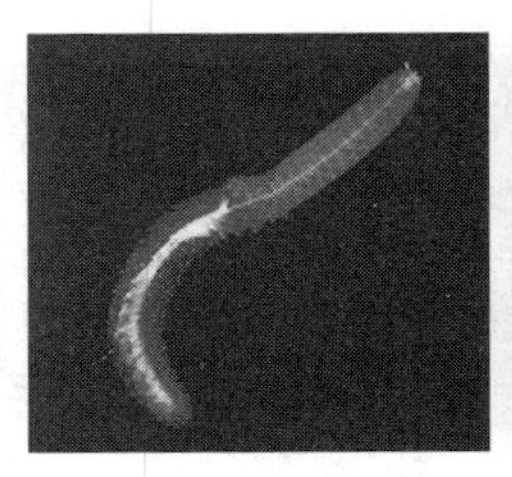 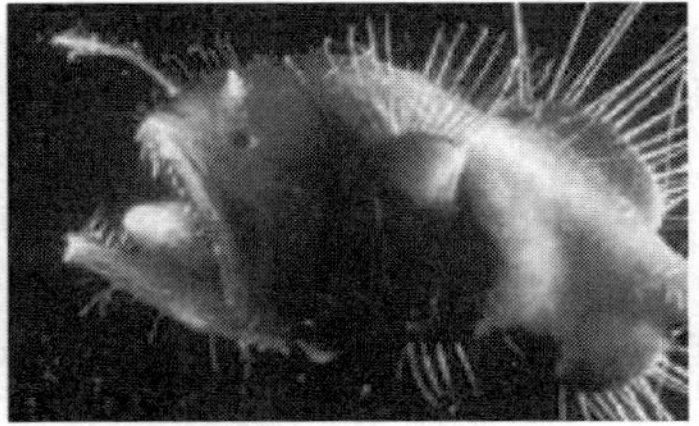 

FIGURE 1.3
Siphonophore, anglerfish, and firefly exhibit bioluminescence.

in early 1950s [14]. The reader is directed to the section on *Terminology Used and Justifications* at the end of this section.

The complexity of neuronal biophysics and electrochemomechanically induced transitions/interactions in biomolecules within neuronal structures is enormous. Therefore, one should be very cautious on the expected progress in ^{BM}PPs and its implication for ^{M}ICs and ^{M}PPs, even though these problems are of an immense importance. However, electrochemomechanically induced transitions/interactions/events and mechanisms are exhibited by living organisms and result in or affect information processing and related tasks.

Terminology Used and Justifications: The electrochemomechanically induced transitions, interactions, and events in the *biomolecular hardware* ultimately result in processing, memory storage, and other related tasks. Do we have an evidence and substantiation of the correctness and accuracy of this hypothesis? With a high degree of confidence, the answer is yes. We do not depart from the generally accepted biophysics, though the conventional neuroscience doctrine on the key role of action potential in the processing and memory, as well as neuron-as-a-device doctrine, are refined. Referencing to the light receptors (photon absorption) and production (photon emission), we recall that:

- A photon-absorbing retinal molecule is bonded to a membrane protein called opsin. The resulting biomolecule is rhodopsin. When rhodopsin absorbs light, the retinal changes shape due to a photochemical reaction. In addition, there are red, green, and blue cones in the retina, each with its own type of opsin associated with retina ensuring high selectivity to the wavelength absorption capabilities of photopsins.
- Bioluminescence is the production and emission of photons (light). This enzyme-catalyzed chemoluminescence reaction was reported.

In both cases, in addition to photochemical reactions, there is the folding (*electromechanical switching*) of biomolecules. A photon is emitted or absorbed as a quantum of electromagnetic energy, and each photon of frequency v ($v = c/\lambda$) has an energy $E = hv$, where h is the Planck constant, $h = 6.62606876 \times 10^{-34}$ J sec. An atom can make a *transition* from one energy level to a lower energy level by emitting a photon of energy equal to the difference between the initial and final levels. When considering transitions between two energy states of an atom (quantized *microscopic* system) in the presence of an electromagnetic field, one studies *stimulated absorption, stimulated emission, spontaneous emission,* and other processes related quantitatively. All these effects are *quantum* and describe the interaction of radiation with the atom or molecule. Considering the emission of electromagnetic energy, assume that the atom is initially in the upper state of energy (excited state) E_2 and decays to the lower state of energy E_1 by the emission of a photon of frequency $v = (E_2 - E_1)/h$. The mean lifetime of an atom in most exited states is $\sim 1 \times 10^{-8}$ sec, but in metastable exited states, the mean lifetime could be as long as 1×10^{-3} sec. From fundamental biophysics and quantum physics, one concludes that absorption and emission processes are: (1) accomplished by biomolecules within biomolecular aggregates that form *biomolecular hardware*, and, (2) are based on quantum effects that describe the electromagnetic transitions and radiation interactions within the *biomolecular hardware*.

There is an overall similarity and correspondence between the aforementioned processes and processing. Therefore, we use the terms:

1. *Biomolecular processing hardware*, which processes the information
2. Electrochemomechanically induced transitions and interactions, reflecting chemical reactions, conformational changes (*electromechanical switching*), and other processes under quantum and electromagnetic effects

Remark: We further briefly discuss the terminology used. The nomenclature "biomolecular electronics" may not be well founded because of the limits to reflect the baseline biophysics. It will be reported in Chapter 3 that DNA- and protein-centered motifs form 3D structures that are not "circuits", and, in general, cannot be defined as "biomolecular electronics." In fact, these "circuits", which ultimately should be based on the electron transport, unlikely exhibit functionality and/or ensure practicality in implementation of even the most simple combinational and/or memory elements. We introduced the term ^{BM}PP, and the device physics of biomolecular devices is different compared with ^{ME}devices. Fluidic-centered molecular processing and ^{M}PPs provide the ability to mimic some features of ^{BM}PPs. Our terminology does not imply that fluidic and biomolecular platforms are predominantly based or centered on electron transport or only on electron-associated transitions. However, synthetic and fluidic ^{M}devices exhibit *specific* electrochemomechanical transitions, and one uses electronic apparatuses to control these transitions.

1.6 Microelectronics and Emerging Paradigms: Challenges, Implications, and Solutions

To design and fabricate planar CMOS ICs, which consist of FETs and BJTs as the major microelectronic devices, processes and design rules have been defined. It should be emphasized that there are various FETs—for example, metal-oxide-semiconductor field-effect transistor (MOSFET), junction field-effect transistor (JFET), metal-semiconductor field-effect transistor (MESFET), heterojunction field-effect transistor (HFET), heterojunction insulated-gate field-effect transistor (HIGFET), and modulation-doped field-effect transistor (MODFET).

Taking note of the topological layout, the physical dimensions and area requirements can be estimated using the CMOS design rules, which are centered on: (1) minimal feature size and minimum allowable separation in terms of absolute dimensional constraints, and, (2) the lambda rule (defined using the length unit λ) that specifies the layout constraints taking note of nonlinear scaling, geometrical constraints, and minimum allowable dimensions (width, spacing, separation, extension, overlap, width/length ratio, and so forth). In general, λ is a function of exposure wavelength, image resolution, depth of focus, processes, materials, device physics, topology, and so forth. For different technology nodes, λ varies from $\sim\frac{1}{2}$ to 1 of the minimal feature size. For the current front-edge 65 nm technology node, introduced in 2005 and deployed by some high-technology companies in 2006, the minimal feature size is 65 nm. It is expected that the feature size could decrease to 18 nm by 2018 [1]. For n-channel MOSFETs (physical cell size is $\sim 10\lambda \times 10\lambda$) and BJTs (physical cell size is $\sim 50\lambda \times 50\lambda$), the *effective* cell areas are in the range of hundreds and thousands of λ^2, respectively. For MOSFETs, the gate length is the distance between the active source and drain regions underneath the gate. This implies that, if the channel length is 30 nm, it does not mean that the gate width or λ is 30 nm. For FETs, the ratio between the effective cell size and minimum feature size will remain $\sim$20.

One cannot define and classify electronic, optical, electrochemomechanical, and other devices, as well as ICs, by taking note of their dimensions (length, area, or volume) or minimal feature size, though device dimensionality is an important feature primarily from the fabrication viewpoint. To classify devices and systems, one examines the device physics, system organization/architecture, and fabrication technologies assessing distinctive features, capabilities, and phenomena utilized. Even if the dimensions of CMOS transistors are scaled down to achieve 100×100 nm *effective* cell size for FETs by late 2020, these solid-state semiconductor devices may not be viewed as nanoelectronic devices because conventional phenomena and evolved technologies are utilized. The fundamental limits on microelectronics and solid-state semiconductor devices were known and have been reported for many years [1].

Although a significant technology progress has been accomplished ensuring high-yield fabrication of ICs, the basic physics of semiconductor devices remains virtually unchanged for decades. Three editions (1969, 1981, and 2007) of a classical textbook *Physics of Semiconductor Devices* [15–17] coherently cover the device physics of various solid-state microelectronic devices. The evolutionary technological developments will continue beyond the current 65 nm technology node. The 45 nm CMOS technology node is expected to emerge in 2007. Let us assume that, by 2018, 18 nm technology node will be deployed with the expected λ of $\sim$18 nm and effective channel length for FETs of $\sim$7–8 nm. This will lead to the estimated footprint area of the interconnected FET to be in the range of tens of thousands of square nanometers because the effective cell area is at least $10\lambda \times 10\lambda$. Sometimes, a questionable size-centered definition of nanotechnology surfaces, ambiguously picking the 100 nm dimensionality to be met. It is uncertain which dimensionality should be used. Also, it is unclear why 100 nm was declared, and not 1 or 999 nm? On other hand, why not to use an area or volumetric measures of 100 nm^2 and 100 nm^3?

CMOS technology provides one with a variety of complimentary deposition and etching processes and techniques that can be used to fabricate two-dimensional (2D) multilayered heterojunctions and structures with nanometer thickness [1,17]. For example, sputtering, low-pressure chemical vapor deposition (LPCVD), plasma-enhanced chemical vapor deposition, thermal evaporation, molecular beam epitaxy, ion-assisted electron beam deposition, thermal evaporation, RF magnetron sputtering, and other processes are used to fabricate uniform 2D structures with $\sim$1 nm thickness, as reported in Figure 1.4a. The deoxyritonucleic acid (DNA) derivatives can be potentially used to form channels in some FETs. In particular, polyG $(dG(C_{10})_2)_n$ is deposited (dropped with subsequent evaporation) on the insulator (silicon oxide, silicon nitride, etc.) between the source and drain, as reported in Figure 1.4b. Although electrons flow through $\sim$100 nm length, 2 nm diameter polyG ropes, this polyG FET cannot be considered to be a nanoelectronic or molecular device. Furthermore, this solution does not ensure better performance, fabrication advantages, or any benefits as compared with the conventional CMOS technology MOSFET, shown in Figure 1.4c, or other CMOS FETs.

An electric current is a flow of charged *microscopic* particles. The current in conductors, semiconductors, and insulators is due to the movement of electrons. In aqueous solutions, the current is due to the movement of charged particles, for example, ions, molecules, and so forth. These devices have not been classified using the dimension of the charged carriers (electrons, ions, or molecules), which are the *microscopic* particles. However, one may compare the device dimensionality with the size of the particles that causes the current flow or transitions. For example, considering a protein as a core component of a biomolecular device and an ion as

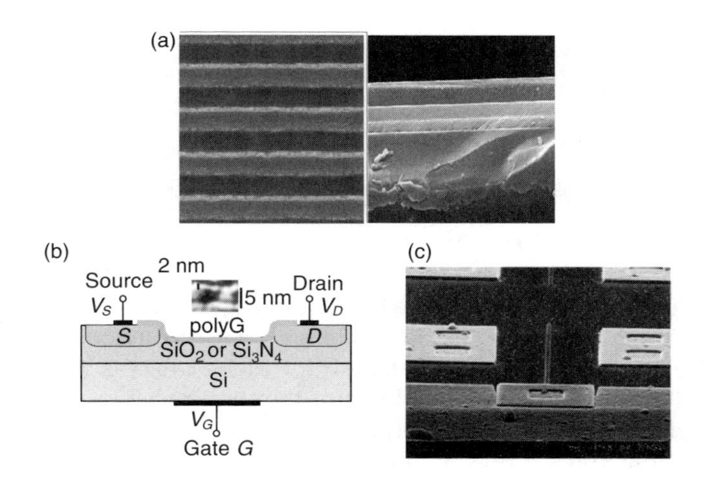

FIGURE 1.4
(a) Multilayered structures: ~1 nm thickness $SiO_2/SiON$ and $SiO_2/Ta_2O_5/SiO_2$; (b) PolyG FET with a channel formed by the adsorbed polyG $(dG(C_{10})_2)_n$; (c) CMOS MOSFET.

a charged carrier that affects the protein transitions, the device/carrier dimensionality ratio would be ~100. The classical electron radius r_0, called the Compton radius, is found by equating the electrostatic potential energy of a sphere with the charge e and radius r_0 to the relativistic rest energy of the electron which is $m_e c^2$. We have $e^2/(4\pi\varepsilon_0 r_0) = m_e c^2$, where e is the charge on the electron, $e = 1.6022 \times 10^{-19}$ C; ε_0 is the permittivity of free space, $\varepsilon_0 = 8.8542 \times 10^{-12}$ F/m; m_e is the mass of electron, $m_e = 9.1095 \times 10^{-31}$ kg; c is the speed of light, and in the vacuum, $c = 299792458$ m/sec. Thus, $r_0 = e^2/(4\pi\varepsilon_0 m_e c^2) = 2.81794 \times 10^{-15}$ m. With the achievable volumetric dimensionality of solid ^{ME}device on the order of $1 \times 1 \times 1$ nm, one finds that the device is much larger than the carrier. Up to 1×10^{18} ^{ME}devices can be placed in 1 mm^3. This upper-limit device density may not be achieved because of the synthesis constraints, technological challenges, expected inconsistency, aggregation/interconnect complexity, and other problems reported in Section 4.6.2 and Chapter 5. The *effective* volumetric dimensionality of interconnected solid ^{ME}devices in ^{M}ICs is expected to be $\sim 10 \times 10 \times 10$ nm. For solid ^{ME}devices, quantum physics is applied to examine the effects, processes, functionality, performance, characteristics, and so forth. The device physics of fluidic and solid ^{M}devices is profoundly different. Solid ^{ME}devices and ^{M}ICs may utilize the so-called *soft materials* such as polymers and biomolecules.

Emphasizing the major premises, nanoelectronics implies the use of:

1. Novel high-performance devices, as devised using new device physics, which exhibit unique phenomena, effects, and capabilities to be exclusively utilized at the gate, module, and system levels.
2. Enabling organizations and advanced architectures that ensure superb performance and superior capabilities. Those developments

are integrated with the device-level solutions, technology-centric SLSI design, and so forth.

3. *Bottom-up* fabrication.

To design ^{M}ICs-comprised processing and memory platforms, one must apply novel paradigms and pioneering developments. Tremendous progress has been accomplished in microelectronics in the last 60 years. For example, from inventions and demonstration of functional solid-state transistors to fabrication of processors that comprise trillions of transistors on a single die. The current high-yield 65 nm CMOS technology node ensures minimal feature sizes of ~65 nm, and the channel length for MOSFETs was scaled down to less than 30 nm. Using this technology for the static random access memory (SRAM) cells, a ~500,000 nm^2 footprint area was achieved by Intel. Optimistic predictions foresee that within 20 years the minimal feature of planar (2D) solid-state CMOS technology transistors may approach ~10 nm, leading to an effective cell size of $\sim 10\lambda \times 10\lambda = 100 \times 100$ nm for FETs. However, the projected scaling trends are based on a number of assumptions and foreseen enhancements [1]. Although the FET cell dimension can reach 100 nm, the overall prospects in microelectronics (technology enhancements, device physics, device/circuits performance, design complexity, cost, and other features) are troubling [1–4]. The near-absolute limits of the CMOS-centered microelectronics can be reached by the next decade. The general trends, prospects and projections are reported in the *International Technology Roadmap for Semiconductors* [1].

The device size- and switching energy-centered version of the first Moore's conjecture for high-yield room-temperature mass-produced microelectronics is reported in Figure 1.5 for past, current (90 and 65 nm), and foreseen (45 and 32 nm) CMOS technology nodes. The minimum switching energy

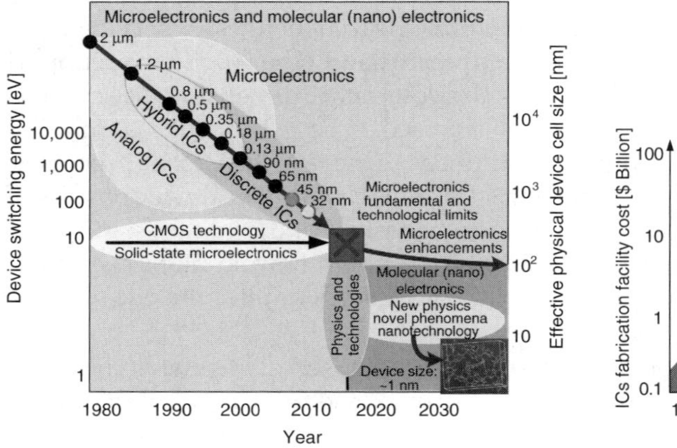

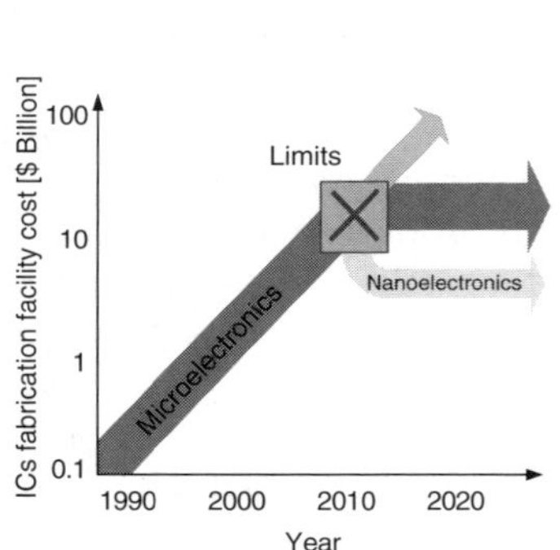

FIGURE 1.5
Microelectronics trends and envisioned molecular (nano) electronics advancements.

is estimated as $1/2CV^2$ J, and 1 eV $= 1.602176462 \times 10^{-19}$ J. Intel expects to introduce 45 nm CMOS technology node in 2007. The envisioned 32 nm technology node is foreseen to emerge in 2010. The expected progress in the baseline characteristics, key performance metrics, and scaling abilities have already been slowed down because of the encountered fundamental and technological challenges and limits. Hence, new solutions and technologies have been sought and assessed [1]. The performance and functionality at the device, module, and system levels can be significantly improved by utilizing novel phenomena, employing innovative topological/organizational/architectural solutions, enhancing device functionality, increasing density, improving utilization, increasing switching speed, and so forth. Molecular electronics (nanoelectronics) is expected to result in departure from the first and second Moore's conjectures. The second conjecture foresees that, to ensure the projected microelectronics scaling trends, the cost of microelectronics facilities can reach hundreds of billion dollars by 2020. High-yield affordable nanoelectronics technologies are expected to emerge and mature ensuring superior performance. Existing superb bimolecular processing/memory platforms and progress in molecular electronics are assured evidence of the fundamental soundness and technological feasibility of ^{M}ICs and ^{M}PPs. Some data and expected developments, reported in Figure 1.5, are subject to adjustments because it is difficult to accurately foresee the fundamental developments and maturity of prospective technologies because of the impact of many factors. However, the overall trends are obvious and likely cannot be rejected. Having emphasized the emerging molecular (nano) electronics, it is obvious that solid-state microelectronics is a core twenty-first century technology. CMOS technology will remain a viable technology for many decades even as the limits will be reached and envisioned nanoelectronics will mature. It may be expected that by 2025–2030 the core modules of super-high-performance processing (computing) platforms may be implemented using ^{M}ICs. However, microelectronics and molecular electronics will be complimentary paradigms, and ^{M}ICs will not diminish the use of ICs. Molecular electronics and ^{M}PPs are impetuous, revolutionary (not evolutionary) changes at the device, system, and technological levels. The foreseen revolutionary changes towards ^{M}devices are analogous to the abrupt change from the vacuum tube to the solid-state transistor.

The fundamental and technological limits are also imposed on molecular electronics and ^{M}PPs. Those limits are defined by the device physics, circuit, system, CAD, and synthesis constraints. Some of these limitations will be examined in Section 2.1. However, there is no end to progress, and new paradigms beyond molecular electronics and processing will be discovered. What lies beyond even molecular innovations and frontiers? The hypothetical answer is provided. In 1993, theoretical physicist G. Hooft proposed the *holographic principle*, which postulates that the information contained in some region of space can be represented as a *hologram*, which gives the bounded region of space that contains at *most* one degree of freedom per the *Planck area*, which is $G\hbar/c^3 = 2.612 \times 10^{-70}$ m^2, where G is the Newtonian constant of gravitation, $G = 6.673 \times 10^{-11}$ m^3/kg sec^2. A *Planck area* is the area enclosed by a square that has the side length equal to the *Planck length* λ_P, where λ_P is defined as the length scale on which the quantized nature of gravity should become evident, that is, $\lambda_P = \sqrt{Gh/c^3} = 4.05096 \times 10^{-35}$ m. Using the modified Planck constant, we obtain $\lambda_P = \sqrt{G\hbar/c^3} = 1.616 \times 10^{-35}$ m. The *Planck time* is the *Planck length* divided by the speed of light, e.g., $t_P = \lambda_P/c = 5.3906 \times 10^{-44}$ sec. For *microscopic* systems, we utilize the so-called *standard model* (particles are considered to be points moving through space and coherently represented by mass, electric charge, interaction, spin, etc.). The *standard model* is consistent within quantum mechanics and relativity theory (the special theory of relativity was originated by Einstein in 1905). Other concepts have been developed, including the *string theory*. In general, one may envision to utilize string vibration, distinct forces, multidimensionality, and so forth.

Atoms ($\sim 1 \times 10^{-10}$ m in diameter) are composed of subatomic *microscopic* particles (protons, neutrons, and electrons). Protons and neutrons form the nucleus with a diameter $\sim 1 \times 10^{-15}$ m. Novel physics and technologies will result in further revolutionary advances in electronics. Figure 1.6 reports the envisioned hypothetical prospects. One should realize the difference between the atom and string. The diameter of the hydrogen atom is 1.058×10^{-10} m, while the size of the string is estimated to be 1.6161×10^{-35} m. Thus,

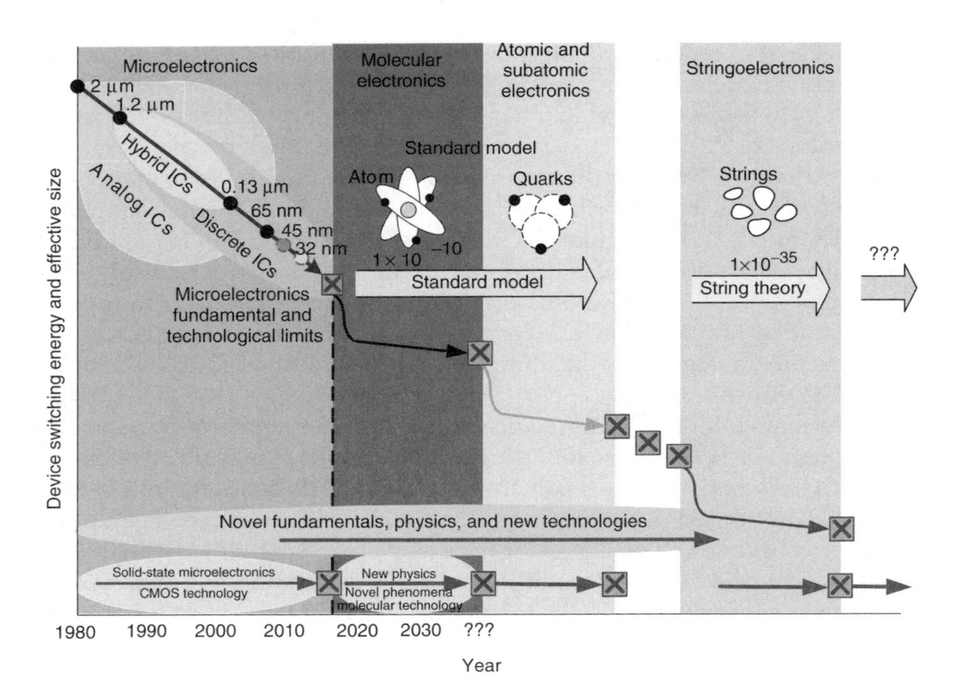

FIGURE 1.6
Envisioned progress in electronics.

the hydrogen atom is larger than the string by 6.56×10^{24}. The molecular electronics will likely progress to *atomic* and *subatomic electronics* (with pico 1×10^{-12} or femto 1×10^{-15} device dimensionality), *stringoelectronics*, and perhaps, further. There is no end to the progress. When will *stringoelectronics* emerge? It may take thousands of years. After Leucippus and Democritus envisioned the atom in 400 BC, it took more than 2400 years to substantiate it and utilize this discovery in electronics. In electronics, it may or may not take 2400 years to utilize the *string theory* to progress to the *Planck length* and *time* ($\lambda_P = 1.616 \times 10^{-35}$ m and $t_P = 5.391 \times 10^{-44}$ sec). Or, perhaps, this is an unreachable target. Within 60 years, tremendous progress in microelectronics has been accomplished. One may estimate that it may take ~50 years for molecular electronics to mature. Further predictions regarding envisioned progress may be quite speculative and exploratory. It is difficult to theorize far-reaching concepts that may or may not emerge. Therefore, we focus on a sound and practical paradigm of molecular electronics.

The key differences between microelectronic and molecular devices were emphasized in Section 1.1. Consider a class of microelectronic, molecular, and atomic devices that operate based on electron transport (flow) or electron-induced transitions/interactions in atomic structures. The microelectronic and $^{\rm ME}$devices cannot be called *femtoelectronic* or subatomic devices by arguing that the electron is the charge carrier and its dimensionality is in femtometers. In fact, though the electron radius is $\sim 2.8 \times 10^{-15}$ m, it is a *microscopic* particle. From the device physics prospective, one may state that:

- In microelectronic devices, individual molecules and atoms do not depict the overall device physics and do not define the device performance, functionality, and capabilities.
- In molecular devices, individual molecules and atoms explicitly define the overall device physics depicting the device performance, functionality, capabilities, and topologies.
- In atomic and subatomic devices, individual atoms and subatomic *microscopic* particles define the overall device physics, depicting the device performance, functionality, capabilities, and topologies.

Commercial high-yield high-performance molecular electronics is expected to emerge after 2015–2020, as shown in Figure 1.5. Molecular devices operate based on ion transport, electron and photon interactions, electrochemomechanical *state* transitions, and so forth. For distinct classes of $^{\rm M}$devices, the basic physics, phenomena exhibited, effects utilized, and fabrication technologies are profoundly different. Molecular electronics can be classified using the following major classes:

1. Solid organic/inorganic molecular electronics
2. Fluidic molecular electronics

3. Synthetic biomolecular electronics
4. Hybrid molecular electronics

Distinct subclasses and classifiers can be developed taking into account the device physics and system features. Biomolecular devices and platforms, which are not within the aforementioned classes, may be classified as well.

As mentioned, the dominating premises of molecular (nano) electronics and ^{M}PPs have a solid bio-association. The device-level biophysics and system-level fundamentals of biomolecular processing are not fully comprehended, but they are fluidic and molecule-centered. From 3D-centered topology and organization standpoints, solid and fluidic molecular electronics mimics superb ^{BM}PPs. Information processing, memory storage, and other relevant tasks, performed by living organisms, are a sound proving evidence of the proposed developments. Molecular electronics will lead to novel ^{M}PPs. Compared with the most advanced CMOS processors, molecular platforms will greatly enhance functionality and processing capabilities, radically decrease latency, power, and execution time, as well as drastically increase device density, utilization, and memory capacity. Many difficult problems at the device and system levels must be addressed, researched, and solved. For example, the following tasks should be carried out: design, analysis, optimization, aggregation, routing, reconfiguration, verification, evaluation, and so forth. Many of the aforementioned problems have not been even addressed yet. Owing to significant challenges, much effort must be focused on solution of these problems. We address and propose solutions to some of the aforementioned fundamental and applied problems. A number of baseline problems are examined progressing from the system-level consideration to the device level, and vice versa. Taking note of the diversity and magnitude of the tasks under consideration, one cannot formulate, examine, and solve simultaneously all of the challenging problems. A gradual step-by-step approach is pursued rather than the attempt to solve abstract problems with a minimal chance to succeed. There is a need to stimulate further developments and foster advanced research focusing on well-defined existing fundamentals and future perspectives emphasizing the near-, medium-, and long-term prospects, vision, solutions, and technologies. Recent fundamental, applied, and experimental developments have notably contributed to progress and have motivated further research, engineering, and technological developments. These prominent trends signify the importance of multidisciplinary science, engineering, and technology, as shown in Figure 1.7.

There is a need for super-high-performance ^{M}ICs to meet the requirements and specifications of envisioned computers, processors, processing platforms, management systems, and so forth. These ^{M}ICs and ^{M}PPs are ultimately expected to guarantee:

- Superior parallel processing capabilities
- Intrinsic data-intensive processing

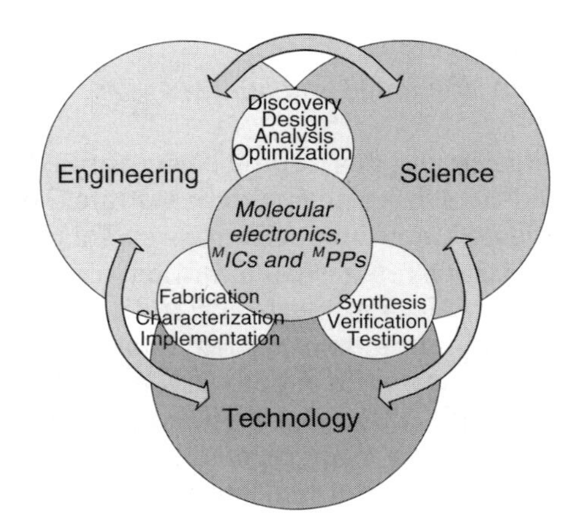

FIGURE 1.7
Multidisciplinary science, engineering and technology in molecular electronics, ^{M}ICs, and ^{M}PPs.

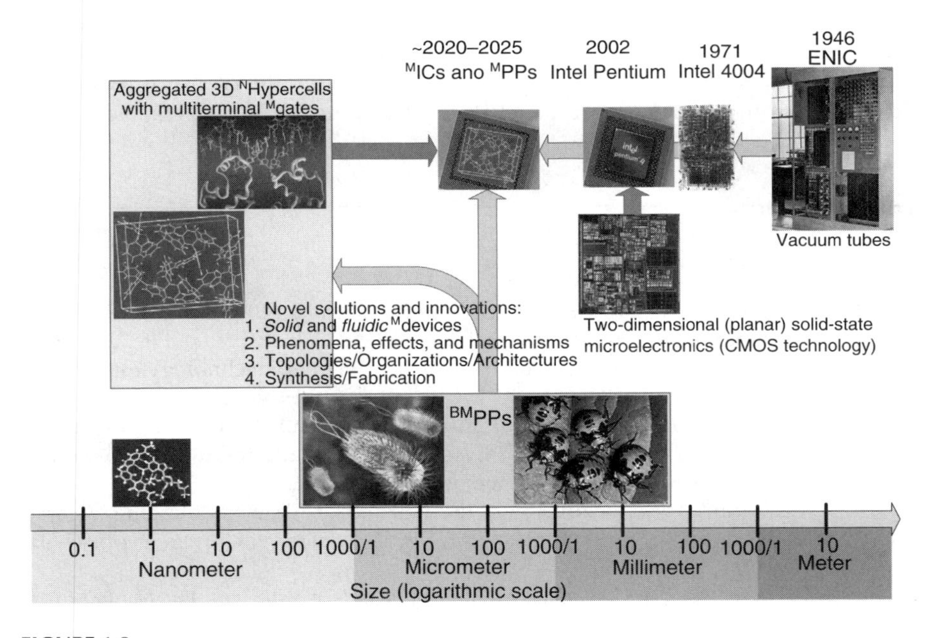

FIGURE 1.8
(See color insert following page 146.) Envisioned roadmap: Toward super-high-performance ^{M}ICs and ^{M}PPs.

- Robust adaptive computing and reconfigurability
- Enhanced functionality
- High reliability and redundancy
- Superior fault tolerance

- Defect-tolerant computing
- High-radix (multiple-valued) capabilities

In addition to facilitating and developing basic, applied, and experimental foundations, our goal is to accelerate and demonstrate the technological and commercial feasibilities of molecular electronics. This is accomplished by pursuing innovative high-payoff research and technological developments in molecular hardware. Molecular electronics ultimately depends on basic theory, pervasive understanding, and implementation of novel concepts, as well as development of *bottom-up* technologies. A roadmap towards super-high-performance ^{M}ICs and ^{M}PPs is documented in Figure 1.8. The proposed ^{M}ICs and ^{M}PPs have an analogy to aggregated brain neurons, which perform information processing, memory storage, and other related tasks. A vertebrate brain is of the most interest. However, as was emphasized, not only vertebrates but even a single-cell bacteria exhibit basic information processing and memory features. Distinct ^{BM}PPs exist in a great variety. Enormous progress has been achieved from the vacuum-tube-based Electronic Numerical Integrator/Computer (1946; 30 t and 150 kW) to Intel's Pentium processors. For breakthrough ^{M}ICs and ^{M}PPs, novel solutions are envisioned and proposed. It may be expected that, under the adequate commitments and resources, commercial ^{M}ICs and ^{M}PPs may emerge by 2020–2025.

References

1. *International Technology Roadmap for Semiconductors*, 2005 Edition, Semiconductor Industry Association, Austin, TX, USA, 2006. http://public.itrs.net/
2. J. E. Brewer, V. V. Zhirnov and J. A. Hutchby, Memory technology for the post CMOS era, *IEEE Circ. Dev. Mag.*, vol. 21, no. 2, pp. 13–20, 2005.
3. D. K. Ferry, R. Akis, A. Cummings, M. J. Gilbert and S. M. Ramey, Semiconductor device scaling: Physics, transport, and the role of nanowires, *Proc. IEEE Conf. Nanotechnol.*, Cincinnati, OH, 2006.
4. V. V. Zhirnov, J. A. Hutchby, G. I. Bourianoff and J. E. Brewer, Emerging research memory and logic technologies, *IEEE Circ. Dev. Mag.*, vol. 21, no. 3, pp. 47–51, 2005.
5. S. E. Lyshevski, *NEMS and MEMS: Fundamentals of Nano- and Microengineering*, CRC Press, Boca Raton, FL, 2005.
6. S. E. Lyshevski, Three-dimensional molecular electronics and integrated circuits for signal and information processing platforms. In *Handbook on Nano and Molecular Electronics*, Ed. S. E. Lyshevski, CRC Press, Boca Raton, FL, pp. 6-1–6-104, 2007.
7. J. C. Ellenbogen and J. C. Love, Architectures for molecular electronic computers: Logic structures and an adder designed from molecular electronic diodes, *Proc. IEEE*, vol. 88, no. 3, pp. 386–426, 2000.
8. J. R. Heath and M. A. Ratner, Molecular electronics, *Physics Today*, no. 1, pp. 43–49, 2003.

9. J. Chen, T. Lee, J. Su, W. Wang, M. A. Reed, A. M. Rawlett, M. Kozaki, et al., Molecular electronic devices. In *Handbook Molecular Nanoelectronics,* Eds. M. A. Reed and L. Lee, American Science Publishers, Stevenson Ranch, CA, 2003.
10. J. M. Tour and D. K. James, Molecular electronic computing architectures. In *Handbook of Nanoscience, Engineering and Technology*, Eds. W. A. Goddard and D. W. Brenner, S. E. Lyshevski and G. J. Iafrate, CRC Press, Boca Raton, FL, pp. 4.1–4.28, 2003.
11. W. Wang, T. Lee, I. Kretzschmar and M. A. Reed, Inelastic electron tunneling spectroscopy of an alkanedithiol self-assembled monolayer, *Nano Lett.*, vol. 4, no. 4, pp. 643–646, 2004.
12. A. Aviram and M. A. Ratner, Molecular rectifiers, *Chem. Phys. Lett.*, vol. 29, pp. 277–283, 1974.
13. H. C. Berg, The rotary motor of bacterial flagella, *J. Annu. Rev. Biochem.*, vol. 72, pp. 19–54, 2003.
14. W. D. McElroy and B. L. Strehler, Bioluminescence, *Bacteriol. Rev.*, vol. 18, no. 3, pp. 177–194, 1954.
15. S. M. Sze, *Physics of Semiconductor Devices*, John Wiley & Sons, NJ, 1969.
16. S. M. Sze, *Physics of Semiconductor Devices*, John Wiley & Sons, NJ, 1981.
17. S. M. Sze and K. K. Ng, *Physics of Semiconductor Devices,* John Wiley & Sons, NJ, 2007.

2

Molecular Electronics: Device-Level and System-Level Considerations

2.1 Performance Estimates

This chapter further introduces the reader to solid and fluidic molecular electronics. Molecular electronics encompasses novel 3D-topology ^{M}devices, new organization, innovative architecture, and *bottom-up* fabrication. The *reachable* volumetric dimensionality of solid ^{ME}device is in the order of $1 \times 1 \times 1$ nm. Due to synthesis, interconnect, and other constraints, the *achievable equivalent* multiterminal device cell volumetric dimensionality is expected to be $\sim 10 \times 10 \times 10$ nm. These multiterminal ^{ME}devices can be synthesized and aggregated as $^{\aleph}$hypercells forming functional ^{M}ICs. New device physics, innovative organization, novel architecture, enabling capabilities, and functionality, but not the dimensionality, are the key features of molecular and nanoelectronics. Solid and fluidic ^{M}devices, as compared to semiconductor devices, are based on new device physics, exhibit exclusive phenomena, provide enabling capabilities, and possess unique functionality, which should be utilized. From the system-level consideration, ^{M}ICs can be designed within novel organization and enabling architecture that guarantee superior performance. The performance estimates are studied in this section.

The designer examines the device, system performance, and capabilities using distinct performance measures, estimates, indexes, and metrics. The combinational and memory ^{M}ICs can be designed as aggregated $^{\aleph}$hypercells comprised of ^{M}gates and molecular memory cells [1]. At the device level, one examines functionality, studies characteristics, and estimates performance of 3D-topology ^{M}devices. The experimental results indicate that transitions in biomolecules and proteins are performed within 1×10^{-6} to 1×10^{-12} sec and require $\sim 1 \times 10^{-19}$ to 1×10^{-18} J of energy. How to find the transition time, derive energy estimates, and evaluate fluidic and *synthetic* ^{M}devices?

To analyze protein energetics, examine the switching energy in microelectronic devices, estimate solid $^{\rm ME}$devices energetics, and so forth, distinct concepts have been applied. Power is defined as the rate at which energy is transformed, that is, $P = E/t$. For solid-state microelectronic devices, the logic signal energy is expected to be reduced to $\sim 1 \times 10^{-16}$ J, and the energy dissipated is

$$E = Pt = IVt = I^2Rt = Q^2R/t,$$

where P is the power dissipation; I and V are the current and voltage along the discharge path; R and Q are the resistance and charge.

The dynamic power dissipation (consumption) in CMOS circuits can be straightforwardly analyzed. Taking note of the equivalent power dissipation capacitance $C_{\rm pd}$ (expected to be reduced to $\sim 1 \times 10^{-12}$ F) and the transition frequency f, one uses the following formula to estimate the device power dissipation due to output transitions

$$P_{\rm T} = C_{\rm pd}V^2f.$$

The energy for one transition can be found using current as a function of transition time, which is found from the equivalent RC models of solid-state transistors. The simplest estimate for the transition time is $-RC\ln(V_{\rm out}/V_{\rm dd})$. During transition, the voltage across the load capacitance $C_{\rm L}$ changes by $\pm V$. The total energy used for a single transition is charge Q times the average voltage change, which is $V/2$. Thus, the total energy per transition is $C_{\rm L}V^2/2$. If there are $2f$ transitions per second, one has

$$P_{\rm L} = C_{\rm L}V^2f.$$

Therefore, the dynamic power dissipation $P_{\rm D}$ is found to be

$$P_{\rm D} = P_{\rm T} + P_{\rm L} = C_{\rm pd}V^2f + C_{\rm L}V^2f = (C_{\rm pd} + C_{\rm L})V^2f.$$

In general, for $^{\rm M}$devices, this analysis cannot be applied.

The term $k_{\rm B}T$ has been used to solve distinct problems. Here, $k_{\rm B}$ is the Boltzmann constant, $k_{\rm B} = 1.3806 \times 10^{-23}$ J/K $= 8.6174 \times 10^{-5}$ eV/K; T is the absolute temperature. For example, the expression $\gamma k_{\rm B}T$ $(\gamma > 0)$ has been used to find the energy (see Section 2.1.1), and $k_{\rm B}T\ln(2)$ was applied with the attempt to assess the lowest energy bound for a binary switching (see Example 2.2). The applicability of distinct equations must be thoroughly examined applying sound concepts. Statistical mechanics and entropy analysis coherently utilize the term $k_{\rm B}T$ within a specific content as reported in the following section, while for other applications and problems, the use of $k_{\rm B}T$ may be impractical.

2.1.1 Entropy and Its Application

For an ideal gas, the kinetic, molecular Newtonian model provides the average translational kinetic energy of a gas molecule:

$$\tfrac{1}{2}m(v^2)_{\text{av}} = \tfrac{3}{2}\text{k}_\text{B}T.$$

One concludes that the average translational kinetic energy per gas molecule depends only on temperature. The most notable equation of statistical thermodynamics is the Boltzmann formula for entropy as a function only the system state, that is,

$$S = k_\text{B} \ln w,$$

where w is the number of possible arrangements of atoms or molecules in the system.

Unlike energy, entropy is a quantitative measure of the system disorder in any specific state and S is not related to each individual atom or particle. At any temperature above absolute zero, the atoms acquire energy, more arrangements become possible, and, because $w > 1$, one has $S > 0$. Entropy and energy are very different quantities. When the interaction between the system and environment involves only reversible processes, the total entropy is constant, and $\Delta S = 0$. When there is any irreversible process, the total entropy increases, and $\Delta S > 0$.

One may derive the entropy *difference* between two distinct states in a system that undergoes a thermodynamic process that takes the system from an initial *macroscopic* state 1, with w_1 possible *microscopic* states, to a final *macroscopic* state 2, with w_2 associated *microscopic* states. The change in entropy is

$$\Delta S = S_2 - S_1 = k_\text{B} \ln w_2 - k_\text{B} \ln w_1 = k_\text{B} \ln(w_2/w_1).$$

Thus, the entropy *difference* between two *macroscopic* states depends on the ratio of the number of possible *microscopic* states. The entropy change for any reversible isothermal process is given using an infinitesimal quantity of heat ΔQ. For initial and final states 1 and 2, one has $\Delta S = \int_1^2 \text{d}Q/T$.

Example 2.1

For a silicon atom, the covalent, atomic, and van der Waals radii are 117, 117, and 200 pm, respectively. The Si—Si and Si—O covalent bonds are 232 and 151 pm, respectively. One can examine the thermodynamics using the enthalpy, Gibbs' function, entropy, and heat capacity of silicon in its solid and gas states. The atomic weight of a silicon atom is 28.0855 amu, where amu denotes the atomic mass unit (1 amu $= 1.66054 \times 10^{-27}$ kg). Hence, the mass of a single Si atom is 28.0855 amu $\times 1.66054 \times 10^{-27}$ kg/amu $= 4.6637 \times 10^{-26}$ kg. Therefore, the number of silicon atoms in, for example, 1×10^{-24} kg of silicon is $1 \times 10^{-24}/4.6637 \times 10^{-26} = 21.44$.

To heat 1 ykg (1×10^{-24} kg) of silicon from 0°C to 100°C, using the constant specific heat capacity $c = 702$ J/kg · K over the specified temperature range, the change of entropy is

$$\Delta S = S_2 - S_1 = \int_1^2 \frac{dQ}{T} = \int_{T_1}^{T_2} mc\frac{dT}{T} = mc \ln \frac{T_2}{T_1}$$

$$= 1 \times 10^{-24} \text{ kg} \times 702 \frac{\text{J}}{\text{kg} \cdot \text{K}} \times \ln \frac{373.15 \text{ K}}{273.15 \text{ K}} = 2.19 \times 10^{-22} \text{ J/K}.$$

From $\Delta S = k_B \ln(w_2/w_1)$, one finds the ratio between *microscopic* states w_2/w_1. For the problem under the consideration, $w_2/w_1 = 7.7078 \times 10^6$ for 21 silicon atoms.

If $w_2/w_1 = 1$, the total entropy is constant, and $\Delta S = 0$.

The energy that must be supplied to heat 1×10^{-24} kg of silicon for $\Delta T = 100$°C is

$$Q = mc\Delta T = 7.02 \times 10^{-20} \text{ J}.$$

To heat 1 g of silicon from 0°C to 100°C, one finds

$$\Delta S = S_2 - S_1 = mc \ln \frac{T_2}{T_1} = 0.219 \text{ J/K} \quad \text{and} \quad Q = mc\Delta T = 70.2 \text{ J}.$$

Taking note of equation $\Delta S = k_B \ln(w_2/w_1)$, it is impossible to derive the numerical value for w_2/w_1.

Consider 2 silicon atoms to be heated from 0°C to 100°C. For $m = 9.3274 \times 10^{-26}$ kg, we have

$$\Delta S = S_2 - S_1 = mc \ln \frac{T_2}{T_1} = 2.04 \times 10^{-23} \text{ J/K}.$$

One obtains an obscure result $w_2/w_1 = 4.39$.

The entropy and *macroscopic*/*microscopic* states analysis is performed for an ideal gas assuming the accuracy of the kinetic, molecular Newtonian model. It should be emphasized again that unlike energy, entropy is a quantitative measure of the system disorder in any specific state, and S is not related to each individual atom or particle. To examine the *microscopic* particle and molecule energetics, quantum physics must be applied.

Example 2.2

Using the results reported, one may carry out similar analysis for other atomic complexes. For example, while carbon has not been widely used in microelectronics, organic molecular electronics is carbon-centered. For a carbon atom, the covalent, atomic, and van der Waals radii are 77, 77, and 185 pm, respectively. Carbon can be in the solid (graphite or diamond) and gaseous states. The atomic weight of a carbon atom is 12.0107 amu, and the mass of a single carbon atom is 12.0107 amu $\times$ 1.66054 $\times$ 10^{-27} kg/amu = 1.9944 $\times$ 10^{-26} kg.

Letting $w = 2$, the entropy is found to be

$$S = k_B \ln 2 = 9.57 \times 10^{-24}\ \text{J/K} = 5.97 \times 10^{-5}\ \text{eV/K}.$$

Having derived S, one cannot conclude that the minimal energy required to ensure the transition (*switching*) between two *microscopic* states or to erase a bit of information (energy dissipation) is $k_B T \ln 2$, which for T = 300 K gives $k_B T \ln 2 = 2.87 \times 10^{-21}$ J = 0.0179 eV. In fact, under this reasoning, one assumes the validity of the *averaging* kinetic, molecular Newtonian model and applies the assumptions of distribution statistics (see Section 2.1.2), at the same time allowing only two distinct *microscopic* system states. The energy estimates should be performed utilizing the quantum mechanics.

2.1.2 Distribution Statistics

Statistical analysis is applicable only to systems with a large number of particles and energy states. The fundamental assumption of statistical mechanics is that in thermal equilibrium every distinct state with the same total energy is equally probable. Random thermal motions constantly transfer energy from one particle to another and from one form of energy to another (kinetic, rotational, vibrational, etc.) obeying the principle of conservation of energy. The absolute temperature T has been used as a measure of the total energy of a system in thermal equilibrium.

In semiconductor devices, enormous number of particles (electrons) are considered using the electrochemical potential $\mu(T)$. The Fermi–Dirac distribution function

$$f(E) = \frac{1}{1 + e^{(E-\mu(T))/k_B T}}$$

gives the average (probable) number of electrons of a system (device) in equilibrium at temperature T in a quantum state of energy E. The electrochemical potential at absolute zero is the Fermi energy E_F, and $\mu(0) = E_F$. The occupation probability that a particle would have the specific energy is not related to quantum indeterminacy.

Electrons in solids obey the Fermi–Dirac statistics. The distribution of electrons, leptons, and baryons (*identical fermions*) over a range of allowable energy levels at thermal equilibrium is

$$f(E) = \frac{1}{1 + e^{(E-E_F)/k_B T}},$$

where T is the equilibrium temperature of the system.

Hence, the Fermi–Dirac distribution function $f(E)$ gives the probability that an allowable energy state at energy E will be occupied by an electron at temperature T.

For *distinguishable* particles, one applies the Maxwell–Boltzmann statistics with a distribution function

$$f(E) = e^{(E-E_F)/k_B T}.$$

The Bose–Einstein statistics is applied to *identical bosons* (photons, mesons, etc.). The Bose–Einstein distribution function is

$$f(E) = \frac{1}{e^{(E-E_F)/k_B T} - 1}.$$

The distribution statistics is applicable to electronic devices, which consist of a great number of particles and interactions can be simplified by deducing the system behavior from statistical considerations. Depending on the device physics, one must coherently apply the appropriate baseline theories and concepts.

Example 2.3

For $T = 100$ K and $T = 300$ K, letting $E_F = 5$ eV, the Fermi–Dirac distribution functions are shown in Figure 2.1a. Figure 2.1b shows the Maxwell–Boltzmann distribution functions $f(E)$.

2.1.3 Energy Levels

In ^{M}devices, one can calculate the energy required to excite the electron, and the allowed energy levels are quantized. In contrast, solids are characterized by energy band structures that define electric characteristics.

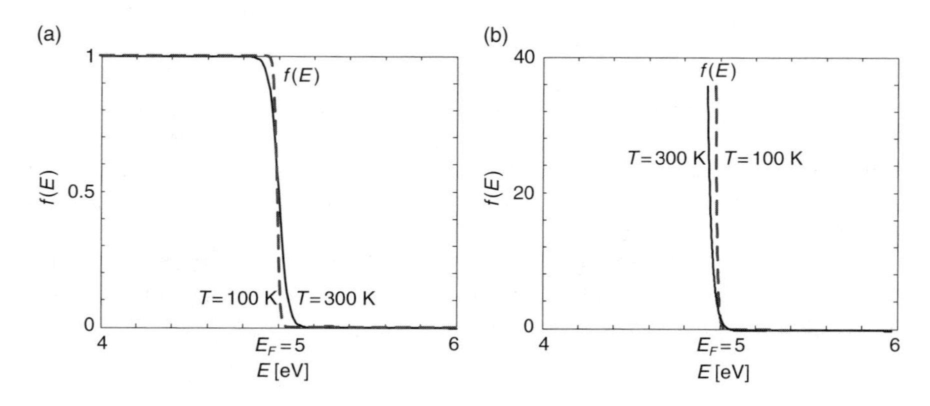

FIGURE 2.1
The distribution functions for $T = 100$ K and $T = 300$ K when $E_F = 5$ eV: (a) Fermi–Dirac distribution functions and; (b) Maxwell–Boltzmann distribution functions.

In semiconductors, the relatively small band gaps allow excitation of electrons from the valence band to conduction band by thermal or optical energy. The application of quantum mechanics allows one to derive the expression for the quantized energy. In Chapter 6 it is derived that for a hydrogen atom

$$E_n = -\frac{m_e e^4}{32\pi^2 \varepsilon_0^2 \hbar^2 n^2},$$

where $\hbar$ is the modified Planck constant, $\hbar = h/2\pi = 1.055 \times 10^{-34}$ J sec = 6.582×10^{-16} eV sec.

The energy levels depend on the quantum number n. As n increases, the total energy of the quantum state becomes less negative, and $E_n \to 0$ if $n \to \infty$. The state of lowest total energy is the most stable state for the electron. The normal state of the electron for a hydrogen (one-electron atom) is at $n = 1$.

Thus, for the hydrogen atom, in the absence of a magnetic field **B**, the energy $E_n = -(m_e e^4)/(32\pi^2 \varepsilon_0^2 \hbar^2 n^2)$ depends only on the principle quantum number n. The conversion 1 eV = $1.602176462 \times 10^{-19}$ J is commonly used, and $E_{n=1} = -2.17 \times 10^{-18}$ J = -13.6 eV. For $n = 2$, $n = 3$, and $n = 4$, we have $E_{n=2} = -5.45 \times 10^{-19}$ J, $E_{n=3} = -2.42 \times 10^{-19}$ J, and $E_{n=4} = -1.36 \times 10^{-19}$ J. When the electron and nucleus are separated by an infinite distance ($n \to \infty$), one has $E_{n\to\infty} \to 0$.

The energy difference between the quantum states n_1 and n_2 is

$$\Delta E = E_{n1} - E_{n2} \quad \text{and} \quad \Delta E = E_{n1} - E_{n2} = \frac{m_e e^4}{32\pi^2 \varepsilon_0^2 \hbar^2}\left(\frac{1}{n_2^2} - \frac{1}{n_1^2}\right),$$

where $(m_e e^4)/(32\pi^2 \varepsilon_0^2 \hbar^2) = 2.17 \times 10^{-18}$ J = 13.6 eV.

The excitation energy of an exited state n is the energy above the ground state, for example, for the hydrogen atom, one has $(E_n - E_{n=1})$. The first exited state ($n = 2$) has the excitation energy $E_{n=2} - E_{n=1} = -3.4 + 13.6 = 10.2$ eV. In atoms, orbits characterized by quantum numbers.

De Broglie's conjecture relates the angular frequency v and energy E. In particular,

$$v = E/h,$$

where h is the Planck constant, $h = 6.626 \times 10^{-34}$ J sec = 4.136×10^{-15} eV sec.

The frequency of a photon electromagnetic radiation is found as $v = \Delta E/h$.

Remark: ***Energy Difference and Energy Uncertainty***: The energy difference between the quantum states ΔE is not the energy uncertainty in the measurement of E, which is commonly denoted in the literature as ΔE. When considering the Heisenberg uncertainty principle, to ensure consistency, we use the notation $\Delta\hat{E}$. Section 6.2 gives the energy–time uncertainty principle as $\sigma_E \sigma_t \geq \frac{1}{2}\hbar$ or $\Delta\hat{E}\Delta t \geq \frac{1}{2}\hbar$, where σ_E and σ_t are the standard deviations, and notations $\Delta\hat{E}$ and Δt are used to define the standard deviations as uncertainties, $\Delta\hat{E} = \sqrt{\langle \hat{E}^2 \rangle - \langle \hat{E} \rangle^2}$. ■

For many-electron atoms, an atom in its normal (electrically neutral) state has Z electrons and Z protons. Here, Z is the atomic number. For boron, carbon, and nitrogen, $Z = 5, 6$, and 7, respectively. The total electric charge of atoms is zero because the neutron has no charge while the proton and electron charges have the same magnitude but opposite sign. For the hydrogen atom, denoting the distance that separates the electron and proton by r, the Coulomb potential is

$$\Pi(r) = -e^2/(4\pi\varepsilon_0 r).$$

The radial attractive Coulomb potential sensed by a single electron due to the nucleus having a charge Ze is

$$\Pi(r) = -Z(r)e^2/(4\pi\varepsilon_0 r),$$

where $Z(r) \to Z$ as $r \to 0$ and $Z(r) \to 1$ as $r \to \infty$.

By evaluating the average value for the radius of the shell, the effective nuclear charge Z_{eff} is found. The common approximation to calculate the total energy of an electron in the outermost populated shell is

$$E_n = -\frac{m_e Z_{\text{eff}}^2 e^4}{32\pi^2\varepsilon_0^2\hbar^2 n^2} \quad \text{and} \quad E_n = -2.17 \times 10^{-18}\frac{Z_{\text{eff}}^2}{n^2}\ \text{J}.$$

The effective nuclear charge Z_{eff} is derived using the electron configuration. For boron, carbon, nitrogen, silicon, and phosphorus, three commonly used Slater, Clementi, and Froese–Fischer Z_{eff} are

2.6, 2.42, and 2.27 for B
3.25, 3.14, and 2.87 for C
3.9, 3.83, and 3.46 for N
4.13, 4.29, and 4.48 for Si
4.8, 4.89, and 5.28 for P

Taking note of the electron configurations for these atoms ($Z/n \approx 1$), one concludes that ΔE is from $\sim 1 \times 10^{-19}$ to 1×10^{-18} J. If one supplies the energy greater than E_n to the electron, the energy excess will appear as kinetic energy of the free electron. The transition energy should be adequate to excite electrons. For different atoms and molecules with different exited states, as prospective solid ^{ME}devices, the transition (switching) energy is estimated to be $\sim 1 \times 10^{-19}$ to 1×10^{-18} J. This energy estimate is in agreement with biomolecular devices, for which we find $\sim 1 \times 10^{-19}$ to 1×10^{-18} J (see Section 2.1.5).

The quantization of the orbital angular momentum of the electron leads to a quantization of the electron total energy. Space quantization permits only quantized values of the angular momentum component in a specific direction. The magnitude L_μ of the angular momentum of an electron in its orbital motion around the center of an atom, as well as the z component L_z, are

$$L_\mu = \sqrt{l(l+1)}\hbar \quad \text{and} \quad L_z = m_l \hbar,$$

where l is the orbital quantum number; m_l is the magnetic quantum number, which is restricted to integer values $-l, -l+1, \ldots, l-1, l$, that is, $|m_l| \leq l$.

If a magnetic field is applied, the energy of the atom will depend on the alignment of its magnetic moment with the external magnetic field. In the presence of a magnetic field $\mathbf{B}$, the energy levels of the hydrogen atom are

$$E_n = -\frac{m_e e^4}{32\pi^2 \varepsilon_0^2 \hbar^2 n^2} - \boldsymbol{\mu}_L \cdot \mathbf{B},$$

where $\boldsymbol{\mu}_L$ is the orbital magnetic dipole moment, $\boldsymbol{\mu}_L = -(e/2m_e)\mathbf{L}$, and $\mathbf{L} = \mathbf{r} \times \mathbf{p}$.

Let $\mathbf{B} = B_z \mathbf{z}$. One finds

$$\begin{aligned} E_n &= -\frac{m_e e^4}{32\pi^2 \varepsilon_0^2 \hbar^2 n^2} + \frac{e}{2m_e}\mathbf{L} \cdot \mathbf{B} = -\frac{m_e e^4}{32\pi^2 \varepsilon_0^2 \hbar^2 n^2} + \frac{e}{2m_e} B_z L_z \\ &= -\frac{m_e e^4}{32\pi^2 \varepsilon_0^2 \hbar^2 n^2} + \frac{e}{2m_e} B_z m_l \hbar. \end{aligned}$$

If the electron is in an $l = 1$ orbit, the orbital magnetic dipole moment is $\boldsymbol{\mu}_L = e\hbar/2m_e = 9.3 \times 10^{-24}$ J/T $= 5.8 \times 10^{-5}$ eV/T. Hence, if the magnetic field is changed by 1 T, an atomic energy level changes by $\sim 10^{-4}$ eV. One concludes that the *switching* energy required to ensure the transitions or interactions between distinct *microscopic* states is straightforwardly derived using the wave function and allowed discrete energies.

Example 2.4

Consider a 1,3-butadiene molecule

```
         H        H
          \      /
     H     C == C
      \   /      \
       C == C      H
      /      \
     H        H
```

The four delocalized π-electrons are assumed to move freely over the four-carbon-atom framework.

Neglecting the 3D configuration, one may perform one-dimensional (1D) analysis. Solving the Schrödinger equation for a particle in the box $-(\hbar^2/2m_e)\nabla^2\Psi(x) + \Pi(x)\Psi(x) = E\Psi(x)$ with an infinite square-well potential

$$\Pi(x) = \begin{cases} 0 & \text{for } 0 \le x \le L, \\ \infty & \text{otherwise,} \end{cases}$$

the wave function $\Psi_n(x)$ and allowed discrete energies are found, as in Chapter 6. In particular,

$$\Psi_n(x) = \sqrt{\frac{2}{L}} \sin\left(\frac{n\pi}{L}x\right) \quad \text{and} \quad E_n = \frac{\hbar^2\pi^2}{2m_e L^2} n^2.$$

The state of the lowest energy is called the *ground state*. The $C_1{=}C_2$, $C_2{-}C_3$, and $C_3{=}C_4$ bond lengths are 0.1467, 0.1349, and 0.1467 nm, respectively. The electron wave function extends beyond the terminal carbons. We add half a bond length at each end. Hence, $L = 0.575$ nm.

The π-electron density is concentrated between carbon atoms C_1 and C_2, as well as C_3 and C_4, because the predominant structure of butadiene has double bonds between these two pairs $C_1{=}C_2$ and $C_3{=}C_4$. Each double bond consists of a π-bond in addition to the underlying σ-bond. One must also consider the residual π-electron density between C_2 and C_3. Thus, butadiene should be described as a resonance hybrid with two contributing structures $CH_2{=}CH{-}CH{=}CH_2$ (dominant structure) and $^\circ CH_2{=}CH{-}CH{=}CH_2^\circ$ (secondary structure).

The lowest unoccupied molecular orbital (LUMO) in butadiene corresponds to the $n = 3$ particle-in-a-box state. Neglecting electron–electron interaction, the longest-wavelength (lowest-energy) electron transition occur from $n = 2$, which is the highest occupied molecular orbital (HOMO). This is visualized as

n = 3 —— LUMO —●—
n = 2 ●● HOMO ●——
n = 1 ●● ——●●

The HOMO $\to$ LUMO transition corresponds to $n \to (n+1)$. The energy difference between HOMO and LUMO is

$$\Delta E = E_3 - E_2 = \frac{\hbar^2\pi^2}{2m_e L^2}(3^2 - 2^2) = \frac{h^2}{8m_e L^2}(3^2 - 2^2) = 9.11 \times 10^{-19}\ \text{J}.$$

From $\Delta E = hc/\lambda$, one finds the Compton wavelength to be $\lambda = 218$ nm. Performing the experiments, it is found that the maximum of the first electronic absorption band occurs at 210 nm. Hence, the use of quantum theory provides one with accurate results.

To enhance accuracy, consider a rectangular ($L_x \times L_y \times L_z$) 3D infinite-well box with

$$\Pi(x, y, z) = \begin{cases} 0 & \text{for } 0 \le x \le L_x,\ 0 \le y \le L_y,\ 0 \le z \le L_z, \\ \infty & \text{otherwise.} \end{cases}$$

One solves a time-independent Schrödinger equation:

$$-\frac{\hbar^2}{2m_e} \nabla^2 \Psi(x, y, z) + \Pi(x, y, z)\Psi(x, y, z) = E\Psi(x, y, z).$$

We apply the separation of variables concept expressing the wave function as

$$\Psi(x, y, z) = X(x)Y(y)Z(z)$$

and let $E = E_x + E_y + E_z$.

One has

$$-\frac{\hbar^2}{2m_e}\frac{d^2X}{dx^2} = E_x X, \quad -\frac{\hbar^2}{2m_e}\frac{d^2Y}{dy^2} = E_y Y \quad \text{and} \quad -\frac{\hbar^2}{2m_e}\frac{d^2Z}{dz^2} = E_z Z.$$

The general solutions are found to be

$$X(x) = A_x \sin k_x x + B_x \cos k_x x, \quad Y(y) = A_y \sin k_y y + B_y \cos k_y y \quad \text{and}$$
$$Z(z) = A_z \sin k_z z + B_z \cos k_z z.$$

Here,

$$k_x^2 = \frac{2m_e}{\hbar^2} E_x, \quad k_y^2 = \frac{2m_e}{\hbar^2} E_y \quad \text{and} \quad k_z^2 = \frac{2m_e}{\hbar^2} E_z.$$

Taking note of the boundary conditions, one finds

$$B_x = 0,\ B_y = 0,\ B_z = 0,\ k_x L_x = n_x \pi,\ k_y L_y = n_y \pi \quad \text{and} \quad k_z L_z = n_z \pi.$$

Normalizing the wave function, we obtain 3D eigenfunctions as

$$\Psi_{n_x,n_y,n_z}(x, y, z) = \sqrt{\frac{8}{L_x L_y L_z}} \sin\left(\frac{n_x \pi}{L_x} x\right) \sin\left(\frac{n_y \pi}{L_y} y\right) \sin\left(\frac{n_z \pi}{L_z} z\right), \quad n_x, n_y, n_z = 1, 2, 3, \ldots.$$

The allowed energies are found to be

$$E_{n_x,n_y,n_z} = \frac{h^2}{8m_e}\left(\frac{n_x^2}{L_x^2} + \frac{n_y^2}{L_y^2} + \frac{n_z^2}{L_z^2}\right) = \frac{\hbar^2 \pi^2}{2m_e}\left(\frac{n_x^2}{L_x^2} + \frac{n_y^2}{L_y^2} + \frac{n_z^2}{L_z^2}\right) = \frac{\hbar^2 k^2}{2m_e},$$

where k is the magnitude of the wave vector $\mathbf{k}, \mathbf{k} = (k_x, k_y, k_z)$, $k_x = n_x\pi/L_x$, $k_y = n_y\pi/L_y$, and $k_z = n_z\pi/L_z$.

Analytic solutions exist for ellipsoidal, spherical, and other 3D wells for infinite and some finite potentials. Numerical solutions can be found for complex potential wells and barriers, as reported in Chapter 6.

Taking note of the wave vector, we conclude that each state occupies a volume $\pi^3/L_xL_yL_z = \pi^3/V$ of a k-space.

Suppose a system consists of Natoms, and each atom contributes M free electrons. The electrons are identical *fermions* that satisfy the Pauli exclusion principle. Thus, only two electrons can occupy any given state. Furthermore, electrons fill one octant of a sphere in k-space, whose radius is $k_R = (3NM\pi^2/V)^{1/3} = (3\rho\pi^2)^{1/3}$. The expression for k_R is derived by making use of $(1/8)(4/3)\pi k_R^3 = (1/2)NM\pi^3/V$. Here, ρ is the *free electron density* ($\rho = NM/V$), that is, ρ is the number of free electrons per unit volume. The boundary separation of occupied and unoccupied states in k-space is called the Fermi surface, and the Fermi energy for a free electron gas is $E_F = \hbar^2/2m_e(3\rho\pi^2)^{2/3}$. The total energy of a free electron gas is

$$E_t = \frac{\hbar^2}{2m_e}\frac{V}{\pi^2}\int_0^{k_R} k^4\,dk = \frac{\hbar^2 V}{10m_e\pi^2}k_R^5 = \frac{\hbar^2(3\pi^2 NM)^{5/3}}{10m_e\pi^2}V^{-2/3}.$$

The expression for E_t is found by taking note of the number of electron states in the shell $2(\frac{1}{2}\pi k^2\,dk)/(\pi^3/V) = (V/\pi^2)k^2\,dk$ and the energy of the shell $dE = (\hbar^2k^2/2m_e)(V/\pi^2)k^2\,dk$ (each state carries the energy $\hbar^2k^2/2m_e$).

2.1.4 Device Transition Speed

The transition (switching) speed of ^{M}devices largely depends on the device physics, phenomena utilized, and other factors. One examines dynamic evolutions and transitions by applying molecular dynamics, Schrödinger equation, time-dependent perturbation theory, and other concepts. The analysis of state transitions and interactions allows one to coherently study the controlled device behavior, evolution, and dynamics. The simplified analysis is also applied to obtain estimates. Considering electron transport, one may assess the device features using the number of electrons. For example, for 1 nA current, the number of electrons that cross the molecule per second is $1 \times 10^{-9}/1.6022 \times 10^{-19} = 6.24 \times 10^9$, which is related to the device state transitions.

The maximum carrier velocity places an upper limit on the frequency response of semiconductor and molecular devices. State transitions can be accomplished by a single photon or electron. Using Bohr's postulates, the average velocity of an optically exited electron is $v = (Ze^2/4\pi\varepsilon_0\hbar n)$. For all atoms $Z/n \approx 1$, one finds the orbital velocity of an optically exited electron to be $v = 2.2 \times 10^6$ m/sec. Hence, $v/c \approx 0.01$.

Considering an electron as a nonrelativistic particle, taking note of $E = mv^2/2$, we obtain the particle velocity as a function of energy as $v(E) = \sqrt{2E/m}$.

Example 2.5

Let $E = 0.1$ eV $= 0.16 \times 10^{-19}$ J. From $v(E) = \sqrt{2E/m}$, one finds $v = 1.88 \times 10^5$ m/sec. Assuming a 1 nm path length, the traversal (*transit*) time is $\tau = L/v = 5.33 \times 10^{-15}$ sec.

We conclude that ^{M}devices can operate at a high switching frequency. However, one may not conclude that the device switching frequency to be utilized is $f = 1/(2\pi\tau)$ because of device physics features (number of electrons, heating, interference, potential, energy, noise, etc.), system-level functionality, circuit specifications, and so forth.

Having estimated the $v(E)$ for ^{M}devices, the comparison to microelectronics devices is of interest. In silicon, at $T = 300$ K, the electron and hole velocities reach 1×10^5 m/sec at a very high electric field with the intensity 1×10^5 V/cm. The reported estimates indicate that particle velocity in ^{M}devices exceeds the carriers saturated drift velocity in semiconductors.

2.1.5 Photon Absorption and Transition Energetics

The reader recalls that covering bioluminescence in Section 1.5 it was reported that fish and firefly emit phonons of wavelength ~500 nm. The energy of a single photon is $E = hc/\lambda$, where λ is the wavelength. Hence, the photon *output* energy is 4×10^{-19} J.

Consider rhodopsin, which is a highly specialized protein-coupled receptor that detects photons in the rod photoreceptor cell. The first event in the monochrome vision process, after a photon (light) hits the rod cell, is the isomerization of the chromophore 11-*cis*-retinal to all-*trans*-retinal. When an atom or molecule absorbs a photon, its electron can move to the higher-energy orbital, and the atom or molecule makes a transition to a higher-energy state. In retinal, absorption of a photon promotes a π-electron to a higher-energy orbital, that is, there is a π–π^* excitation. This excitation breaks the π component of the double bond allowing free rotation about the bond between carbon-11 and carbon-12. This isomerization, which corresponds to switching, occurs in picoseconds.

The energy of a single photon is given by $E = hc/\lambda$. The maximum absorbance for rhodopsin is 498 nm. For this wavelength, one finds $E = 4 \times 10^{-19}$ J. This energy is sufficient to ensure transitions and functionality.

The photochemical reaction, which should be referenced as the electrochemomechanical transitions, changes the shape of retinal, causing a conformational change in the opsin protein, which consists of 348 amino acids covalently linked together to form a single chain. The sensitivity of the photoreceptor in the eye is one photon. Thus, the energy of a single photon, which is $E = 4 \times 10^{-19}$ J, ensures the functionality of a molecular complex of 348 amino acids (~5000 atoms). We derived the excitation energy (signal energy) sufficient to ensure electrochemomechanically induced state transitions and interactions leading to processing in ^{M}device. This provides a conclusive evidence that ~1×10^{-19} to 1×10^{-18} J of energy is required to guarantee state transitions for molecular aggregates in the *biomolecular processing hardware*.

2.1.6 System-Level Performance Estimates

Aggregated brain neurons perform information processing, perception, learning, robust reconfigurable networking, memory storage, and other functions. The number of neurons in the human brain is estimated to be ~100 billion, mice and rats have ~100 millions of neurons, while honeybees and ants have ~1 million neurons. Bats use echolocation sensors for navigation, obstacle avoidance, and hunting. By processing the sensory data, bats can detect 0.1% frequency shifts, and they distinguish echoes received ~100 μsec apart. To accomplish these tasks, as well as to perform shift compensation and transmitter/receiver isolation, real-time signal/data processing should be accomplished within at least microseconds. Flies accomplish a real-time precisely coordinated motion by means of remarkable actuation, adaptive control, and a visual system that maps the relative motion using retinal photodetector arrays. The information from the visual system, sensors, and actuators is transmitted and processed within nanoseconds requiring microwatts of power.

The biophysics of biomolecular information and signal/data processing are not fully understood. State transitions are accomplished by specific biophysical phenomena, effects, and mechanisms. The electrochemomechanically induced transitions and interactions in biomolecules (propagation of molecules and ions through the synaptic cleft and membrane channels, protein folding, binding/unbinding, etc.) may require microseconds. In contrast, photon- and electron-induced transitions can be performed within femtoseconds. The energy estimates were made obtaining the transition energy requirements ~1×10^{-19} to 1×10^{-18} J.

To process enormous amounts information and to perform related tasks with immense performance capability, which are far beyond foreseen capabilities of envisioned parallel vector processors (which perform signal/data processing), the human brain consumes only ~20 W. Only some of this power is required to accomplish information and signal/data processing. This contradicts some postulates of slow processing, immense delays, high energy and power requirements, low switching speed, and other hypotheses reported in [2–5]. The review of electrical excitability of neurons is reported in [6].

The human retina has ~125 million rod cells and ~6 million cone cells. An enormous amount of data, among other tasks, is processed by the visual system and brain in real time. Real-time 3D image processing, ordinarily accomplished even by primitive vertebrates and insects that consume less than 1 μW cannot be performed by envisioned processors with trillions of transistors, device switching speed of 1 THz, circuit speed of 10 GHz, device switching energy of 1×10^{-16} J, writing energy of 1×10^{-16} J/bit, read time of 10 nsec, and so forth.

Molecular devices can operate with the estimated transition energy of $\sim 1 \times 10^{-18}$ J, discrete energy levels (ensuring multiple-valued logics and memory), and femtosecond transition dynamics. These guarantee exceptional device transition (switching) speed, low losses, unique functionality, and other features ensuring superior overall performance. Furthermore, 3D-topology ^{M}devices give the ability to design super-high-performance processing and memory platforms within novel organizations and enabling architectures, ensuring unprecedented capabilities including massive parallelism, robustness, reconfigurability, and so forth.

Departing from a conventional neuroscience a neuron-as-a-device doctrine, the neuron as a biomolecular information processing/memory module (system) can be examined. The dimension of the brain neuron is ~10 μm, and their density is ~50,000 neurons/mm^3. The neuron has thousands of synaptic inputs and outputs. Chapter 3 describes *biomolecular processing hardware* and suggests that the reconfigurable networking is accomplished by means of a reconfigurable *axo-dendritic, dendro-axonic, axo-axonic,* and *dendro-dendritic* mapping. Each neuron consists of neuronal processing-and-memory primitives within a neuronal reconfigurable networked organization. The immense m-input and z-output vector capability ($m \approx 10,000$ and $z \approx 1,000$) per neuron is accomplished by ~10,000 input and ~1,000 output synapses, as well as various neurotransmitters with specific receptors (see Figures 3.3 and 3.5). These neurotransmitters may serve as: (1) *information* carriers, ensuring processing and memory storage at very high radix due to electrochemomechanically induced transitions and interactions in neuronal processing-and-memory primitives, and, (2) *routing* carriers, enhancing the reconfiguration capabilities. Under some assumptions, the processing capabilities of a single neuron can be estimated assigning the number of inputs, outputs, *equivalent* processing-and-memory primitives per neuron, radix, and other data. One can imagine the system's processing capabilities if 50,000 reconfigurable high-performance processing modules (neurons) with parallel capabilities housed in 1 mm^3 and consuming less that 20 μW.

Distinct performance measures, estimates, indexes, and metrics are used. For profoundly different paradigms (microelectronics versus molecular electronics), Figure 2.2 shows some baseline performance estimates, such as transition (switching) energy, delay time, dimension, and number of modules/gates. It was emphasized that the device physics and system organization/architecture are the dominating features rather than the dimensionality or number of devices. Owing to limited basic/applied/experimental results, as well as the attempts to use four performance estimates, as in Figure 2.2, some projected performance measures are expected to be refined. Furthermore, molecular electronics and ^{M}ICs can utilize diverse molecular primitives and

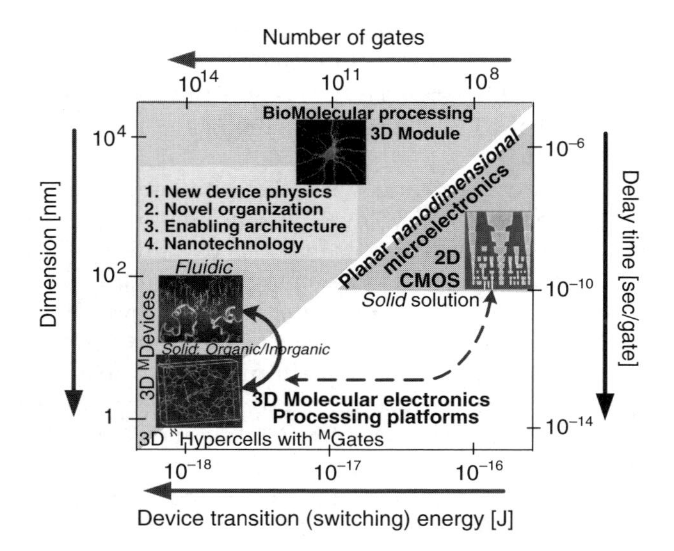

FIGURE 2.2
(See color insert following page 146.) Toward molecular electronics and processing/memory platforms. Revolutionary advancements: From two-dimensional (2D) microelectronics to 3D molecular electronics. Evolutionary developments: From ^{BM}PPs to solid and fluidic molecular electronics and processing.

^{M}devices that:

1. Operate based on different physics, such as electron transport, electrostatic transitions, photon emission, conformational changes, quantum interactions, and so forth
2. Exhibit and utilize distinct phenomena and effects
3. Possess different functionality and capabilities

Therefore, biomolecular, fluidic, and solid ^{M}devices and systems will exhibit distinct performance and capabilities. As demonstrated in Figure 2.2, advancements are envisioned towards 3D solid molecular electronics mimicking ^{BM}PPs, which can resemble a familiar solid-state microelectronics solution. It should be emphasized that solid ^{ME}devices and ^{M}ICs may utilize the so-called soft materials such as polymers and biomolecules. In Figure 2.2, a neuron is represented as a biomolecular information processing/memory module (system).

2.2 Topologies, Organizations, and Architectures

Molecular devices inherently possess 3D topology because they are made of atoms. The 3D-centered topology and organization of ^{M}devices and envisioned ^{M}PPs resemble the device topology and system organization of ^{BM}PPs. The 3D-topology solid and fluidic ^{M}devices are covered in detail in Chapters 3, 5, and 6. Diverse organizations can be implemented. For example, linear, star, ring, and hypercube organizations are possible, as shown in Figure 2.3.

More complex organizations can be designed. A 3D-meshed hypercube organization with hypercube-connected nodes, cyclic-connected subnodes, and radial subnode–module connection are illustrated in Figure 2.4.

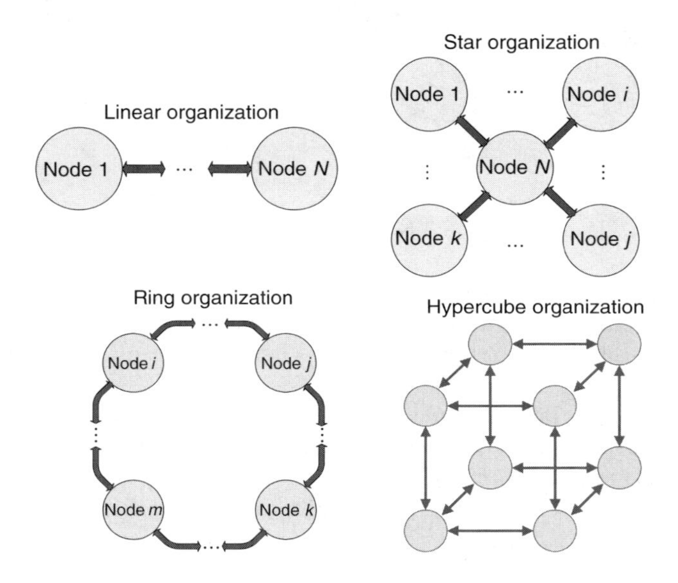

FIGURE 2.3
Linear, star, ring, and hypercube 3D-centered organizations.

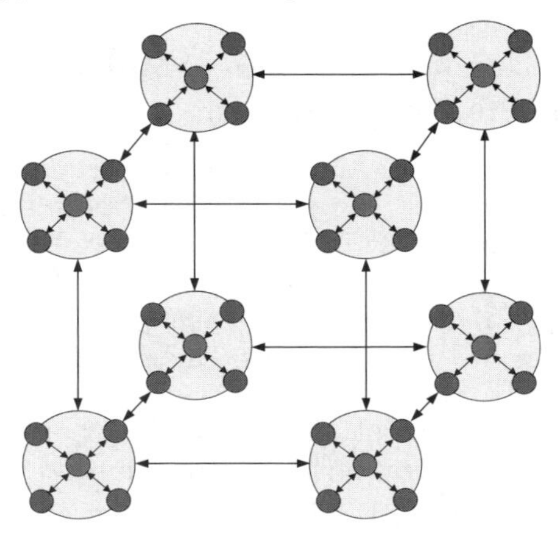

FIGURE 2.4
Three-dimensional hypercube organization.

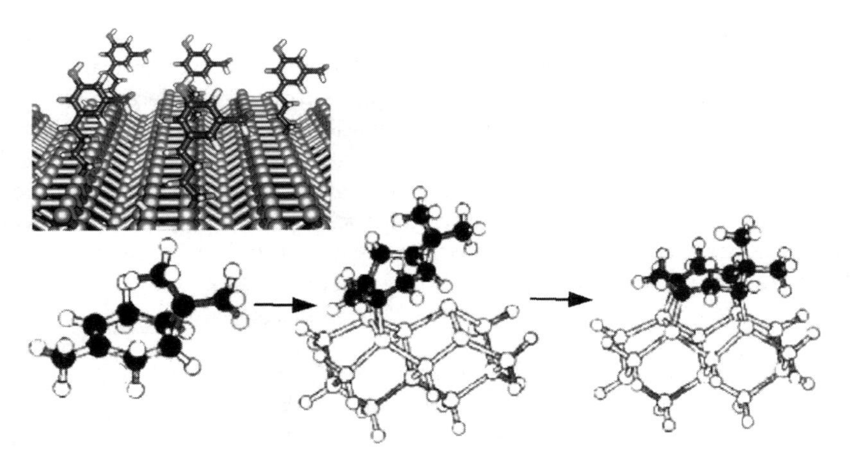

FIGURE 2.5
Bonding and assembling of organic molecules forming a molecular hardware.

These 3D-centered topologies and organizations are fully supported by solid and fluidic molecular electronics hardware. Figure 2.5 illustrates bonding and assembling of organic molecules, and the *design rules* are discussed in Section 4.6.2.

The ^{M}PPs can be designed within enabling hierarchical architectures (neuronal, processor-and-memory, fused memory-in-processor, etc.) utilizing novel organization of molecular processing and memory hardware. For example, neuromorphological reconfigurable solid and fluidic ^{M}PPs are devised by utilizing a ^{3D}networking-and-processing paradigm, see Sections 3.7 and 3.8.

2.3 Synthesis Taxonomy in Design of Molecular Integrated Circuits and Processing Platforms

Devising, design, optimization, and analysis are sequential activities. Devising starts with the discovery of new and/or application of existing sound concepts, developing and applying basic physics, examining phenomena and mechanisms, analyzing specifications, assessing performance, and so forth. We propose to apply a *molecular architectronics* paradigm in order to devise and design pre-eminent ^{M}ICs and ^{M}PPs. This paradigm is based on

1. Discovery of novel topological/organizational/architectural solutions, as well as utilization of new phenomena and capabilities of 3D molecular electronics at the device and system levels
2. Development and implementation of sound methods, technology-centric CAD, and super-large-scale-integration (SLSI) design concurrently associated with bottom-up fabrication

In the design of ^{M}ICs, one faces a number of challenging tasks such as analysis, optimization, aggregation, verification, reconfiguration, validation, evaluation, and so forth. Technology-centric synthesis and design at the device and system levels must be addressed, researched, and solved by making use of a CAD-supported SLSI design of complex ^{M}ICs. Molecular electronics provides a unique ability to implement signal/data processing hardware within 3D-centered novel organizations and enabling architectures. This guarantees massive parallel distributed computations, reconfigurability, and large-scale data manipulation thereby ensuring super-high-performance computing and processing. Combinational and memory ^{M}ICs should be designed using a hypercube concept (reported in Chapter 4) and implemented as aggregated *modular* processing and memory ${}^{\aleph}$hypercells. The device physics of solid and fluidic ^{M}devices is described in Chapters 3 and 6. Those ^{M}devices are aggregated as ^{M}gates that must guarantee the desired performance and functionality of ${}^{\aleph}$hypercells.

Various design tasks for 3D ^{M}ICs are not analogous to the CMOS-centered design, planar layout, placement, routing, interconnect and other tasks that were successfully solved. Conventional very-large-scale-integration (VLSI)/ultra-large-scale-integration (ULSI) design flow is based on the well-established system specifications, design rules, functional design, conventional architecture, verification (functional, logic, circuit, and layout), as well as CMOS fabrication technology. The CMOS technology utilizes 2D topology of gates with FETs and BJTs. For ^{M}ICs, device- and system-level technology-centric design must be performed using novel methods, which are discussed in Chapter 4. Figure 2.2 illustrates the proposed 3D-centered solid and fluidic molecular electronics hardware departing from 2D multilayer CMOS microelectronics. For ^{M}ICs, we propose to utilize a unified top-down (system level)

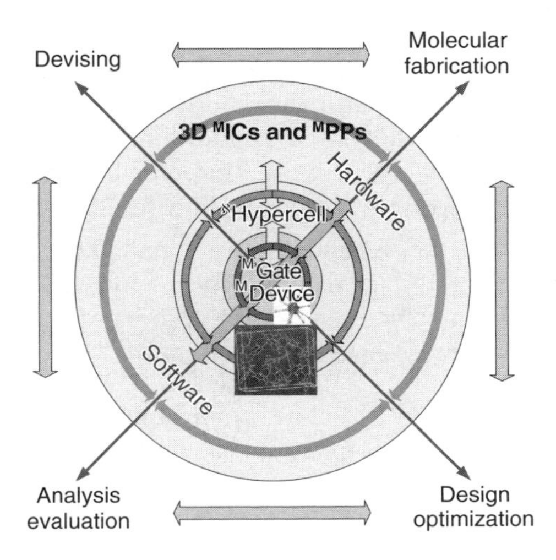

FIGURE 2.6
Top-down and bottom-up synthesis taxonomy within an x-domain flow map.

and bottom-up (device/gate level) synthesis taxonomy within an x-domain flow map, as shown in Figure 2.6. The core 3D design themes are integrated within four domains:

- Devising with validation
- Analysis and evaluation
- Design and optimization
- Molecular fabrication

As shown in Figure 2.6, the synthesis and design of ^{M}ICs and ^{M}PPs should be performed by utilizing a bidirectional flow map. Novel design, analysis, and evaluation methods must be developed. Design in 3D space is radically different compared with VLSI/ULSI because of novel topology/organization, enabling architectures, new devices, enhanced functionality, enabling capabilities, complexity, technology dependence, and so forth. The unified top-down/bottom-up synthesis taxonomy should be coherently supported by developing innovative solutions to carry out a number of major tasks such as:

1. Devise and design ^{M}devices, ^{M}gates, $^{\aleph}$hypercells, and networked $^{\aleph}$hypercells aggregates that form ^{M}ICs
2. Develop new methods in design and verification of ^{M}ICs
3. Analyze and evaluate performance characteristics, estimates, measures, and metrics at the device and system levels
4. Develop technology-centric CAD to concurrently support design at the device and system levels

The reported unified synthesis taxonomy integrates the following:

1. *Top-Down Synthesis*: Devise super-high-performance molecular processing and memory platforms implemented by designed ^{M}ICs within new organizations and enabling architectures. These ^{M}ICs are implemented as aggregated $^{\aleph}$hypercells composed of ^{M}gates that are engineered from ^{M}devices (Figures 2.7a and b).
2. *Bottom-Up Synthesis*: Engineer functional 3D-topology ^{M}devices that compose ^{M}gates in order to form $^{\aleph}$hypercells (for example, multiterminal solid ^{ME}devices can be engineered as cyclic molecules arranged from atoms ensuring functionality).

Super-high-performance ^{M}PPs can be synthesized using $^{\aleph}$hypercells D_{ijk}. The proposed synthesis taxonomy utilizes a number of innovations at the system and device levels such as:

1. Innovative architecture, organization, topology, aggregation, and networking

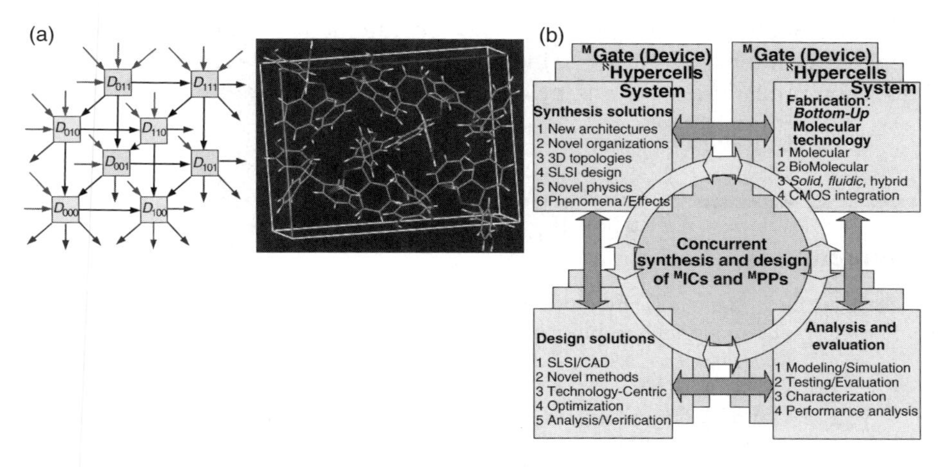

FIGURE 2.7
(a) Molecular electronics: Aggregated $^{\aleph}$hypercells D_{ijk} composed of ^{M}gates that integrate multi-terminal solid ^{ME}devices engineered from atomic complexes; (b) concurrent synthesis and design at system, module and gate (device) levels.

2. Novel enabling ^{M}devices that form ^{M}gates, $^{\aleph}$hypercells, and ^{M}ICs
3. Unique phenomena, effects, and solutions (quantum interaction, parallelism, etc.)
4. Bottom-up fabrication
5. CAD-supported technology-centric SLSI design

Biomolecular and fluidic ^{M}devices, which operate utilizing different phenomena as compared with solid ^{ME}devices, are covered in Chapter 3. Chapters 5 and 6 focus on the synthesis, analysis, and basic fundamentals of solid ^{ME}devices. Performance and baseline characteristics of solid ^{ME}devices are drastically affected by the molecular structures, aggregation, bonds, atomic orbitals, electron affinity, ionization potential, arrangement, sequence, assembly, folding, side groups, and other features. Molecular devices and ^{M}gates must ensure desired functionality, transitions, switching, logics, electronic characteristics, performance, and so forth. Enabling capabilities, functionality, high switching frequency, superior density, expanded utilization, low power, low voltage, desired I–V characteristics, noise immunity, robustness, integration, and other characteristics can be ensured by a coherent design. In ^{M}devices, performance and characteristics can be changed and optimized by utilizing and controlling state transitions and parameters. For example, the number of quantum wells/barriers, their width, energy profile, tunneling length, dielectric constant, and other key features of solid ^{ME}devices can be adjusted and optimized by engineering molecules with specific atomic sequences, bonds, side groups, and so forth. The goal is to achieve functionality ensuring the best achievable performance at the device, module, and system levels. The reported interactive synthesis taxonomy coherently integrates all tasks, including devising of ^{M}devices, discovering novel organization, synthesizing enabling architectures, designing ^{M}ICs.

References

1. S. E. Lyshevski, Three-dimensional molecular electronics and integrated circuits for signal and information processing platforms. In *Handbook on Nano and Molecular Electronics*, Ed. S. E. Lyshevski, CRC Press, Boca Raton, FL, pp. 6-1–6-100, 2007.
2. P. S. Churchland and T. J. Sejnowski, *The Computational Brain*, The MIT Press, Cambridge, MA, 1992.
3. W. Freeman, *Mass Action in the Nervous System*, Academic Press, New York, 1975.
4. W. Freeman, Tutorial on neurobiology from single neurons to brain chaos, *Int. J. Biforcation Chaos*, vol. 2, no. 3, pp. 451–482, 1992.
5. S. Laughlin, R. van Steveninck and J. C. Anderson, The metabolic cost of neural computation, *Nature Neurosci.*, vol. 1, pp. 36–41, 1998.
6. U. B. Kaupp and A. Baumann, Neurons—The molecular basis of their electrical excitability. In *Handbook of Nanoelectronics and Information Technology*, Ed. R. Waser, Wiley-VCH, Darmstadt, Germany, pp. 147–164, 2005.

3

Biomolecular Processing and Molecular Electronics

PREAMBLE Heated debates concerning consciousness, information processing, and intelligence have emerged on the cornerstone fundamentals and developments of neuroscience, neurobiology, biophysics, and other life sciences. Focusing on processing platforms, some of the basics, current status, envisions, and hypotheses are reported. The author, to his best perception and knowledge, solicits various questions with the attempt to answer them under large uncertainties. This chapter introduces new concepts such as *biomolecular processing hardware*, *biomolecular processing software*, *information*, and *routing* carriers. Certain postulates, hypotheses, assessments, and solutions may be viewed from different perspectives, and their broad society engagement is expected in the various multidisciplinary topics covered in this chapter. It also should be emphasized that the terminology used in science, medicine, and engineering are different, and correspondingly, consistency in definitions and meanings cannot be ensured.

3.1 Neuroscience: Conventional Outlook and Brainstorming Rational

Following the accepted theory, we first introduce some concepts of life science disciplines related to the scope of the book. Appreciating neuroscience, neurophysiology, neurobiology and other disciplines, this section covers conventional views and addresses open-ended problems from engineering and technology reflecting some of the author's inclinations and hypothesis. The reader is aware that there are virtually unlimited literature resources that cover conventional neuroscience, neurophysiology, neurobiology, and so forth. The author does not aim to cover the material from thousands of books and journals available. Rather, problems directly related to the molecular electronics and molecular processing are emphasized.

The human central nervous system that includes the brain and spinal cord, performs information processing, adaptive evolutionary learning, memory

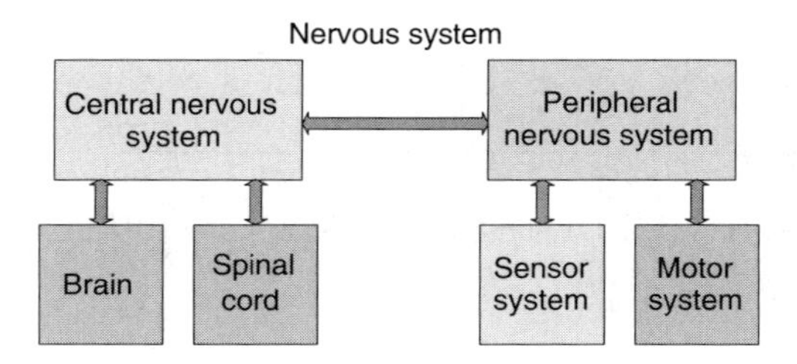

FIGURE 3.1
Vertebrate nervous system: high-level diagram.

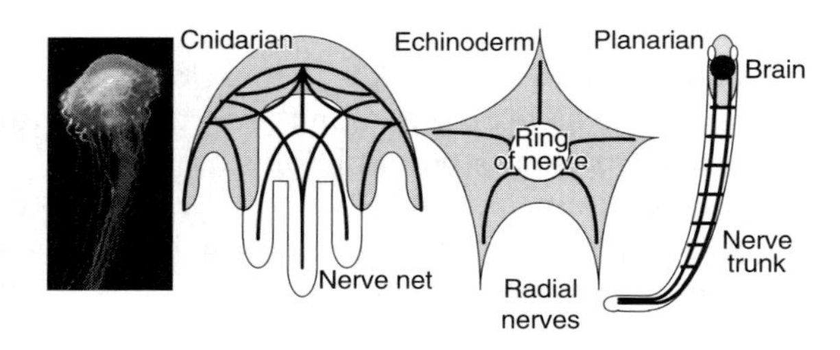

FIGURE 3.2
Invertebrate nervous systems.

storage, and so forth. The human brain consists of the hindbrain (controls homeostasis and coordinates movement), midbrain (receives, integrates, and processes the sensory information), and forebrain (information processing, integration, image processing, short- and long-term memories, learning, decision making, and motor control). The peripheral nervous system consists of the sensory system. Sensory neurons transmit information from internal and external environments to the central nervous system, and motor neurons carry information from the brain or spinal cord to the motor peripheral nervous system (effectors, muscles, and gland cells). The nervous system schematics is depicted in Figure 3.1.

There is a great diversity in the organization of different nervous systems. The cnidarian (*hydra*) nerve net is an organized system of simple nerves (with no central control), which performs elementary tasks such as propulsion. Echinoderms have a central nerve ring with radial nerves. For example, sea stars have central and radial nerves with a nerve net. Planarians have small brains that send information through two or more nerve trunks as illustrated in Figure 3.2. Jellyfish have been on the earth for over 650 million years, and have no heart, bones, brain, or eyes. A network of nerve cells allows the jelly fish to move and react to food, danger, light, temperature, and so forth. Sensors around the bell rim provide information as to whether they are heading up or down, toward light or away from it, and so forth. The basic information processing, actuation (propulsion mechanism and stinging cells, cnidocytes, which contain tiny harpoons called nematocyst triggered by the contact), and sensing are performed by the simplest invertebrates.

The anatomist Heinrich Wilhelm Gottfried Waldeyer-Hartz found that the nervous system consists of nerve cells in which there are no mechanical joints. In 1891, he used the word *neuron*. The cell body of a typical vertebrate neuron consists of the nucleus (soma) and other cellular organelles. The human brain is a network of $\sim 1 \times 10^{11}$ aggregated neurons with more than 1×10^{14} synapses. Action potentials, and likely other information-containing signals, are generated and transmitted to other neurons by means of complex and not fully comprehended electrochemomechanical transitions, interactions and mechanisms within *axo-dendritic, dendro-axonic, axo-axonic,* and *dendro-dendritic* mapping formed by biomolecules within axonic and dendritic structures. Branched projections of neurons (axons and dendrites), shown in Figure 3.3, are packed with ~25 nm diameter microtubules that may play a significant role in signal/data transmission, communication, and networking, ultimately affecting information processing and memory storage. The cylindrical wall of each microtubule is formed by 13 longitudinal protofilaments of tubuline molecules, for example, altering α- and β-heterodimers. Within a complex microtubule network, there are also nucleus-associated microtubules. The cross-sectional representation of a microtubule is a ring of 13 distinct tubuline molecules. Numerous and extensively branched dendrite structures

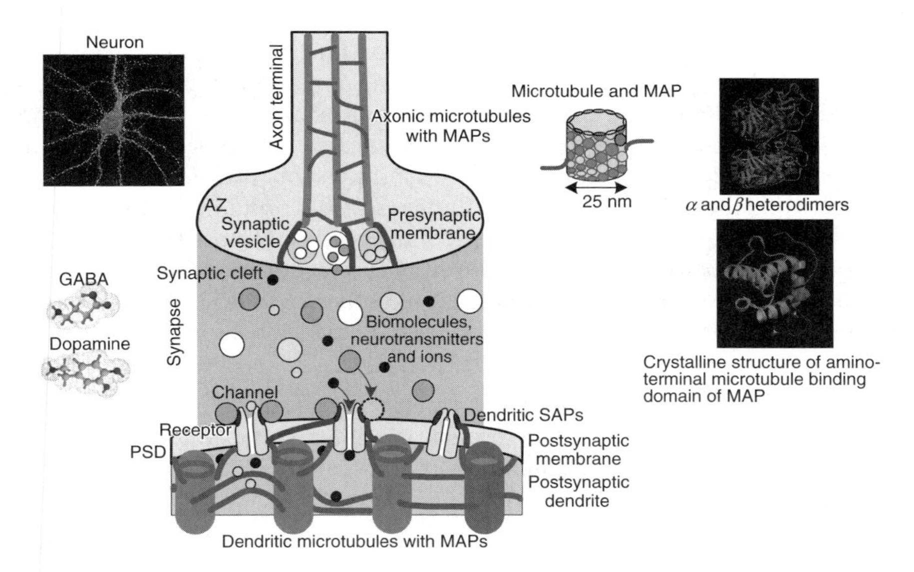

FIGURE 3.3
(See color insert following page 146.) Schematic representation of the *axo-dendritic* organelles with presynaptic active zones (AZ) and postsynaptic density (PSD) protein assemblies. Electrochemomechanically induced transitions, interactions, and events in a *biomolecular processing hardware* may result to information processing and memory storage. For example, the *state* transitions results due to binding/unbinding of the *information carriers* (biomolecules and ions). 3D-topology lattice of synapse-associated proteins (SAPs) and microtubules with microtubule-associated proteins (MAPs) may ensure reconfigurable networking, and biomolecules can be utilized as the *routing carriers*.

are believed to transmit information to the cell body. The information is transmitted from the cell body through the axon structures. The axon originates from the cell body and terminates in numerous terminal branches. Each axon terminal branch may have thousands of synaptic axon terminals. These presynaptic axon terminals and postsynaptic dendrites establish the interface between neurons or between a neuron and target cells. As shown in Figure 3.3, various neurotransmitters are released into the synaptic cleft and propagate to the postsynaptic membrane binding to the receptor.

The conventional neuroscience theory postulates that in neurons the information is transmitted by action potentials, which result due to ionic fluxes that are controlled by the cellular mechanisms. The ionic channels are opened and closed by binding and unbinding of neurotransmitters that are released from the synaptic vesicles (located at the presynaptic axon sites). Neurotransmitters propagate through the synaptic cleft to the receptors at the postsynaptic dendrite (see Figure 3.3). According to conventional theory, binding/unbinding of neurotransmitters in multiple synaptic terminals result in selective opening/closing of membrane ionic channels, and the flux of ions causes the action potential which is believed to contain and carry-out information. The membrane potential is due to the unequal distribution of ions across the membrane, and the membrane can be depolarized or hyperpolarized.

Axo-dendritic organelles with microtubules and microtubule-associated proteins (MAPs), as well as the propagating ions and neurotransmitters in synapse, are schematically depicted in Figure 3.3. There are axonic and dendritic microtubules, MAPs, synapse-associated proteins (SAPs), endocytic proteins, and so forth. Distinct pre- and postsynaptic SAPs have been identified and examined. Large multidomain scaffold proteins, including SAP and MAP families, form the framework of the presynaptic active zones (AZ), postsynaptic density (PSD), endocytic zone (EnZ), and exocytic zone (ExZ) assemblies. There are numerous interactions between AZ, PSD, EnX, and ExZ proteins. With a high degree of confidence, one may conclude that:

- There could exist processing- and memory-associated, electrochemomechanically induced *state* transitions, interactions, and events in extracellular and intracellular protein assemblies.
- Extracellular and intracellular protein assemblies form a *biomolecular processing hardware* that possesses *biomolecular processing software*. This built-in *biomolecular processing software* functions by inherent events and mechanisms.

In a microtubule, each tubulin dimer ($\sim 8 \times 4 \times 4$ nm) consists of positively and negatively charged α-tubulin and β-tubulin (see Figure 3.3). Each heterodimer is made of ~450 amino acids, and each amino acid contains ~15–20 atoms. Tubulin molecules exhibit different geometrical conformations (states). Tubulin dimer subunits are arranged in a hexagonal lattice with different chirality. The interacting negatively charged C-terminals extend outward from each monomer (protrude perpendicularly to the microtubule surface), attracting positive ions from the cytoplasm. The intratubulin relative permittivity ε_r is $\sim$2, while outside the microtubule ε_r is $\sim$80.

MAPs are proteins that interact with the microtubules of the cellular cytoskeleton. A large variety of MAPs have been identified. MAPs accomplish different functions such as stabilization/destabilization of microtubules, guiding microtubules toward specific cellular locations, interfacing of microtubules and proteins. MAPs bind directly to the tubulin monomers. Usually, the MAPs carboxyl-terminus —COOH (C-terminal domain) interacts with tubulin, while the amine-terminus, $-NH_2$ (N-terminal domain) binds to organelles, intermediate filaments, and other microtubules. Microtubule–MAPs binding is regulated by phosphorylation. This is accomplished through the function of the microtubule-affinity-regulating kinase protein. Phosphorylation of MAP by microtubule-affinity-regulating kinase protein causes MAP to detach from any bound microtubules. By utilizing the charge-induced interactions, MAP1a and MAP1b, found in axons and dendrites, bind to microtubules differently than other MAPs. While the C-terminals of MAPs bind the microtubules, the N-terminals bind other parts of the cytoskeleton or the plasma membrane. MAP2 is found mostly in dendrites, while tau-MAP is located in the axon. These MAPs have a C-terminal microtubule-binding domain and variable N-terminal domains projecting outward and interacting with other proteins. In addition to MAPs, there are many other proteins that affect microtubule behavior. These proteins are not considered to be MAPs because they do not bind directly to tubulin monomers, but they affect the functionality of microtubules and MAPs. The mechanism of the so-called *synaptic plasticity* and the role of proteins, neurotransmitters, and ions, which likely affect processing, learning and memory, are not fully understood.

Biomolecular processing hardware, which is formed by biomolecular assemblies, built-in *biomolecular processing software*, and information and routing carriers, are introduced. Some neurotransmitters can be utilized as information carriers. Neurotransmitters are: (1) synthesized (reprocessed) and stored into vesicles in the presynaptic cell, (2) released from the presynaptic cell, propagate, and bind to receptors on one or more postsynaptic cells, and, (3) removed and/or degraded. There are more than 100 known neurotransmitters. In general, neurotransmitters are classified as (i) small-molecule neurotransmitters, and, (ii) neuropeptides (composed of 3–36 amino acids). It is reported in the literature that small-molecule neurotransmitters mediate rapid synaptic actions, while neuropeptides tend to modulate slower ongoing synaptic functions. As an illustrative example, the structure and three-dimensional (3D) configuration of the γ-aminobutyric acid (GABA) and dopamine neurotransmitters are illustrated in Figures 3.4a and b.

Some Conclusions, Open Problems, and Possible Directions: At the cellular level, a wide spectrum of biophysical phenomena and mechanisms

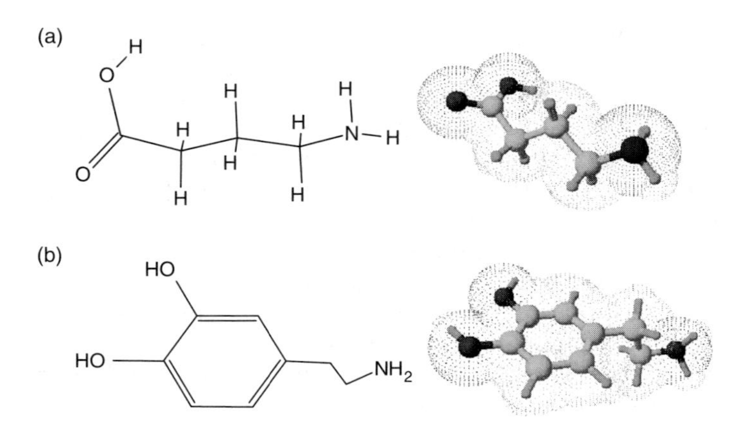

FIGURE 3.4
γ-Aminobutyric acid and dopamine neurotransmitters.

remain unknown or are not sufficiently studied. For example, the production, activation, reprocessing, binding, unbinding, and propagation of neurotransmitters, although studied for decades, are not adequately comprehended. There are debates on the role of microtubules, MAPs, and SAPs, which are establish, or at least affect, a *biomolecular processing hardware*. *Biomolecular processing hardware*, as well as the built-in software, can be questioned. The open problems and major questions are

1. Which biomolecular assemblies form the *biomolecular processing hardware* within a reconfigurable networked organization?
2. Which phenomena/effects/mechanisms within *biomolecular processing hardware* execute the software tasks?
3. Which electrochemomechanically induced transition/interactions/events are utilized guarantying information processing, memory storage and other related tasks?
4. Which biomolecules and ions are the *information* and potentially *routing* carriers?

With a limited knowledge even on signal transmission and communication in neurons, as well as the role of action potentials, one can conclude that other stimuli of different origin likely exist and should be examined. Some results, experimental data, and estimates are reported in Sections 3.2, 3.3, and 3.4. From the conventional action potential doctrine, the required high energy to ensure transitions, very-low-potential propagation velocity ~100 m/sec, and other performance estimate assessments are inadequate to conclude that the conventional postulates are undisputable at the device and system levels. We reported the possible role of biomolecules (neurotransmitters, enzymes, etc.) and ions as *information* (to accomplish processing and memory storage) and *routing* (to accomplish interconnect and reconfiguration) carriers. Though there is a lack of experimental verification on the utilization of biomolecules as *information* and *routing* carriers, there is no explicit proof on the role of action potentials even as communication signals. In synapse, there could exist stimuli of electromagnetic, mechanical, thermodynamic, or other origins that may ultimately result in communication and data exchange between neurons, coding, activation, and so forth. In general, information processing, with all related tasks, is accomplished utilizing electrochemomechanically induced transitions/interaction/events due to biophysical phenomena in biomolecules. With uncertainties in the cornerstone biophysics of processing by biomolecules, one cannot explicitly specify the biophysical phenomena, effects, and mechanisms utilized. However, for *synthetic* ^{M}devices and ^{M}ICs, specific biophysical phenomena can be defined and uniquely employed to ensure their soundness. At the system level, for ^{BM}PPs, the situation is even more disturbing as compared with that at the device level. In particular, there are no sound explanation, justification, and validation of information processing, signal/data processing, memory storage, and other related processes within virtually unknown system organization and architecture. New concepts and hypotheses, which may, or may not, be sound or adequate according to conventional doctrines, have emerged. Though these concepts could be questionable for ^{BM}PPs, they have significant merit for envisioned *synthetic* ^{M}PPs, for which one departs from ultimate prototyping of bioprocessing in order to devise fundamentally and technologically sound practical paradigms. In fact, ^{M}PPs are not expected to coherently prototype or mimic ^{BM}PPs.

3.2 Processing in Neurons

3.2.1 Introduction and Discussions

Biosystems detect various stimuli, and the information is processed through complex biophysical phenomena and mechanisms at the molecular and cellular levels. Biosystems accomplish cognition, learning, perception, knowledge generation, memory storage, coding, transmission, communication, adaptation, and other tasks related to the information processing. Owing to a lack of conclusive evidence, there are disagreements on baseline biophysical phenomena (electrochemical, thermodynamic, etc.) and mechanisms that ultimately result in signal/data and information processing. The baseball, football, basketball, and tennis players perform information processing and other tasks in real time, that is, 3D image processing, estimation of the ball velocity and trajectory, analysis of the situation under uncertainties, decision making based upon the possessed situation awareness, and so forth. They respond accordingly by running, jumping, throwing or hitting the ball, and so forth. The reader can imagine the overall performance of the ^{BM}PP to coherently execute the coordinated response (action through sensing–processing–control–actuation by means of various processing tasks) within 0.2 sec. This "slow" ~0.2 sec response is largely due to the slow *torsional-mechanical* dynamics of actuation system (Newtons' second law gives $\Sigma F = ma$ or $\Sigma T = J\alpha$), although the information processing is performed much faster.

The human retina has ~125 million rod cells and ~6 million cone cells. Assuming the conventional postulates, the resulting communication delay $\tau_{\text{data transmission}}$ of the action potential, which is believed establishing the communication (or data transmission and exchange from the engineering viewpoint) between neurons, can be estimated as

$$\tau_{\text{communication}} = \tau_{\text{spike propagation}} + \tau_{\text{spike generation}} + \tau_{\text{spike transit}} + \Delta\tau_{\Sigma},$$

where $\tau_{\text{spike propagation}}$ and $\tau_{\text{spike transit}}$ are the spike propagation and transit delays to and from the neuron, as estimated by $\tau_i = L_i/v$; L_i is the path length (form 100 μm to centimeters); v is the propagation velocity which is ~100 m/sec; $\tau_{\text{spike generation}}$ is the spike generation delay that can be estimated by taking note that ~10 to ~1000 spikes are induced per second, and hence $\tau_{\text{spike generation}}$ is ~0.01 sec; $\Delta\tau_{\Sigma}$ denote other delays, which could be considered to be negligibly small.

We found that the communication per each neuron results in a ~0.01 sec data transmission delay. One can postulate the number of associated neurons to perform image processing tasks from each rod cell and human retina. Assume that a single spike contains adequate data, there is no communication redundancy and error assessment, no feedback, no synchronization and protocols, and so forth. If there are only five neurons dynamically processing the data in series from each rod, the communication (data transmission) takes ~0.05 sec. The processing capabilities of the visual system are considerably higher, and less than 0.05 sec is needed to cognitively processes images. As was emphasized, the communication is a "low-end" task as compared with the processing and memories. Therefore, other possible solutions should be researched deriving sound concepts refining conventional postulates.

An innovative hypothesis on the microtubule-assisted quantum information processing is reported in [1]. The authors consider microtubules as assemblies of oriented dipoles and postulate that: (1) Conformational states of individual tubulins within neuronal microtubules are determined by the induced dipole interactions (mechanical London forces) that may lead to changes via electromechanical coupling. These mechanical forces can induce a conformational quantum superposition; (2) In superposition, tubulins communicate/compute with entangled tubulins in the same microtubule, with other microtubules in the same neuron, with microtubules in neighboring neurons, and through macroscopic regions of brain by tunneling through gap junctions; (3) Quantum states of tubulins/microtubules are isolated from environmental decoherence by biological mechanisms, such as quantum isolation, ordered water, Debye layering, coherent pumping, and quantum error correction; (4) Microtubule quantum computations/superpositions are tuned by MAPs during a classical liquid phase that alternates with a quantum solid-state phase of actin gelation; (5) Following periods of preconscious quantum computation, tubulin superpositions reduce or collapse by Penrose quantum gravity *objective reduction*; (6) The output states, which result from the *objective reduction* process, are nonalgorithmic (noncomputable) and govern neural events of binding of MAPs, regulating synapses and membrane functions; (7) The reduction or self-collapse in the *orchestrated objective reduction* model is a *conscious moment*, related to Penrose's quantum gravity mechanism, which relates the process to the space–time geometry. The results reported in [1] suggest that tubulins can exist in quantum superposition of two or more possible states until the threshold for quantum state reduction (quantum gravity mediated by *objective reduction*) is reached. A double-well potential, according to [1], enables the interwell quantum tunneling of a single electron and spin states because its energy is greater than the thermal fluctuations. The debate continues on the soundness of this concept examining the feasibility of utilization of quantum effects in tubulin dimers, analyzing the tunneling in quantum wells with relatively high width (the separation is $\sim$2 nm), studying decoherence, assessing noise, and so forth.

There are ongoing debates on devices and systems, neuronal organization, as well as fundamental biophysical phenomena observed, utilized, embedded, and exhibited by neurons and their organelles. There is no agreement on whether or not a neuron is a device (according to a conventional neuroscience postulate) or a module (system), or on how the information is processed, encoded, controlled, transmitted, routed, and so forth. Information processing is a far more complex task compared with signal/data processing, data transmission, routing and communication [2]. Under these uncertainties, new theories, paradigms, and concepts have emerged. As reported in Section 3.1, there are electrochemomechanically induced transitions and interactions in biomolecules, which may result in processing, memory storage, and other directly related tasks such as signal/data processing and coding. For example, binding/unbinding of neurotransmitters cause these electrochemomechanical transitions and interactions by means of: (1) charge variation, (2) force generation, (3) moment transformation, (4) potential change, (5) electromagnetic radiation, (6) orbital overlap variation, (7) vibration, (8) resonance, (9) folding. Distinct biophysical phenomena and effects can be utilized. For example, a biomolecule (protein) can be used as a *biomolecular electrochemomechanical switch* utilizing the conformational changes, or as a *biomolecular electromechanical switch* using the charge changes that affect the potential or charge distribution.

Reference [3] examines the subneuronal processing integrating quantum, electromagnetic, and mechanical stochastic phenomena. The interaction between the cytosolic water electric dipole field and the quantized electromagnetic field, induced by transmembrane neuronal currents, as well as vibrationally assisted tunneling, are researched. Soliton electromagnetically induced collisions are viewed as phenomena that may ultimately result in bioelectromechanical logic gates, functionality of which can be due to the interaction of the soliton with *C-terminal tails* projecting from the cytoskeletal microtubules. In particular, the *C-terminal tails energase* action of vibrationally assisted tunneling affects the conformational dynamics of the neuronal cytoskeletal protein network by facilitating the mechanisms leading to neuronal neurite outgrowth, synaptogenesis, and membrane fusion. The nonlinear time-dependent Schrödinger and Klein–Gordon equations are applied to model the dynamics of bioenergetics, solitary and propagating electromagnetic waves, mechanical–stochastic processes, and other mechanisms of protein (AZ, PSD, membrane, microtubules, microtubular surfaces, tubulines, etc.) biodynamics [3]. The complexity in fundamental, applied, and experimental analysis of dynamic electrochemomechanically induced transitions, interactions, and effects is overwhelming. For example, coherent high-fidelity modeling, heterogeneous simulations and verifications have not been performed for a simple stand-alone microtubule, MAP and SAP because of enormous computational and experimental complexity, multidisciplinary constraints, and inadequate attention. It can be expected that *in vivo* input–output synaptic activities will become available, allowing one to examine these activities within a *black box* concept. It is likely that electrochemomechanically induced transitions/interactions/effects at least can be modeled omitting the details of biophysics, which result in enormous complexity. It can be expected that the number of *state* transitions in the baseline electrochemomechanical processing processes is not high because in performing specific tasks (processing, sensing, actuation, etc.) living systems strive to minimize losses (associated with processing, sensing, actuation, etc.) from the bioenergetics viewpoint. Correspondingly, we do not specify the electrochemomechanically induced transitions, interactions, and effects.

3.2.2 Processing and Networking in Neurons: Proposed Outlook

It is the author's belief that a neuron, as a complex system, performs information processing, memory storage, and other tasks utilizing biophysical phenomena and effects of the electrochemomechanically induced *state* transitions and interactions between biomolecules. Microtubules, MAPs, SAPs, and other proteins establish a *biomolecular processing hardware*, which possesses software capabilities utilizing the inherent events and mechanisms. In neurons, some biomolecules (neurotransmitters and enzymes) and ions function as the *information* and *routing* carriers. Signal and data processing (computing, logics, coding, and other tasks), memory storage, memory retrieval, and information processing may be accomplished by utilizing a *biomolecular processing hardware* (implemented by neuronal processing-and-memory primitives within a neuronal reconfigurable networked organization) with possessed *biomolecular processing software*. There are distinct *information* and *routing* carriers, for example, *activating*, *regulating*, and *executing*. The information carriers are released into and propagate in the synaptic cleft, membrane channels, and cytoplasm. Control of released specific neurotransmitters in a particular synapse and their binding to the receptors result in *state* transitions (charge distribution, bonding, switching, folding, etc.), interactions, and events (release, binding, unbinding, etc.). These *state* transitions, interactions, and events ensure signal/data/information processing and memory by the utilization of the possessed *biomolecular processing software* due to the inherent events and mechanisms. Control of *information* and *routing* carriers under the electrostatic, magnetic, hydrodynamic, thermal and other fields (forces) are described in Sections 3.3 and 3.6. Complex biomolecular *electromechanical* logic gates and combinational and memory platforms can be designed utilizing the proposed concept. Robust reconfiguration is accomplished by the *routing* carriers enhancing reconfigurable networking attained by microtubules, MAPs, SAPs, and other protein aggregates. We originate the following major postulates:

1. Microtubules, SAPs, MAPs, and other cellular proteins form *biomolecular processing hardware,* which consists of processing-and-memory primitives (biomolecules and biomolecular aggregates) within a neuronal, reconfigurable, networked organization engineered by the protein aggregates. The 3D-topology neuronal processing-and-memory primitives exhibit electrochemomechanically induced transitions, interactions, and mechanisms resulting in information processing and related tasks. *Axo-dendritic, dendro-axonic, axo-axonic,* and *dendro-dendritic* mapping establishes neuronal processing organization and architecture.
2. Certain biomolecules and ions are the *activating, regulating,* and *executing information* carriers that interact with SAPs, MAPs, and other cellular proteins; see Figure 3.3.
    a. The *activating information* carriers activate and trigger the regulating and executing processes and events—for example, the membrane channels are open or closed, the proteins and specific biomolecule sites steered, proteins fold, and so forth.
    b. The *regulating information* carriers control the feedback mechanisms at the primitive (device/gate) and system levels. The control and adaptation of activation and execution processes is accomplished utilizing the *biomolecular processing software* by means of events and mechanisms.
    c. The *executing information* carriers accomplish the processing and memory functions and tasks.

    Controlled binding/unbinding of *information* carriers lead to the biomolecule-assisted electrochemomechanically induced *state* transitions (folding, bonding, chirality changes, etc.) and interactions affecting the direct (by *executing information* carriers) and indirect (by *activating* and *regulating information* carriers) processing- and memory-associated transitions and interaction in biomolecules and protein assemblies. This ultimately results in processing and memory storage. As the simple typifying examples

i. Binding/unbinding of *information* carriers ensures a combinational logics equivalent to *on* and *off* switching analogous to the AND- and OR-centered electronic gates, mechanical switches and logics. For example, the AND and OR biomolecular logic gates, equivalent to gates documented in Figure 4.6, result.
ii. Charge variation is analogous to the functionality of the molecular storage capacitor in the memory cell, see Figure 4.1.
iii. Processing and memories may be accomplished on a high radix because of the multiple-conformation by distinct *executing information* carriers, which is equivalent to the multiple-valued logics and memories.

3. Reconfigurable networking is accomplished by microtubules, MAPs, SAPs, and other protein aggregates that form a reconfigurable *biomolecular processing hardware*. Specific biomolecules and ions are the *activating, regulating*, and *executing routing* carriers that bind to and unbind from the specific sites of SAPs, MAPs, and other proteins. The binding/unbinding leads to the electrochemomechanical *state* transitions and events. The *activating routing* carriers activate the routing processes and events. The feedback mechanisms, adaptation, and control of the routing processes are performed by the *regulating routing* carriers. The electrochemomechanical *state* transitions (folding, bonding, chirality changes, etc.) in biomolecules with the resulting events, accomplished (induced) by the *executing routing* carriers, ensure robust reconfiguration, adaptation, and interconnect.
4. Presynaptic AZ and PSD (composed of SAPs, MAPs, and other proteins), as well as microtubules, form a biomolecular assembly (organization) within a reconfigurable processing-and-memory neuronal architecture.

Biomolecular processing includes various tasks, such as communication, signaling, routing, reconfiguration, coding, and so forth. Consider biomolecular processing between neurons using the axo-dendritic inputs and dendro-axonic outputs. We do not specify the information-containing signals (action potential, polarization vector, phase shifting, folding, modulation, vibration, switching, etc.), cellular mechanisms, as well as electrochemomechanically induced transitions/interactions/events in biomolecules. Each neuron consists of m_i neuronal processing-and-memory primitives. The transitions result in the axo-dendritic input vectors $\mathbf{x}_i$; see Figure 3.5. Hence, the inputs to neuron $\mathbf{N}_0$ are m vectors to m primitives, that is,

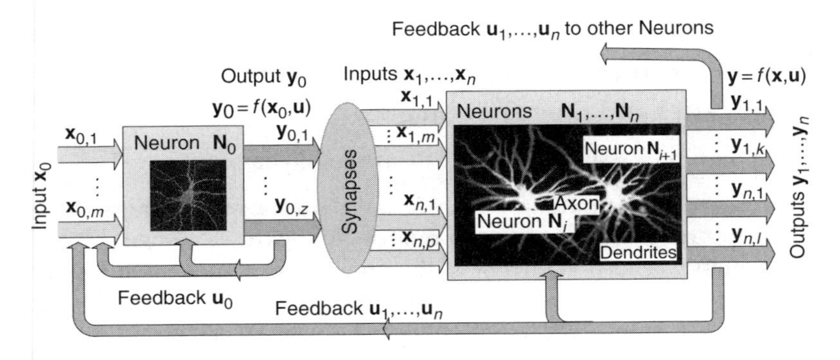

FIGURE 3.5
Input–output representation of $(n + 1)$ aggregated neurons with *axo-dendritic* inputs and *dendro-axonic* outputs.

$\mathbf{x}_0 = [\mathbf{x}_{0,1}, \mathbf{x}_{0,2}, \ldots, \mathbf{x}_{0,m-1}, \mathbf{x}_{0,m}]$. The first neuron $\mathbf{N}_0$ has the z dendro-axonic output vectors, and $\mathbf{y}_0 = [\mathbf{y}_{0,1}, \mathbf{y}_{0,2}, \ldots, \mathbf{y}_{0,z-1}, \mathbf{y}_{0,z}]^{\mathrm{T}}$. Spatially distributed $\mathbf{y}_0$ furnishes the inputs to neurons $\mathbf{N}_1, \mathbf{N}_2, \ldots, \mathbf{N}_{n-1}, \mathbf{N}_n$. For all neurons, $\mathbf{x}_{i,j} \in \mathbb{R}^{\mathrm{b}}$ and $\mathbf{y}_{i,j} \in \mathbb{R}^{\mathrm{c}}$ are the vectors (not variables) due to multiple neurotransmitters, binding cites, receptors, as well as *discrete* biomolecule-assisted transitions, interactions, and events. For the first neuronal processing-and-memory primitive, one has $\mathbf{x}_{0,1} = [x_{0,1\ 1}, \ldots, x_{0,1\ m}]^{\mathrm{T}}$ and $\mathbf{y}_{0,1} = [y_{0,1\ 1}, \ldots, y_{0,1\ z}]^{\mathrm{T}}$. Hence, $\mathbf{x}_{i,1} = [x_{i,1\ 1}, \ldots, x_{i,1\ m}]^{\mathrm{T}}$ and $\mathbf{y}_{i,1} = [y_{i,1\ 1}, \ldots, y_{i,1\ z}]$.

The aggregated neurons $\mathbf{N}_0, \mathbf{N}_1, \ldots, \mathbf{N}_{n-1}, \mathbf{N}_n$ *process* the information utilizing electrochemomechanically induced transitions, interactions, events, and mechanisms. The output vector is $\mathbf{y} = f(\mathbf{x})$, where f is the nonlinear function. In the logic design of ICs, f is called the *switching* function.

To ensure robustness, reconfigurability, and adaptiveness, we consider the feedback vector $\mathbf{u}$. The output of the neuron $\mathbf{N}_0$ is a nonlinear function of the input vector $\mathbf{x}_0$ and feedback vector $\mathbf{u} = [\mathbf{u}_0, \mathbf{u}_1, \ldots, \mathbf{u}_{n-1}, \mathbf{u}_n]^{\mathrm{T}}$, that is, $\mathbf{y}_0 = f(\mathbf{x}_0, \mathbf{u})$. As the information is processed by $\mathbf{N}_0$, it is fed to the aggreagated neurons $\mathbf{N}_1$, $\mathbf{N}_2, \ldots, \mathbf{N}_{n-1}$, $\mathbf{N}_n$. The neurotransmitters release, binding/unbinding, conformations, as well as other electrochemomechanically induced transitions performed by all neurons, are the dendro-axonic output $\mathbf{y}_i$. Neurons have a branched dendritic tree with ending axo-dendritic synapses.

We have an immense m-input and z-output vector capability ($m \approx 10{,}000$ and $z \approx 1{,}000$) per neuron, and a single neuron has thousands of neuronal processing-and-memory primitives. Figure 3.5 illustrates the 3D aggregation of $(n + 1)$ neurons with the resulting input–output maps $\mathbf{y}_i = f(\mathbf{x}_i, \mathbf{u})$. Dendrites may form dendro-dendritic interconnects, while in axo-axonic connects, one axon may terminate on the terminal of another axon modifying its neurotransmitter release as well as accomplishing other transitions. The analysis of the considered neuronal topology and organization is reported in Section 3.8.

3.3 Biomolecules and Ions Transport: Communication Energetics and Energy Estimates

3.3.1 Active and Passive Transport: Brownian Dynamics

The analysis of propagation of biomolecules and ions is of a great importance. Kinetic energy is the energy of motion, while the stored energy is called potential energy. Thermal energy is the energy associated with the random motion of molecules and ions, and therefore can be examined in terms of kinetic energy. Chemical reaction energy changes are expressed in calories, that is, 1 cal = 4.184 J.

In cells, the directional motion of biomolecules and ions results in *active* and *passive* transport. One studies the *active* transport of *information* and *routing* carriers. The controlled Brownian dynamics of molecules and ions in the synaptic cleft, channels, and fluidic cavity should be examined. It is feasible to control the propagation (motion) of biomolecules by changing the force $F_n(t, \mathbf{r}, \mathbf{u})$ or varying the asymmetric potential $V_k(\mathbf{r}, \mathbf{u})$. The high-fidelity mathematical model is given as

$$m_i \frac{d^2\mathbf{r}_i}{dt^2} = -F_{v_i}\left(\frac{d\mathbf{r}_i}{dt}\right) + \sum_{i,j,n} F_n(t, \mathbf{r}_{ij}, \mathbf{u}) + \sum_{i,k} q_i \frac{\partial V_k(\mathbf{r}_i, \mathbf{u})}{\partial \mathbf{r}_i}$$

$$+ \sum_{i,j,k} \frac{\partial V_k(\mathbf{r}_{ij}, \mathbf{u})}{\partial \mathbf{r}_{ij}} + f_r(t, \mathbf{r}, \mathbf{q}) + \boldsymbol{\xi}_{ri}, \quad i = 1, 2, \ldots, N-1, N, \tag{3.1}$$

$$\frac{d\mathbf{q}_i}{dt} = f_q(t, \mathbf{r}, \mathbf{q}) + \boldsymbol{\xi}_{qi},$$

where $\mathbf{r}_i$ and $\mathbf{q}_i$ are the displacement and extended state vectors; $\mathbf{u}$ is the control vector; $\boldsymbol{\xi}_r(t)$ and $\boldsymbol{\xi}_q(t)$ are the Gaussian white noise vectors; F_v is the viscous friction force; m_i and q_i are the mass and charge; $f_r(t, \mathbf{r}, \mathbf{q})$ and $f_q(t, \mathbf{r}, \mathbf{q})$ are the nonlinear maps.

The Brownian particle velocity vector $\mathbf{v}$ is $\mathbf{v} = d\mathbf{r}/dt$. The Lorenz force on a particle possessing the charge q is $\mathbf{F} = q(\mathbf{E} + \mathbf{v} \times \mathbf{B})$, while using the surface charge density ρ_v one obtains $\mathbf{F} = \rho_v(\mathbf{E} + \mathbf{v} \times \mathbf{B})$. The released carriers propagate in the fluidic cavity and can be controlled or uncontrolled. For the controlled particle, $F_n(t, \mathbf{r}, \mathbf{u})$ and $V_k(\mathbf{r}, \mathbf{u})$ vary.

Example 3.1

Consider a Brownian particle with mass m, under the external time-varying force $F(t, x)$, in a one-dimensional (1D) spatially periodic potential $V(x)$ with a period l, $V(x) = V(x + l)$. This particle dynamics is usually modeled by using the displacement x and thermal fluctuations $\xi(t)$. The Langevin stochastic equation is

$$m\frac{\mathrm{d}^2x}{\mathrm{d}t^2} = -\eta\frac{\mathrm{d}x}{\mathrm{d}t} + F(t,x) - \frac{\partial V(x)}{\partial x} + \xi_x(t),$$

where η is the viscous friction coefficient, and $F_v = \eta v$.

The force term $\partial V(x)/\partial x$ results due to electromagnetic, hydrodynamic, thermal, and other effects. Taking into account bistable modes and using the control variable u, from Equation 3.1, under some assumptions on the extended variable of dynamics, one finds a set of first-order stochastic differential equations as

$$\frac{\mathrm{d}v}{\mathrm{d}t} = \frac{1}{m}\left[-\eta v + \sum_n F_n(t,x,u) - \sum_k \frac{\partial V_k(x,u)}{\partial x} + \xi_x(t)\right],$$

$$\frac{\mathrm{d}x}{\mathrm{d}t} = v,$$

$$\frac{\mathrm{d}q}{\mathrm{d}t} = a_{qx}x + a_{qq}q + \xi_q(t).$$

As the electromagnetic, hydrodynamic, thermal, and other effects are examined, and control inputs are defined, the explicit equations for forces $F_n(t,x,u)$ and potentials $V_k(x,u)$ should be used to solve analysis and control problems. For example, letting $n = 1$, $k = 1$, $F(t,x,u) = -e^t \tanh x$ and $V(x,u) = -xu$, we have

$$\frac{\mathrm{d}v}{\mathrm{d}t} = \frac{1}{m}[-\eta v + \mathrm{e}^{-t}\tanh x + u + \xi_x(t)],$$

$$\frac{\mathrm{d}x}{\mathrm{d}t} = v, \quad \frac{\mathrm{d}q}{\mathrm{d}t} = a_{qx}x + a_{qq}q + \xi_q(t).$$

To describe $V(\mathbf{r},\mathbf{u})$, one may use the strength $\boldsymbol{\varphi}(\cdot)$, asymmetry $\boldsymbol{\phi}(\cdot)$, and decline-displacement $\mathbf{f}(\cdot)$ functions. We have,

$$V(\mathbf{r},\mathbf{u}) = \sum_k \boldsymbol{\varphi}_k(\mathbf{r},\mathbf{u})\boldsymbol{\varphi}_k(\mathbf{r},\mathbf{u})\mathbf{f}_k(\mathbf{r},\mathbf{u}). \tag{3.2}$$

Example 3.2

For a 1D case, from Equation 3.2, one finds $V(x,u) = \varphi(x,u)\phi(x,u)f(x,u)$. Using the strength, asymmetry and decline magnitudes (A_{0n}, A_{1n}, and A_{2n}) that can vary, as well as the strength, asymmetry and decline constants (a_n, b_n, and d_n), the uncontrolled potential is defined as

$$V(x) = \sum_{n=1}^{K} \underbrace{A_{0n}\,\mathrm{e}^{-a_n(x/l)}\sin\frac{2n\pi x}{l}}_{\text{Strength } \varphi(x)} \underbrace{A_{1n}\,\mathrm{e}^{-c_n|x/l|}}_{\text{Asymmetry } \phi(x)} \underbrace{A_{2n}\tanh\left(d_n\frac{x^g}{l^g} + d_{n0}\right)}_{\text{Decline-Displacement } f(x)},$$

where g is the integer.

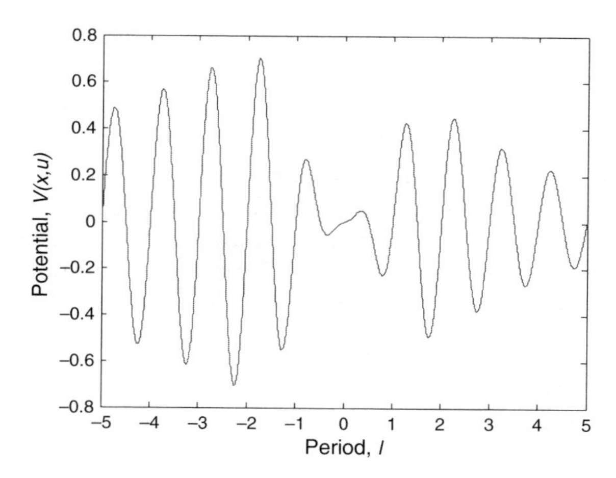

FIGURE 3.6
Potential $V(x,u) = u\,\mathrm{e}^{-0.1(x/l)}\sin(2n\pi x)/(l)\,\mathrm{e}^{-0.25|x/l|}\tanh(0.5(x^g/l^g)+0.01)$ if $u=1$.

One can express the controlled potential $V(x,u)$ as

$$V(x,u) = \sum_{n=1}^{K} f_n(u) \underbrace{A_{0n}\,\mathrm{e}^{-a_n(x/l)} \sin\frac{2n\pi x}{l}}_{\text{Strength } \varphi(x)} \underbrace{A_{1n}\,\mathrm{e}^{-c_n|x/l|}}_{\text{Asymmetry } \phi(x)} \underbrace{A_{2n}\tanh\left(d_n\frac{x^g}{l^g} + d_{n0}\right)}_{\text{Decline-Displacement } f(x)}.$$

For $K=1$, $l = 2.5\times10^{-8}$ m, $f_1(u) = u$, $A_{01} = 0.5$, $A_{11} = 1$, $A_{21} = 2$, $a_1 = 0.1$, $c_1 = 0.25$, $d_1 = 0.5$, $d_{10} = 0.01$, and $g = 2$, one finds $V(x,u) = u\,\mathrm{e}^{-a_1(x/l)}\sin((2n\pi x)/l)\mathrm{e}^{-c_1|x/l|}\tanh(d_1(x^g/l^g)+d_{10})$. For $u=1$, the resulting potential is documented in Figure 3.6. The MATLAB statement to calculate and plot $V(x,u)$ is

```
l=25e-9; x=-125e-9:1e-10:125e-9;
V=0.5.*exp(-0.1*x/l).*sin(2*pi*x/l),*...
  exp(-0.25*abs(x/l)).*2.*tanh(0.5*(x/l).^2+0.01);
plot(x/l,V);
xlabel('Period, \itl', 'FontSize',14);
ylabel('Potential, \itV(x,u)','FontSize',14);
```

Example 3.3

Using continuous differentiable functions, the two-dimensional (2D) asymmetric potential $V(\mathbf{r},u)$, $\mathbf{r} = [x\ \ y]^{\mathrm{T}}$ is given as $V(x,y,u) = u(\sin\sqrt{x^2+y^2+2}/\sqrt{x^2+y^2+2})\mathrm{e}^{-|0.05\sqrt{x^2+y^2+2}|}$. The resulting potential is shown in Figure 3.7 for $u=1$. To calculate and plot $V(\mathbf{r},u)$, we use the following MATLAB statement

```
u=1; d=2; a=0.05; [x,y]=meshgrid([-9:0.2:9]);
  xy=sqrt(x.^2+y.^2+d);
```

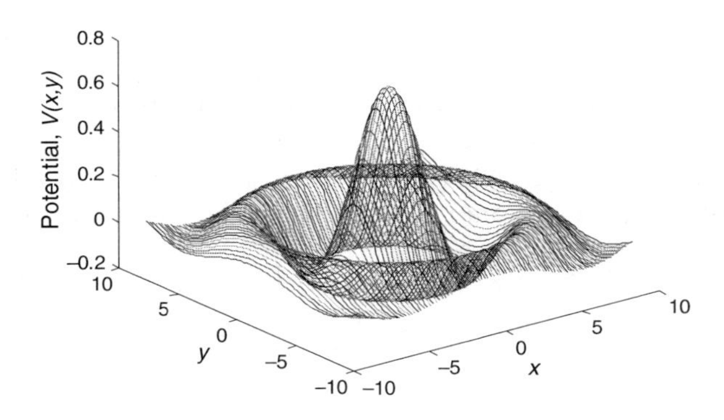

FIGURE 3.7
Two-dimensional potential.

```
v=u.*sin(xy).*exp(-abs(a*xy))./xy; plot3(x,y,v);
xlabel('\itx','FontSize',14);
ylabel('\ity', 'FontSize',14);
zlabel('Potential, \itV(x,y)','FontSize',14);
```

Example 3.4

Consider the GABA neurotransmitter in the synaptic cleft. The GABA receptor complex contains specific binding sites. The number of ions in the synaptic cleft is defined by their concentration. Let the ionic concentration for Na^+, Cl^-, K^+, and Ca^{2+} be 140, 100, 5, and 2 mM. In the synaptic cleft, with the 25 nm ($L = 25$ nm) separation between membranes, we study 20, 15, 1, and 1 Na^+, Cl^-, K^+, and Ca^{2+} ions. These ions interact with a polar neurotransmitter. We examine the motion of 38 particles in 3D space. This results in 342 first-order stochastic differential equations (see Equation 3.1), and $i = 1, 2, \ldots, 37, 38$.

The electric dipole moment for GABA is 4.8×10^{-29} C m. The length of GABA is 0.91 nm, and the neurotransmitter mass and diffusion coefficient are: $m_{\text{GABA}} = 1.71 \times 10^{-25}$ kg and $D_{\text{GABA}} = 4 \times 10^{-11}$ m^2/sec.

The masses, diffusion coefficients at 37°C and ionic radii of Na^+, Cl^-, K^+, and Ca^{2+} ions are

$$m_{\text{Na}} = 3.81 \times 10^{-26}\ \text{kg}, \quad m_{\text{Cl}} = 5.89 \times 10^{-26}\ \text{kg},$$

$$m_{\text{K}} = 6.49 \times 10^{-26}\ \text{kg}, \quad m_{\text{Ca}} = 6.66 \times 10^{-26}\ \text{kg},$$

$$r_{\text{Na}} = 0.95 \times 10^{-10}\ \text{m}, \quad r_{\text{Cl}} = 1.81 \times 10^{-10}\ \text{m},$$

$$r_{\text{K}} = 1.33 \times 10^{-10}\ \text{m}, \quad r_{\text{Ca}} = 1 \times 10^{-10}\ \text{m},$$

$$D_{\text{Na}} = 1.33 \times 10^{-9}\ \text{m}^2/\text{sec}, \quad D_{\text{Cl}} = 2 \times 10^{-9}\ \text{m}^2/\text{sec},$$

$$D_{\text{K}} = 1.96 \times 10^{-9}\ \text{m}^2/\text{sec}, \quad \text{and} \quad D_{\text{Ca}} = 0.71 \times 10^{-9}\ \text{m}^2/\text{sec}.$$

The relative permittivities of presynaptic and postsynaptic membranes are $\varepsilon_{rp} = 2.3$ and $\varepsilon_{rP} = 2$. The numerical solution of a set of the stochastic differential equations (3.1) defines the position and velocity of each Brownian particle at time t. Thus, solving (3.1) one obtains the motion dynamics of *microscopic* particles in the synaptic cleft. The neurotransmitter is released at the origin (presynaptic membrane). Hence, the initial conditions for GABA are

$$\mathbf{r}_{0\mathrm{GABA}} = \begin{bmatrix} x_{0\mathrm{GABA}} \\ y_{0\mathrm{GABA}} \\ z_{0\mathrm{GABA}} \end{bmatrix} = \begin{bmatrix} 0 \\ 0 \\ 0 \end{bmatrix}$$

and

$$\mathbf{v}_{0\mathrm{GABA}} = \begin{bmatrix} v_{0\mathrm{GABA}} \\ v_{0\mathrm{GABA}} \\ v_{0\mathrm{GABA}} \end{bmatrix} = \begin{bmatrix} 0 \\ 0 \\ 0 \end{bmatrix}.$$

The neurotransmitter should reach the receptor site, and *microscopic* particles propagate in the 3D (x, y, z) space. Assume that the receptor is at (0, 0, 25) nm, that is,

$$\mathbf{r}_f = \begin{bmatrix} \mathbf{r}_{f,x} \\ \mathbf{r}_{f,y} \\ \mathbf{r}_{f,z} \end{bmatrix} = \begin{bmatrix} 0 \\ 0 \\ 25 \times 10^{-9} \end{bmatrix}.$$

The initial positions of ions were assigned randomly with equal probability within the synaptic cleft.

The electric field intensity is related to the potential as $\mathbf{E} = -\nabla V$. The particle and particle–membrane interactions are studied. For neurotransmitters and ions, the potential is derived using the superposition of the point, line, surface and volume charges. Hence

$$V_i = V_{ai} + V_{ei} + \sum_{j}^{2} V_{mij} + \sum_{j,i \neq j}^{N} V_{Eij} + \sum_{j,i \neq j}^{N} V_{Cij},$$

where V_{ai} is the asymmetric periodic potential with period $l \neq \mathrm{const}$ (V_{ai} results due to the temperature gradient and hydrodynamic field); V_{ei} is the external potential including the effective membrane potentials; V_{mij} is the interacting Brownian particle–membrane potential; V_{Eij} is the electrostatic potential due to the charge; V_{Cij} is the Coulomb potential due to jth Brownian particle, $V_{Cij} = (1/4\pi\varepsilon)(q_j/|\mathbf{r}_i - \mathbf{r}_j|)$, and for point charges one has $F_{Cij} = (1/4\pi\varepsilon)(q_i q_j/|\mathbf{r}_i - \mathbf{r}_j|^2)$.

The membrane potential is $V_{\mathrm{m}}(\mathbf{r}) = (1/4\pi\varepsilon) \int_S (\rho_S(\mathbf{r}')/|\mathbf{r} - \mathbf{r}'|)\mathrm{d}\mathbf{S}$. The field intensity of the charged disc-shaped membrane with radius a and surface charge ρ_S at a distance z from its center is $E_z = (\rho_S/2\varepsilon)(1 - 1/\sqrt{(a^2/z^2) + 1})\mathbf{a}_z$. The force on a charged particle is $\mathbf{F}_z = (\rho_S q/2\varepsilon)(1 - 1/\sqrt{(a^2/z^2) + 1})\mathbf{a}_z$, $z > 0$.

In the synaptic cleft the magnetic field force is negligible due to small **B**. Correspondingly, the dominant electrostatic force is under our consideration. The Poisson's equation $\nabla^2 V_{\mathrm{m}}(\mathbf{r}) = -(\rho_{\mathrm{S}}(\mathbf{r})/\varepsilon)$ is solved to define the electrostatic forces on Brownian particles due to the membrane surface charges $\rho_{\mathrm{S}i}(\mathbf{r})$. Here, $i = 1, 2$, that is, pre- and postsynaptic membranes are considered. Owing to the fast dynamics of *microscopic* particle motion, the membrane charges are time-invariant. Thus, one can assume that the membrane potential is constant during the neurotransmitter transients as it moves through the synaptic cleft reaching the receptor site. Using the strength $\varphi(\mathbf{r})$, asymmetry $\phi(\mathbf{r})$, and decline-displacement $f(\mathbf{r})$ nonlinear maps, the potential $V(\mathbf{r}, u)$ is given as

$$V(\mathbf{r}) = \mathrm{e}^{-z/L} \cos^4\left(\frac{2n\pi}{L} z\right) \times \left(\frac{\sin\sqrt{2\times 10^{19}x^2 + 2\times 10^{19}y^2 + 2\times 10^{-5}}}{\sqrt{2\times 10^{19}x^2 + 2\times 10^{19}y^2 + 2\times 10^{-5}}} + \frac{1}{4}\right), \quad n = 5.$$

The analysis of the directed Brownian motion is performed if the external force $F(t, \mathbf{r}_{ij}) = 0$. Only the electrostatic, hydrodynamic, and thermal forces are considered. For different initial conditions, the numerical solution of 342 coupled highly nonlinear first-order stochastic differential equations (3.1) is obtained in the MATLAB environment using the differential equation solver. Figure 3.8 shows the dynamics for neurotransmitter displacement $z(t)$. The velocity and forces in the z axis are reported in the arbitrary units. The Einstein equation for the thermal *passive* diffusion gives the average neurotransmitter diffusion time of 7.81 μsec (the details are covered in Section 3.3.2). Under the electrostatic force, the neurotransmitter reaches the receptor within 1 μsec. The neurotransmitter evolution in **r** is documented in Figure 3.8.

3.3.2 Ionic Transport

For years the analysis of neuronal activities has largely been focused on action potentials. Conventional neuroscience postulates that the neuronal communication is established by means of action potentials. There is a potential difference across the axonal membrane, and the resting potential is $V_0 = -0.07$ V. The voltage-gated sodium and potassium channels in the membrane result in propagation of action potential with a speed ~100 m/sec, and the membrane potential changes from $V_0 = -0.07$ V to $V_{\mathrm{A}} = +0.03$ V. The ATP-driven pump restores the Na^+ and K^+ concentration to their initial values within $\sim 1 \times 10^{-3}$ sec, making the neuron ready to fire again, if triggered. Neurons can fire more than 1×10^3 times per second.

Consider a membrane with the uniform thickness h. For the voltage difference $\Delta V = (V_{\mathrm{A}} - V_0)$ across the membrane, the surface charge density $\pm\rho_{\mathrm{S}}$

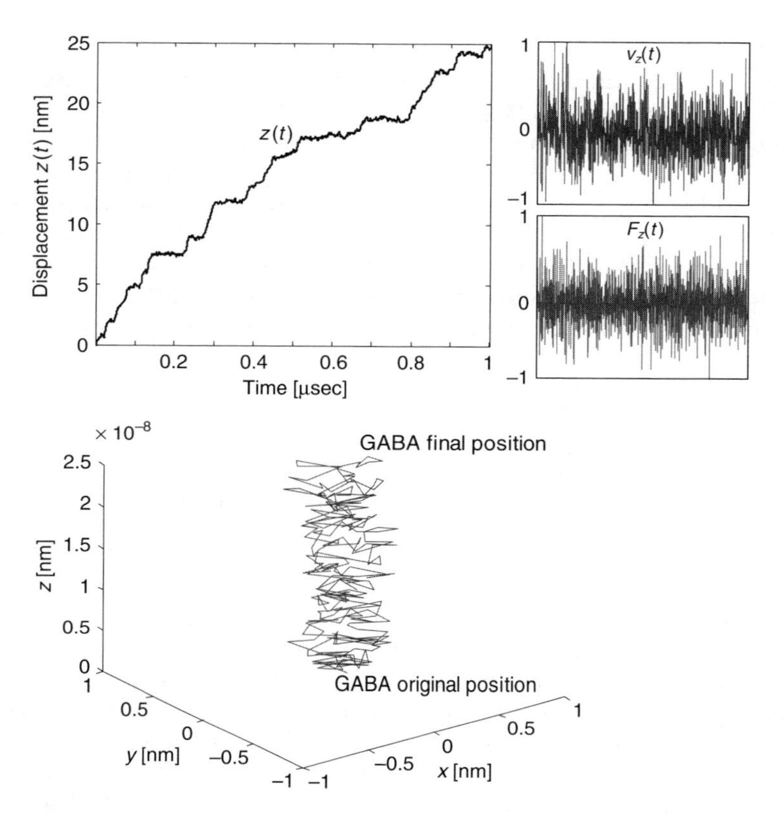

FIGURE 3.8
Controlled motion of neurotransmitter in the synaptic cleft: Neurotransmitter displacement z, velocity v_z and force F_z.

inside/outside membrane is $\rho_S = \varepsilon E_E$. Here, ε is the membrane permittivity, $\varepsilon = \varepsilon_0 \varepsilon_r$; E_E is the electric field intensity, $E_E = \Delta V/h$.

The total active surface area is estimated as $A = \pi d L_A$, where d is the diameter; L_A is the active length.

The total number of ions that should propagate to ensure a single action potential is

$$n_I = \frac{A\rho_S}{q_I} = \frac{\pi \mathrm{d} L_A \varepsilon_0 \varepsilon_r (V_A - V_0)}{q_I h},$$

where q_I is the ionic charge.

One recalls that for a parallel-plate capacitors, the capacitance is given as $C = \varepsilon_0 \varepsilon_r A/h$, and the number of ions that flow per action potential is

$$n_I = Q/q_I,$$

where $Q = C\Delta V$.

Example 3.5

Let $d = 1 \times 10^{-5}$ m, $L_A = 1 \times 10^{-4}$ m, $\varepsilon_r = 2$, $V_A = 0.03$ V, $V_0 = -0.07$ V, and $h = 8 \times 10^{-9}$ m. The charge of Na^+ ion is $q_{Na} = e$. We have $n_{Na} = 4.34 \times 10^5$. By making use of $n_I = A\rho_S/q_I = \pi d L_A \varepsilon_0 \varepsilon_r (V_A - V_0)/q_I h$, one finds that 4.34×10^5 ions are needed to ensure $\Delta V = 0.1$ V. The synapses separation is ~1 μm. The single sodium and potassium pump maximum transport rates are ~200 Na^+ ions/sec and ~100 K^+ ions/sec, respectively. Hence, one finds that for the assigned L_A the firing rate is ~1 spike/sec.

The instantaneous power is $P = dW/dt$.

Using the force F and liner velocity v, one finds $P = Fv$.

The output power, as well as power required, can be found using the kinetic energy $\Gamma = \frac{1}{2}mv^2$, and $W = \Delta\Gamma$.

Consider a spherical particle with radius r that moves at velocity v in the liquid with viscosity μ. For the laminar flow, the Stokes's law gives the viscous friction (drag) force as

$$F_v = \eta v,$$

where η is the viscous friction (drag) coefficient, $\eta = 6\pi\mu r$.

The inverse of the drag coefficient is called the mobility, $\mu_B = 1/\eta = 1/(6\pi\mu r)$.

The diffusion constant D for a particle is related to the mobility and the absolute temperature. The Einstein fluctuation dissipation theorem gives $D = k_B T \mu_B$.

Using the ionic radii of Na^+ and K^+ ions, for $\mu = 9.5 \times 10^{-4}$ N sec/m^2 at 37°C ($T = 310.15$ K), one calculates $D_{Na} = 2.52 \times 10^{-9}$ m^2/sec and $D_K = 1.8 \times 10^{-9}$ m^2/sec, which agree with the experimental values $D_{Na} = 1.33 \times 10^{-9}$ m^2/sec and $D_K = 1.96 \times 10^{-9}$ m^2/sec, reported in Example 3.4.

By regulating the ionic channels, the cell controls the ionic flow across the membrane. The membrane conductance g has been experimentally measured using an expression $I = gV$. It is found that, for the sodium and potassium open channels, the conductance g is $\sim 2 \times 10^{-11}$ A/V.

The ionic current through the channel is estimated as

$$I = q_I J_I A_c,$$

where J_I is the ionic flux, $J_I = cv$; c is the ionic concentration; A_c is the channel cross-sectional area.

The average velocity of ion under the electrostatic field is estimated using the mobility and force as $v = \mu_B F = \mu_B q_I E = \mu_B q_I V/x$. Hence, one has $v = (D/k_B T) q_I (V/x)$. One finds the values for the velocity, force, and power. To transport a single ion, by using the data reported, the estimated power is $\sim 1 \times 10^{-13}$ W. Taking note of the number of ions required to produce (induce) and amplify the action potential for the firing rate 100 spike/sec and letting the instantaneous neuron utilization to be 1%, hundreds of watts are required to ensure communication only. It must be emphasized that binding/unbinding, production (reprocessing) of biomolecules, controlled propagation, and other cellular mechanisms require additional power. These also result in numerous losses.

The neuron energetics is covered in [4]. Using the longitudinal current, the intracellular longitudinal resistivity is found to be from 1×10^3 to 3×10^3 Ω mm, while the channel conductance is 25 pS or $g = 2.5 \times 10^{-11}$ A/V. For a 100 μm segment with 2 μm radius, the longitudinal resistance is found to be 8×10^6 Ω [4]. The membrane resistivity is 1×10^6 $\Omega\,\text{mm}^2$. To cross the ionic channel, the energy is $q_I V$. Taking note of the number of ions to generate a spike in neuron, the switching energy can be estimated to be $\sim 1 \times 10^{-14}$ J/spike. This section reports the cellular energetics by taking note of the conventional consideration. The action potentials, ionic transport, spike generation, and other cellular mechanisms exist, guarantying the functionality of cellular processes. However, the role and specificity of some biophysical phenomena, effects, and mechanisms may be revisited and coherently examined from the communication energetics, coding, and other perspectives. Recently, the research in *synaptic plasticity* has culminated in results departing from the past oversimplified analysis. However, the complexity of the processes and mechanisms is overwhelming. Correspondingly, novel postulates were proposed in Section 3.2.

3.4 Applied Information Theory and Information Estimates

Considering a neuron as a *switching* device, which is an oversimplified hypothesis, the interconnected neurons are postulated to be exited only by the action potential I_i. Neurons were attempted to be modeled as a spatio-temporal lattice of aggregated processing elements (neurons) by the second-order linear differential equation [5]:

$$\frac{1}{ab}\left(\frac{\mathrm{d}^2 x_i}{\mathrm{d}t^2} + (a+b)\frac{\mathrm{d}x_i}{\mathrm{d}t} + abx_i\right)$$

$$= \sum_{j \neq i}^{N} [w_{1ij} Q(x_j, q_j) + w_{2ij} f_j(t, Q(x_j, q_j))] + I_i(t), \quad i = 1, 2, \ldots, N-1, N,$$

$$Q(x,q) = \begin{cases} q(1 - \mathrm{e}^{(\mathrm{e}^x - 1)/q}) & \text{if } x > \ln[1 - q\ln(1+q^{-1})], \\ -1 & \text{if } x < \ln[1 - q\ln(1+q^{-1})], \end{cases}$$

where a, b, and q are the constants; w_1 and w_2 are the topological maps. This model, according to [5], is an extension of the results reported in [6,7] by taking into consideration the independent dynamics of the dendrites wave density and the pulse density for the parallel axons action.

Synaptic transmission has been researched by examining action potentials and the activity of the pre- and postsynaptic neurons [8–10] in order to understand communication, learning, cognition, perception, and so forth. Reference [11] proposes the learning equation for a synaptic adaptive weight $z(t)$ associated with a long-term memory as $dz/dt = f(x)[-Az + g(y)]$, where x is the activity of a presynaptic (postsynaptic) cell; y is the activity of a postsynaptic (presynaptic) cell; $f(x)$ and $g(y)$ are the nonlinear functions; A is the matrix. Reference [8–10] suggest that matching the action potential generation in the pre- and postsynaptic neurons equivalent to the condition of associative (Hebbian) learning that results in a dynamic change in synaptic efficacy. The excitatory postsynaptic potential results because of presynaptic action potentials. After matching, the excitatory postsynaptic potential changes. Neurons are firing irregularly at distinct frequencies. The changes in the dynamics of synaptic connections, resulting from Hebbian-type pairing, lead to significant modification of the temporal structure of excitatory postsynaptic potentials generated by irregular presynaptic action potentials [4]. The changes that occur in synaptic efficacy because of the Hebbian pairing of pre- and postsynaptic activity substantially change the dynamics of the synaptic connection. The long-term changes in synaptic efficacy (long-term potentiation or long-term depression) are believed to be dependent on the relative timing of the onset of the excitatory postsynaptic potential generated by the pre- and postsynaptic action potentials [8–10]. These, as well as other numerous concepts, have given rise to many debates. Furthermore, the cellular mechanisms responsible for the induction of long-term potentiation or long-term depression are not known.

Analysis of distinct cellular mechanisms and even unverified hypotheses that exhibit sound merits have a direct application to molecular electronics, envisioned bio-inspired processing, and so forth. The design of processing and memory platforms may be performed by examining baseline fundamentals at the device and system levels by making use of or prototyping (assuming that this task is achievable) cellular organization, phenomena, and mechanisms. Based on the inherent phenomena and mechanisms, distinct processing paradigms, different reconfiguration concepts, and various networking approaches can be envisioned for fluidic and solid electronics. This networking and interconnect, however, can unlikely be based on the semiconductor-centered interfacing as reported in [12]. Biomolecular, as against envisioned solid/fluidic ^{M}PPs, can be profoundly different from the device- and system-level standpoints.

Intelligent biosystems exhibit goal-driven behavior, evolutionary intelligence, learning, perception, and knowledge generation functioning in a non-Gaussian, nonstationary, rapidly changing dynamic environment. There does not exist a generally accepted concept for a great number of key open problems such as biocentered processing, memory, coding, and so forth. Attempts have been pursued to perform bioinspired symbolic, analog, digital (discrete-state and discrete-time), and hybrid processing by applying stochastic and deterministic concepts. To date, those attempts have not been culminated in feasible and sound solutions. At the device/module level, utilizing electrochemomechanical transitions and using biomolecules as the *information* and *routing* carriers, novel devices and modules have been proposed for the envisioned fluidic molecular electronics [2]. These carriers (intra- and extracellular ions and biomolecules) are controlled in cytoplasm, synaptic cleft, membrane channels, and so forth. The information processing platforms use stimuli and capture the goal-relevant information into the cognitive information processing, perception, learning, and knowledge generation. In bio-inspired fluidic devices, to ensure data/signal processing, one utilizes electrochemomechanically induced transitions that result, for example, from binding/unbinding of biomolecules. These biomolecules could be in *active*, *available*, *reprocessing*, and other states, and there are controlled propagation, production, activation, and other processes in neurons. Unfortunately, there is a significant gap between basic, applied, and experimental research as well as consequent engineering practice and technologies. Owing to technological and fundamental challenges and limits, this gap may not be overcome in the near future despite its tremendous importance.

Neurons in the brain, among various information processing and memory tasks, code and generate signals (stimuli) that are transmitted to other neurons through axon–synapse–dendrite *biomolecular hardware*. Unfortunately, we may not be able to coherently answer fundamental questions, including how neurons process (compute, store, code, extract, filter, execute, retrieve, exchange, etc.) information. Even the communication in neurons is a disputed topic. The central assumption is that the information is transmitted and possibly processed by means of an action potential–spikes mechanism. Unsolved problems exist in other critical areas, including information theory. Consider a series connection of processing primitives ($^{\rm ME}$device, biomolecule, or protein). The input signal is denoted as x, while the outputs of the first and second processing elements are y_1 and y_2. Even simplifying the data processing to a Markov chain $x \rightarrow y_1(x) \rightarrow y_2(y_1(x))$, cornerstone methods and information measures used in communication theory can be applied only to a very limited class of problems. One may not be able to explicitly, quantitatively, and qualitatively examine the information-theoretic measures beyond communication and coding problems. The information-theoretic estimates in neurons and molecular aggregates, shown in Figure 3.5, can be applied to the communication-centered analysis assuming the soundness of various postulates and availability of a great number of relevant data. Performing the communication and coding analysis, one examines the entropies of the variable x_i and y, denoted as $H(x_i)$ and $H(y)$. The probability distribution functions, conditional entropies $H(y|x_i)$ and $H(x_i|y)$, relative information $I(y|x_i)$ and $I(x_i|y)$, mutual information $I(y, x_i)$, and joint entropy $H(y, x_i)$ are of interest.

In a neuron and its intracellular structures and organelles, baseline processes, mechanisms, and phenomena are not fully understood. The lack of ability to soundly examine and coherently explain the basic biophysics has resulted in numerous hypotheses and postulates. From the signal/data processing standpoints, neurons are commonly studied as *switching* devices, while networked neuron ensembles have been considered assuming *stimulus-induced, connection-induced, adaptive,* and other *correlations.* Conventional neuroscience postulates that networked neurons transmit data, perform processing, accomplish communication, as well as perform other functions by means of sequence of spikes that are the propagating time-varying action potentials. Consider communication and coding in networked neurons assuming the validity of conventional hypotheses. Each neuron usually receives inputs from many neurons. Depending on whether the input produces a spike (excitatory or inhibitory) and how the neuron processes inputs, the neuron's functionality is determined. Excitatory inputs cause spikes, while inhibitory inputs suppress them. The rate at which spikes occur is believed to be changed by stimulus variations. Though the spike waveform (magnitude, width, and profile) vary, these changes usually considered to be irrelevant. In addition, the probability distribution function of the interspike intervals varies. Thus, input stimuli, as processed through a sequence of processes, result in outputs that are encoded as the pattern of action potentials (spikes). The spike duration is $\sim$1 msec, and the spike rate varies from one to thousands spikes per second. The doctrine that the spike occurrence, timing, frequency, and its probability distribution encode the information has been extensively studied. It is found that the same stimulus does not result in the same pattern, and debates continue, with an alarming number of recently proposed hypotheses.

Let us discuss the relevant issues applying the information-theoretic approach. In general, one cannot determine if a signal (neuronal spike, voltage pulse in ICs, electromagnetic wave, etc.) is carrying information or not. There are no coherent information measures and concepts beyond communication- and coding-centered analysis. One of the open problems is to qualitatively and quantitatively define what information is. It is not fully understood how neurons perform signal/data processing, not to mention information processing, but it is obvious that networked neurons are not analogous to combinational and memory ICs. Most importantly, by examining any signal, it is impossible to determine if it is carrying information or not and to coherently assess the signal/data processing, information processing, coding, or communication features. It is evident that there is a need to further develop the information theory. Those meaningful developments, as succeeded, can be applied in the analysis of neurophysiological signal/data and information processing.

Entropy, which is the Shannon quantity of information, measures the complexity of the set. That is, sets having larger entropies require more bits to represent them. For M objects (symbols) X_i that have probability distribution functions $p(X_i)$, the entropy is given as

$$H(X) = -\sum_{i=1}^{M} p(X_i)\log_2 p(X_i), \quad i = 1, 2, \ldots, M-1, M.$$

Example 3.6
Let

$$X = \begin{cases} a & \text{with probability } \frac{1}{2}, \\ b & \text{with probability } \frac{1}{4}, \\ c & \text{with probability } \frac{1}{8}, \\ d & \text{with probability } \frac{1}{8}. \end{cases}$$

The entropy $H(X)$ is found to be

$$\begin{aligned} H(X) &= -\sum_{i=1}^{M} p(X_i)\log_2 p(X_i) \\ &= -\tfrac{1}{2}\log_2\tfrac{1}{2} - \tfrac{1}{4}\log_2\tfrac{1}{4} - \tfrac{1}{8}\log_2\tfrac{1}{8} - \tfrac{1}{8}\log_2\tfrac{1}{8} = 1.75 \text{ bit}. \end{aligned}$$

The MATLAB statement to carry out the calculation of $H(X)$ is

```
H=-log2(1/2)/2-log2(1/4)/4-log2(1/8)/8-log2(1/8)/8
resulting in
H = 1.7500
```

One finds that the entropy of a fair coin toss is 1 bit, that is, $H(X) = 1$ bit.

Example 3.7
Let

$$X = \begin{cases} 1 & \text{with probability } p, \\ 0 & \text{with probability } 1-p. \end{cases}$$

The entropy $H(X)$ is a function of p, and

$$H(X) = -\sum_{i=1}^{M} p(X_i)\log_2 p(X_i) = -p\log_2 p - (1-p)\log_2(1-p).$$

One finds that $H(X) = 1$ bit when $p = \frac{1}{2}$. We have $H(X) = 0$ bit when $p = 0$ and $p = 1$. The entropy $H(X) = 0$ corresponds to the case when the variables are not random and there is no uncertainty. In general, $H(X)$ if p varies is of

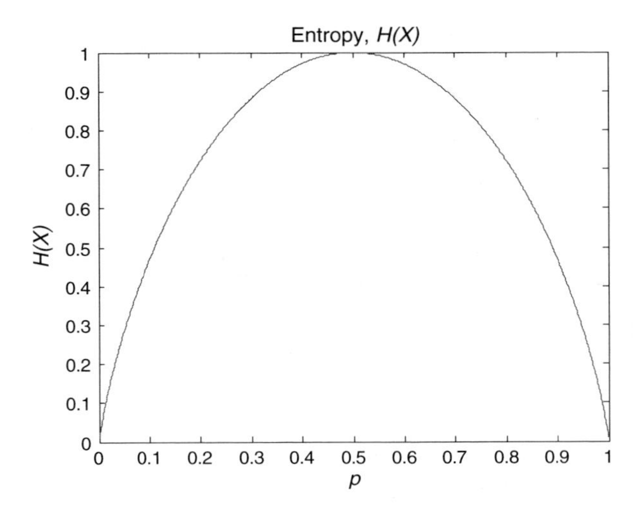

FIGURE 3.9
Entropy as a function of probability.

our interest. From the derived equation for $H(X)$, a plot $H(X)$ as a function of p is given in Figure 3.9. One concludes that the uncertainty is maximum when $p = \frac{1}{2}$, for which the entropy is maximum, that is, $H(X) = 1$ bit.

The MATLAB statement to perform the calculation and plotting is

```
p=0:0.001:1; H=-p.*log2(p)-(1-p).*log2(1-p);
plot(p,H); title('Entropy, \itH(X)','FontSize',14);
xlabel('\itp','FontSize',14);
ylabel('\itH(X)','FontSize',14);
```

We have $H \geq 0$, and, hence the number of bits required by the source coding Theorem is positive. Examining analog action potentials and considering spike trains, a *differential entropy* can be applied. For a continuous-time random variable X, the *differential entropy* is

$$H(X) = -\int p_X(x) \log_2 p_X(x) \mathrm{d}x,$$

where $p_X(x)$ is a 1D probability distribution function of x, $\int p_X(x)\mathrm{d}x = 1$.

In general,

$$H(X_1, X_2, \ldots, X_{n-1}, X_n) = -\int p_X(\mathbf{x}) \log_2 p_X(\mathbf{x}) \mathrm{d}\mathbf{x}.$$

The relative entropy between probability density functions $p_X(\mathbf{x})$ and $g_X(\mathbf{x})$ is

$$H_R(p_X \| g_X) = \int p_X(\mathbf{x}) \log_2 \frac{p_X(\mathbf{x})}{g_X(\mathbf{x})} \, \mathrm{d}\mathbf{x}$$

The *differential entropy* for common densities are easy to derive and well known. For example, for Cauchy, exponential, Laplace, Maxwell–Boltzman, normal, and uniform probability distribution functions $p_X(x)$, we have

$$p_X(x) = \frac{a}{\pi} \frac{1}{a^2 + x^2}, \quad -\infty < x < \infty, \; a > 0, \quad \text{and} \quad H(X) = \ln(4\pi a),$$

$$p_X(x) = \frac{1}{a} \mathrm{e}^{-(x/a)}, \; x > 0, \; a > 0, \quad \text{and} \quad H(X) = 1 + \ln a,$$

$$p_X(x) = \frac{1}{2a} \mathrm{e}^{-(|x-b|/a)}, \; -\infty < x < \infty, \; a > 0, \; -\infty < b < \infty, \quad \text{and}$$
$$H(X) = 1 + \ln(2a),$$

$$p_X(x) = 4\pi^{-(1/2)} a^{(3/2)} x^2 \, \mathrm{e}^{-ax^2}, \; x > 0, \; a > 0, \quad \text{and}$$
$$H(X) = -0.0772 + 0.5 \ln(\pi/a),$$

$$p_X(x) = \frac{1}{\sqrt{2\pi\sigma^2}} \mathrm{e}^{-((x-a)^2/2\sigma^2)}, \; -\infty < x < \infty, \; -\infty < a < \infty, \; \sigma > 0, \quad \text{and}$$
$$H(X) = 0.5 \ln(2\pi e \sigma^2),$$

$$p_X(x) = \frac{1}{b-a}, \; a \le x \le b, \quad \text{and} \quad H(X) = \ln(b-a),$$

respectively. We conclude that the *differential entropy* can be negative. For example, the *differential entropy* of a Gaussian random variable is $H(X) = 0.5 \ln(2\pi e \sigma^2)$, and $H(X)$ can be positive, negative, or zero depending on the variance. Furthermore, *differential entropy* depends on scaling. For example, if $Z = kX$, one has $H(Z) = H(X) + \log_2 |k|$, where k is the scaling constant.

To avoid the aforementioned problems, from the entropy analysis standpoints, continuous signals are discretized. Let X_n denotes a discretized continuous random variable with a binwidth ΔT. We have, $\lim_{\Delta T \to 0} H(X_n) + \log_2 \Delta T = H(X)$. The problem is to identify the information-carrying signals for which ΔT should be obtained.

One may use the a-order Renyi entropy measure as given by [13]:

$$R^a(X) = \frac{1}{1-a} \log_2 \int p_X^a(x) \, \mathrm{d}x,$$

where a is the integer, $a \ge 1$.

The first-order Renyi information ($a = 1$) leads to the Shannon quantity of information. However, Shannon and Renyi's quantities measure the complexity of the set, and even for this specific problem, the unknown probability distribution function should be obtained.

The Fisher information $I_F = \int ((\mathrm{d}p(x)/\mathrm{d}x)^2/p(x))\mathrm{d}x$ is a metric for the estimations and measurements. In particular, I_F measures an adequate change in knowledge about the parameter of interest.

Entropy does not measure the complexity of a random variable, which could be voltage pulses in ICs, neuron inputs or outputs (response) such as spikes, or any other signals. Entropy can be used to determine whether random variables are statistically independent or not. Having a set of random variables denoted by $\mathbf{X} = \{X_1, X_2, \ldots, X_{M-1}, X_M\}$, the entropy of their joint probability function equals the sum of their individual entropies, $H(\mathbf{X}) = \sum_{i=1}^{M} H(X_i)$, only if they are statistically independent.

One may examine the mutual information between the stimulus and the response in order to measure how similar the input and output are. We have

$$
\begin{aligned}
I(X,Y) &= H(X) + H(Y) - H(X,Y), \\
I(X,Y) &= \int p_{X,Y}(x,y) \log_2 \frac{p_{X,Y}(x,y)}{p_X(x)p_Y(y)} \,\mathrm{d}x\,\mathrm{d}y \\
&= \int p_{Y|X}(y|x)p_X(x) \log_2 \frac{p_{Y|X}(y|x)}{p_Y(y)} \,\mathrm{d}x\,\mathrm{d}y.
\end{aligned}
$$

Thus, $I(X,Y) = 0$ when $p_{X,Y}(x,y) = p_X(x)p_Y(y)$ or $p_{Y|X}(y|x) = p_Y(y)$. For example, $I(X,Y) = 0$ when the input and output are statistically independent random variables of each other. When the output depends on the input, one has $I(X,Y) > 0$. The more the output reflects the input, the greater is the mutual information. The maximum (infinity) occurs when $Y = X$. From a communications viewpoint, the mutual information expresses how much the output resembles the input.

Taking note that for discrete random variables

$$
I(X,Y) = H(X) + H(Y) - H(X,Y) \quad \text{or} \quad I(X,Y) = H(Y) - H(Y|X),
$$

one may utilize the conditional entropy

$$
H(Y|X) = -\sum_{x,y} p_{X,Y}(x,y) \log_2 \times p_{Y|X}(y|x).
$$

Here, $H(Y|X)$ measures how random the *conditional* probability distribution of the output is, on average, given a specific input. The more random it is, the larger the entropy, reducing the mutual information and $I(X,Y) \leq H(X)$ because $H(Y|X) \geq 0$. The less random it is, the smaller the entropy until it equals zero when $Y = X$. The maximum value of mutual information is the entropy of the input (stimulus).

The random variables X, Y, and Z are said form a Markov chain as $X \rightarrow Y \rightarrow Z$ if the conditional distribution of Z depends only on Y and is conditionally independent of X. For a Markov chain, the joint probability is $p(x, y, z) = p(x)p(y|x)p(z|y)$. If $X \to Y \to Z$, the data processing inequality $I(X, Y) \geq I(X, Z)$ ensures the qualitative and quantitative analysis features, imposes the limits on the data manipulations, provides the possibility to evaluate hardware solutions, and so forth.

The channel capacity is found by maximizing the mutual information subject to the input probabilities:

$$C = \max_{p_X(\cdot)} I(X, Y) \text{ (bit/symbol)}.$$

Thus, the analysis of mutual information results in the estimation of the channel capacity C, which depends on $p_{Y|X}(y|x)$, which defines how the output changes with the input. In general, it is very difficult to obtain or estimate the probability distribution functions. Using conventional neuroscience hypotheses, the neuronal communication, to some extent is equivalent to the communication in the *point process channel* [14]. The *instantaneous* rate at which spikes occur cannot be lower than $r_{\min}$ and greater than $r_{\max}$, which are related to the discharge rate. Let the average sustainable spike rate is r_0. For a Poisson process, the channel capacity of the point processes when $r_{\min} \leq r \leq r_{\max}$ is derived in [14] as

$$C = r_{\min}\left[e^{-1}\left(1 + \frac{r_{\max} - r_{\min}}{r_{\min}}\right)^{(1+r_{\min})/(r_{\max}-r_{\min})} - \left(1 + \frac{r_{\min}}{r_{\max} - r_{\min}}\right)\ln\left(1 + \frac{r_{\max} - r_{\min}}{r_{\min}}\right)\right],$$

which can be expressed as [15]:

$$C = \begin{cases} \dfrac{r_{\min}}{\ln 2}\left(e^{-1}\left(\dfrac{r_{\max}}{r_{\min}}\right)^{r_{\max}/(r_{\max}-r_{\min})} - \ln\left(\dfrac{r_{\max}}{r_{\min}}\right)^{r_{\max}/(r_{\max}-r_{\min})}\right) \text{ for} \\ \quad r_0 > e^{-1} r_{\min}\left(\dfrac{r_{\max}}{r_{\min}}\right)^{r_{\max}/(r_{\max}-r_{\min})}, \\ \dfrac{1}{\ln 2}\left((r_0 - r_{\min})\ln\left(\dfrac{r_{\max}}{r_{\min}}\right)^{r_{\max}/(r_{\max}-r_{\min})} - r_0 \ln\left(\dfrac{r_0}{r_{\min}}\right)\right) \text{ for} \\ \quad r_0 < e^{-1} r_{\min}\left(\dfrac{r_{\max}}{r_{\min}}\right)^{r_{\max}/(r_{\max}-r_{\min})}. \end{cases}$$

Let the minimum rate is zero. For $r_{\min} = 0$, the expression for a channel capacity is simplified to be

$$C = \begin{cases} \dfrac{r_{\max}}{e \ln 2}, & r_0 > \dfrac{r_{\max}}{e}, \\ \dfrac{r_0}{\ln 2}\ln\left(\dfrac{r_{\max}}{r_0}\right), & r_0 < \dfrac{r_{\max}}{e}. \end{cases}$$

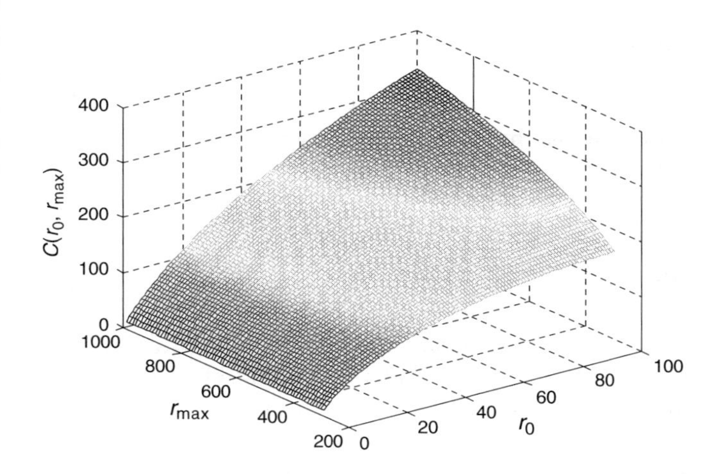

FIGURE 3.10
Channel capacity.

Example 3.8

Assume that the maximum rate varies from 300 to 1000 pulse/sec (or spike/sec), and the average rate changes from 1 to 100 pulse/sec. Taking note of $r_{max}/e = 0.3679r_{max}$, one obtains $r_0 < r_{max}/e$, and the channel capacity is given as $C = (r_0/\ln 2)\ln(r_{max}/r_0)$. The channel capacitance $C(r_0, r_{max})$ is documented in Figure 3.10. The MATLAB statement to calculate C and perform 3D plotting is

```
[r0,rmax]=meshgrid(1:1:100, 300:10:1000);
C=r0.*log(rmax./r0)/log(2);
plot3(r0,rmax,C); mesh(r0,rmax,C);
xlabel('r_0','FontSize',14);
ylabel('r_m_a_x', 'FontSize',14);
zlabel('C(r_0, r_m_a_x)', 'FontSize',14);
```

For $r_0 = 50$ and $r_{max} = 1000$, one finds $C = 216.1$ bits, or $C = 2.16$ bits/pulse.

Entropy is a function of the window size T and the time binwidth ΔT. For $\Delta T = 3 \times 10^{-3}$ sec and $18 \times 10^{-3} < T < 60 \times 10^{-3}$ sec, the entropy limit is found to be 157 ± 3 bit/sec [16]. For the spike rate $r_0 = 40$ spike/sec, $\Delta T = 3 \times 10^{-3}$ sec and $T = 0.1$ sec, the entropy is 17.8 bits [17]. This data agrees with these calculations for the capacity of the *point process channel* (see Figure 3.10).

For $r_0 = 50$ and $r_{max} = 1000$, one finds that $C = 216.1$ bits ($C = 2.16$ bit/pulse). However, this does not mean that each pulse (spike) represents 2.16 bits or any other number of bits of information. In fact, the capacity is derived for digital communication. In particular, for a Poisson process, using r_{min}, r_{max}, and r_0, we found specific rates with which digital signals (data) can be sent by a *point process channel* without incurring massive transmission errors.

For analog channels, the channel capacity is

$$C = \lim_{T \to \infty} \frac{1}{T} \max_{p_X(\cdot)} I(X, Y) \text{ (bit/sec)},$$

where T is the time interval during which communication occurs.

In general, analog communication cannot be achieved through a noisy channel without incurring error. Furthermore, the probability distribution functions as well as the distortion function must be known so as to perform the analysis. Probability distributions and distortion functions are not available, and processes are non-Poisson. In studying various communication problems one considers many senders and receivers applying the network information theory taking into account communication, channel transition matrices, interference, noise, feedback, distributed source coding (data compression), distributed communication, capacity, and other issues. Under many assumptions, for Gaussian relay, Gaussian interference, multiple access, and other channels, some estimates may be found. A great number of required details, statistics, and postulates are phenomena-, organization- and hardware-specific. This makes it virtually impossible to coherently apply information theory to ^{BM}PPs. For ^{M}PPs, information theory can be utilized with a great degree of caution. The focus of research should be directed on the verification of biomimetics-centered hypothesis and analysis of various molecular solutions utilizing sound fundamentals and technologies developed.

Other critical assumption commonly applied in the attempt to analyze bioprocessing features is a binary-centered hypothesis. Binary logics has a radix of two, meaning that it has two logic levels -0 and 1. The radix r can be increased by utilizing r states (logic levels). Three- and four-valued logics are called ternary and quaternary [18]. The number of unique permutations of the truth table for r-valued logic is r^{r^2}. Hence, for two-, three-, and four-valued logic, we have 2^4 (16), 3^9 (19,683), and 4^{16} (4,294,967,296) unique permutations, respectively. The use of multiple-valued logic significantly reduces circuitry complexity, device number, power dissipation, and improves interconnect, efficiency, speed, latency, packaging, and other features. However, sensitivity, robustness, noise immunity, and other challenging problems arise. A r-valued system has r possible outputs for r possible input values, and one obtains r^r outputs of a single r-valued variable [18]. For the radix $r = 2$ (binary logic), the number of possible output functions is $2^2 = 4$ for a single variable x. In particular, for $x = 0$ or $x = 1$, the output f can be 0 or 1, that is, the output can be the same as the input (identity function), reversed (complement), or constant (either 0 or 1). With a radix of $r = 4$ for quaternary logic, the number of output functions is $4^4 = 256$. The number of functions of two r-valued variables is r^{r^2}, and for the two-valued case $2^{2^2} = 16$. The larger the radix, the smaller is the number of digits necessary to express a given quantity.

The radix (base) number can be derived from optimization standpoints. For example, mechanical calculators, including Babbage's calculator, mainly utilize a 10-valued design. Though the design of multiple-valued memories is similar to the binary systems, multistate elements are used. A T-gate can be viewed as a universal primitive. It has $(r + 1)$ inputs, one of which is an r-valued control input whose value determines which of the other r (r-valued) inputs is selected for output. Because of quantum effects in solid ^{ME}devices, or controlled release-and-binding/unbinding of specific *information* carriers in the fluidic ^{M}devices, it is possible to employ enabling multiple-valued logics and memories.

Kolmogorov Complexity: Kolmogorov defined the algorithmic (descriptive) complexity of an object to be the length of the shortest computer program that describes the object. The *Kolmogorov complexity* for a finite-length binary string x with length $l(x)$ is

$$K_{\mathbf{U}}(x) = \min_{\mathbf{P}:\mathbf{U}(\mathbf{P})=x} l(\mathbf{P}),$$

where $\mathbf{U}(\mathbf{P})$ is the output of the computer $\mathbf{U}$ with a program $\mathbf{P}$. The Kolmogorov complexity provides the shortest descriptive length of x over all descriptions interpreted by a computer $\mathbf{U}$. One can derive the *Kolmogorov complexity* of a number ($\pi, \pi^2, \pi^{1/2}$, or any other), text, painting, image, and so forth. For example, in MATLAB, we compute π^2 as

```
» format long; pi^2
9.86960440108936
```

For an 8-bit-character, 73-symbol program, we calculate the descriptive complexity to be $8 \times 73 = 584$ bits.

The application of the *Kolmogorov complexity* for an *arbitrary* computer $\mathbf{U}_\mathrm{A}$ with $K_\mathrm{A}(x)$ is straightforward. The *universality Kolmogorov complexity* theorem states that there exists a constant c_A, which does not depend on x, such that

$$K_{\mathbf{U}}(x) \leq K_{\mathbf{A}}(x) + c_{\mathbf{A}}, \forall x \in [0, 1].$$

One can apply the conditional complexity, calculate the upper and lower bounds on *Kolmogorov complexity*, prove that $|K_{\mathbf{U}}(x) - K_{\mathbf{A}}(x)| < c$ (c is the constant $c \gg 1$), relate the *Kolmogorov complexity* to entropy, and so forth. However, the *Kolmogorov complexity*, as well as entropy, cannot be applied to perform the information estimates related to cognition, intelligence, and so forth. One can derive the *Kolmogorov complexity* of any image such as a painting using the number of pixels. For example, for Rembrandt and Rubens portraits $K(\text{Renbrandt}|n) \leq c + n/3$ and $K(\text{Rubens}|n) \leq c + n/3$. For biosystems, where the string, binary and other hypothesis most likely cannot be applied, it is difficult to imagine that *minimal* and *universal programs* are utilized. The *Kolmogorov complexity* can be enhanced to the high radix, strings of vectors $\mathbf{x} \in \mathbb{R}^n$ and other premises as applied to emerging ^{M}ICs and ^{M}PPs explicitly defining the hardware solutions and baseline processing features.

3.5 Biomimetics and Bioprototyping

In theory, the fundamentals, operation, functionality, and organization/architecture of ^{M}PPs can be devised through *biomimetics*. The theories of computing, computer architecture, information processing, and networking have been focused on the study of efficient robust processing, communication, networking, and other problems of hardware and software design. One can address and study fundamental problems of molecular electronics and ^{M}PPs utilizing advanced architectures/organizations, 3D topologies, multiterminal ^{M}devices, multithreading, error recovery, massive parallel processing, shared memory, message passing parallelism, and so forth. These features are likely possessed by living systems.

Novel devices and systems can be designed utilizing the phenomena and effects observed in biosystems. The necessary condition is the comprehension of various biophysical phenomena, effects, and mechanisms. The sufficient conditions include: (1) utilization of these phenomena, effects, and mechanisms, and, (2) ability to fabricate devices and systems. Prototyping must guarantee an eventual consensus and coherence between

- Basic phenomena and their utilization
- Device and system functionality, and capabilities
- System organization and architecture

ensuring descriptive and integrative features to carry out various design tasks. To acquire and expand the engineering–science–technology core, there is a need to integrate interdisciplinary areas as well as to link and place the synergetic perspectives integrating hardware with the discovery, optimization, and synthesis tasks. Biomimetics, as an envisioned paradigm in the design of ^{M}ICs and ^{M}PPs, is introduced to attack, integrate, and solve a great variety of emerging problems. Through biomimetics, fundamentals of engineering and science can be utilized with the ultimate objective of guarantying the synergistic combination of systems design, basic physics, fundamental theory, and fabrication technologies. In biosystems, the principles of matching and compliance are the general inherent design principles that require the system organization/architectures be synthesized by integrating all subsystems, modules, and components. The matching conditions, functionality, operationability, and systems compliance have to be examined and guaranteed. For example, the device–gate–hypercell compliance and operating functionality must be satisfied. We define biomimetics, as applied to ^{M}ICs and ^{M}PPs, as: Biomimetics is the coherent abstraction and sound practice in devising, prototyping, and design of molecular devices, modules, and systems using biological analogies through cornerstone fundamentals, bioinformatics, bioarchitectronics, and synthesis.

One can attempt to apply complex biological patterns to devise, analyze, and examine distinct devices and systems. However, biosystems cannot be copied because many phenomena, effects, and mechanisms have not been comprehended. For example, bioarchitectures have not been assessed, reconfiguration and control mechanisms are unknown, functionality of biomolecular primitives is not comprehended, and so forth. There is a need for focused studies and application of biomimetics to ensure the systematic design. Biological systems perform the following tasks of our specific interest:

1. Establish highly hierarchical multifunctional complex modules and subsystems, aggregating primitives (devices)–modules–systems within unified optimal topologies/organization/architectures
2. Information processing with various related tasks (coding, communication, etc.)
3. Evolutionary learning, real-time reconfiguration, and adaptive networking under rapidly evolving dynamic internal/external/environment changes
4. Robust self-assembling and self-organization (thousands of individual components are precisely defined in complex functional and operational devices and systems)
5. Adaptation, optimization, self-diagnostics, reconfiguration, and repairing (for example, biosystems can identify damage, perform adaptation, execute optimal decisions, and repair themselves to the functional state and operational level)
6. Precise bottom-up molecular assembly of very complex and multifunctional modules (for example, proteins and protein complexes, which integrate thousands of atoms, become dysfunctional by the change of a single atom)

Biomolecular processing platforms provide evidence of the soundness and achievable super-high-performance of envisioned ${}^{\mathrm{M}}$PPs. The engineering biomimetics paradigm potentially provides a meaningful conceptual tool to understand how ${}^{\mathrm{M}}$devices and ${}^{\mathrm{M}}$PPs can be devised and designed. For example, neurons, which perform various processing and memory tasks, are examined with the attempt to design integrated processor-and-memory bioinspired ${}^{\mathrm{M}}$PPs. Using the axo-dendritic input and dendro-axonic output vectors $\mathbf{x}$ and $\mathbf{y}$, the input–output mapping is schematically represented in Figure 3.11 for two neurons, as well as for a molecular processing primitive and ${}^{\mathrm{M}}$ICs. Taking note of the feedback vector $\mathbf{u}$, which is an important feature for robust reconfigurable (adaptive) processing, we have $\mathbf{y} = f(\mathbf{x}, \mathbf{u})$. It is possible to describe the information processing and transfer by bidirectional stream-oriented input–output operators, for

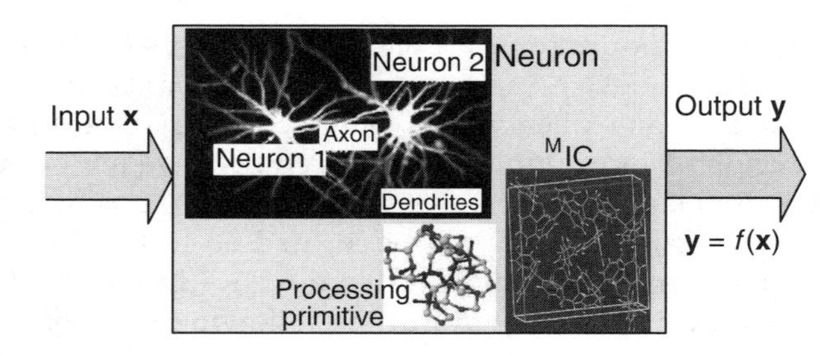

FIGURE 3.11
Input–output representation of aggregated neurons, processing molecular primitive and ^{M}IC.

example,

$$\textit{receive} \leftrightarrow \textit{process} \leftrightarrow \textit{send}.$$

By applying the possessed knowledge, it is a question *whether is it possible to accomplish a coherent biomimetics/bioprototyping and devise (discover and design) bio-identical or bi-centered processing and memory platforms*. Unfortunately, even for signal/data processing, it seems unlikely that those objectives could be achieved in the near future. There are a great number of unsolved fundamental, applied, and technological problems. To some extent, a number of problems can be approached by examining and utilizing different biomolecular-centered processing features and postulates. Our goal is to develop general and application-centric fundamentals, molecular hardware, and foundations that do not rely on hypotheses, postulates, assumptions, exclusive solutions, and hardware. Sound technology-centric solid and fluidic molecular electronics are prioritized.

It will be covered in Section 3.9 that the *information coding* is accomplished by the DNA, and the application of transcription–translation mechanism (DNA $\rightarrow$ RNA $\rightarrow$ protein) may result in the synthesis of *biomolecular processing hardware* with embedded *software*. Among the major tasks of an interest to be accomplished, one may emphasize the following:

1. Identify and comprehend the phenomena, effects, mechanisms, transitions, and interactions in biomolecular processing hardware (biomolecules, proteins, etc.) that ultimately accomplish processing in living organisms
2. Examine and comprehend the *instructional codes* to perform *bottom-up* synthesis of the *biomolecular processing hardware*

The successful solution of these formidable problems will enable one to design and synthesize

- Biomolecular processing and memory primitives
- ^{BM}Modules and ^{BM}PPs

by fully utilizing biomimetics applying bioinformatics. However, these are long-term goals.

3.6 Fluidic Molecular Platforms

The activity of brain neurons has been extensively studied using single microelectrodes as well as microelectrode arrays to probe and attempt to influence the activity of a single neuron or assembly of neurons in brain and neural culture. The integration of neurons and microelectronics is covered in [12,19–21]. We propose a fluidic molecular processing device/module that mimics, to some extent, a neuronal processing-and-memory primitive or a brain neuron. In general, signal/data processing and memory storage can be accomplished by various electrochemomechanically induced transitions, interactions, and events. For example, release, propagation, and binding/unbinding of movable molecules result in the state transitions to be utilized. Owing to fundamental complexity and technological limits, one may not coherently mimic and prototype bioinformation processing. Therefore, we mimic 3D topologies/organizations of biosystems, utilize molecular hardware, and employee molecular transitions. These innovations imply novel synthesis, design, aggregation, utilization, functionalization, and other features. Using the specific electromechanical transitions (*electromechanical switching* in biomolecules or electron tunneling in organic molecules) and molecules/ions as *information/routing* carriers, we propose a novel concept to perform signal/data processing and memory storage.

Utilizing 3D topology/organization, observed in ^{BM}PPs, a *synthetic fluidic* device/module is illustrated in Figure 3.12. The inner enclosure can be made of proteins, porous silicon, or polymers to form membranes with fluidic channels that should ensure the selectivity. The information and

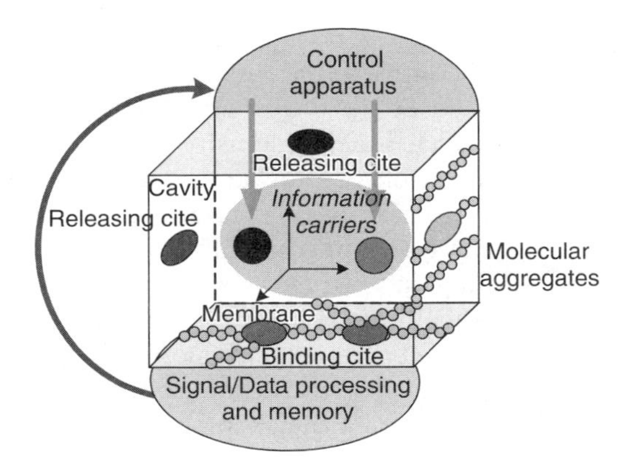

FIGURE 3.12
Synthetic fluidic molecular processing module.

routing carriers are encapsulated in the outer enclosure or cavities. The release and steering of different carriers are controlled by the control apparatus. Selective information carriers are used as logic and memory inputs ensuring the multiple-valued solution. Hence, computing, processing, and memory storage can be performed on the high radix. Using routing carriers, persistent and robust morphological reconfigurable networking can be achieved. The proposed fluidic ^{M}PPs can be designed within a reconfigurable networking processing-and-memory organization. In particular, *routing* carriers are steered in the fluidic cavity to the binding sites resulting in the binding/unbinding of routers to the stationary molecules. The binding/unbinding events lead to reconfigurable routing, for example, reconfiguration. Independent control of *information* and *routing* carriers cannot be accomplished through preassigned steady-state conditional logics, synchronization, timing protocols, and other conventional concepts. The motion and dynamics of the carrier release, propagation, binding/unbinding, and other events should be examined.

The proposed device/module mimics to some extent a neuron with synapses, membranes, channels, cytoplasm, and other components. Specific ions, molecules, and enzymes can pass through the porous membranes. These passed molecules (*information* and *routing* carriers) bind to the specific receptor sites, while enzymes free molecules from binding sites. Binding and unbinding of molecules result in the state transitions. The carriers that are released pass through selective fluidic channels and propagate through cavity and are controlled by changing the electrostatic potential or thermal gradient. The goal is to achieve a controlled Brownian motion of *microscopic* carriers. Distinct control mechanisms (electrostatic, electromagnetic, thermal, hydrodynamic, etc.) may allow one to uniquely utilize selective control ensuring super-high performance and enabling functionality. The controlled Brownian dynamics (*active* transport) of molecules and ions in the fluidic cavity and channels can be examined by applying the concept reported in Section 3.3. Having developed a hardware solution, the carrier displacement $\mathbf{r}_i$ is controlled by the control vector $\mathbf{u}$. The released carriers propagate in the fluidic cavity and controlled by a control apparatus varying $F_n(t, \mathbf{r}, \mathbf{u})$ and $V_k(\mathbf{r}, \mathbf{u})$. This apparatus is comprised of molecular structures that change the temperature gradient or the electric field intensity. The *state* transitions occur in the anchored processing polypeptide or organic molecular complexes as *information* and *routing* carriers bind/unbind. For example, conformational *switching*, charge changes, electron transport, and other phenomena can be utilized. The settling time of electronic, photoelectric, and electrochemomechanical *state* transitions is from picoseconds to microseconds. The molecular hardware predifines the phenomena and effects utilized. For example, *electromechanical switching* can be accomplished by biomolecules, while electron tunneling can be of a major interest as organic molecules are used as ^{ME}devices.

In general, it is possible to design, and potentially synthesize, aggregated networks of reconfigurable fluidic modules. These modules can be characterized in terms of input–output activity. However, the complexity to fabricate the reported fluidic ^{M}PPs is enormous, imposing significant challenges. Reported in this section solution is an elegant theoretical concept that will hold until the available fabrication technologies emerge. To overcome the fabrication deficiencies, one may utilize the cultured neurons.

3.7 Neuromorphological Reconfigurable Molecular Processing Platforms and Cognitive Processing

Reconfigurable computing is a well-established general term that applies to any device or primitive that can be configured, at run time, to implement a function utilizing a specific hardware solution. A reconfigurable device should possess adequate logic, reprogramming and routing capabilities to ensure reconfiguration features, as well as to compute a large set of functions. The reconfigured ^{M}device performs a different set of functions. Consider a gate with binary inputs A and B. Using the outputs to be generated by the universal logic gate, one has the following 16 functions: 0, 1, A, B, $\bar{A}$, $\bar{B}$, A + B, A + $\bar{B}$, $\bar{A}$ + B, $\bar{A}$ + $\bar{B}$, AB, A$\bar{B}$, $\bar{A}$B, $\overline{AB}$, A$\bar{B}$ + $\bar{A}$B, and AB + $\overline{AB}$. The standard logic primitives (AND, NAND, NOT, OR, NOR, etc.) can be implemented using a Fredkin gate, which performs conditional permutations. Consider a gate with a *switched* input A and a *control* input B. As illustrated in Figure 3.13, the input A is routed to one of two outputs, conditional on the state of B. The routing events change the output switching function, which is AB or A$\bar{B}$.

Utilizing the proposed fluidic molecular processing paradigm, *routable* molecular universal logic gates (^{M}ULG) can be designed and implemented. We define a ^{M}ULG as a reconfigurable combinational gate that can be reconfigured to realize specified functions of its input variables. The use of specific multi-input ^{M}ULGs is defined by the technology soundness, specifications, and achievable performance. These ^{M}ULGs can realize logic functions using multi-input variables with the same delay as a two-input ^{M}gate. Logic functions can be efficiently factored and decomposed using ^{M}ULGs.

Figure 3.14 schematically depicts the proposed routing concepts for reconfigurable logics. The typified 3D-topologically reconfigurable routing is accomplished through the binding/unbinding of *routing* carriers to the stationary molecules, which perform processing and memory storage. For illustrative purposes, Figure 3.14 documents reconfiguration of 5 ^{M}gates

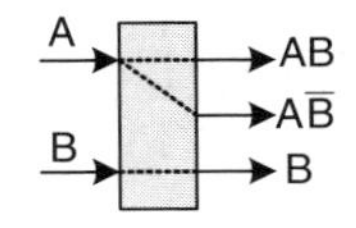

FIGURE 3.13
Gate schematic.

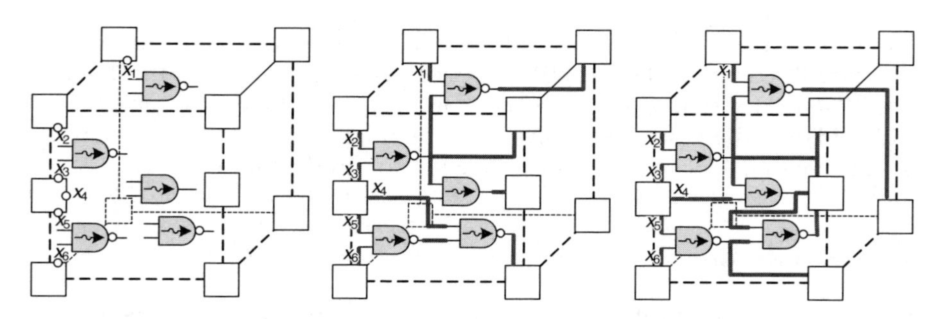

FIGURE 3.14
Reconfigurable routing and networking.

within 10 ℵhypercells depicting a reconfigurable networking and processing in 3D. The state transitions, caused by the *information* carriers, are represented as x_1, x_2, x_3, x_4, x_5, and x_6. The *routing* carriers ensure a reconfigurable routing and networking of Mgates and hypercells, uniquely enhancing and complementing the capabilities of the ℵhypercell solution. In general, one may not be able to route any output of any gate/hypercell/module to any input of any other gate/hypercell/module. There are synthesis constraints, selectivity limits, complexity to control the spatial motion of *routers* and other limits that should be integrated in the design.

The fluidic module can perform computations, implement complex logics, ensure memory storage, guarantee memory retrieval, and ensure other tasks through electrochemomechanically induced transitions/interactions/events. Sequences of conditional aggregation, carriers steering, 3D directed routing, and spatial networking events form the basis of the logic gates and memory retrieval in the proposed neuromorphological reconfigurable fluidic MPPs. In Section 3.3 we documented how to integrate the Brownian dynamics in the performance analysis and design. The transit time of *information* and *routing* carriers depends on the steering mechanism, control apparatus, carriers used, sizing features, and so forth. From the design prospective, one applies the state-space paradigm using the processing and routing transition functions F_p and F_r, which describe previous states to the resulting new states in $[t, t_+]$, $t_+ > t$. The output evolution is

$$\mathbf{y}(t_+) = F_i[t, \mathbf{x}(t), \mathbf{y}(t), \mathbf{u}(t)],$$

where $\mathbf{x}$ and $\mathbf{u}$ are the state and control vectors.

For example, $\mathbf{u}$ leads to the release and steering of the *routing* carriers with the resulting routing and reconfigurable networking. The reconfiguration is described as $P \subset X \times Y \times U$, where X, Y, and U are the input, output, and control sets.

The proposed neuromorphological reconfigurable fluidic MPPs surpass the overall performance, functionality, and capabilities of envisioned microelectronic solutions and ICs. However, the theoretical and technological foundations of neuromorphological reconfigurable ^{3D}networking-processing-and-memory ^{M}PPs remain to be developed and implemented. It is expected that cognition can be achieved utilizing neuromorphological reconfigurable processing. The issues of cognitive processing are briefly examined.

Information (***I***) causes changes either in the whole system (***S***) that receives information or in an information processing logical subsystem ($\boldsymbol{S_I}$) of this system. There are different types of information measures, estimates, and indexes. For example, potential or perspective measures of information should determine (reflect) what changes may be caused by ***I*** in ***S***. Existential or synchronic measures of information should determine (reflect) what changes ***S*** experiences during a fixed time interval after receiving ***I***. Actual or retrospective measures of information should determine (reflect) what changes were actually caused by ***I*** in ***S***. For example, synchronic measures reflect the changes of the short-term memory, while retrospective measures represent transformations in the long-term memory.

Consider the system mapping tuple (***S***, ***L***, ***E***), where ***E*** denotes the environment and ***L*** represents the linkages between ***S*** and ***E***. There are three structural types of measure of information, for example, internal, integral and external. The internal information measure should reflect the extent of inner changes in ***S*** caused by ***I***. The integral information measure should reflect the extent of changes caused by ***I*** on ***S*** due to ***L*** between ***S*** and ***E***. Finally, the external information measure should reflect the extent of outer changes in ***E*** caused by ***I*** and ***S***. One can define three constructive types of measures of information—abstract, realistic, and experimental. The abstract information measure should be determined theoretically under general assumptions, while a realistic information measure must be determined theoretically subject to realistic conditions applying sound information-theoretic concepts. Finally, the experimental information measure should be obtained through experiments.

The information can be measured, estimated, or evaluated only for simple systems examining a limited number of problems (communication and coding) for which the information measures exist. Any ***S*** has many quantities, parameters, stimuli, states, events, and outputs that evolve. Different measures are needed to be used in order to reflect variations, functionality, performance, capabilities, efficiency, and so forth. It seems that currently the prospect of finding and using a universal information measure is unrealistic. The structural-attributive interpretation of information does not represent information itself but may relate ***I*** to the information measures (for some problems), events, information carriers, and communication in ***S***. In contrast, the functional-cybernetic consideration is aimed to explicitly or implicitly examine information from the functional viewpoint descriptively studying *state* transitions in systems that include information processing logical subsystems.

Cognitive systems are envisioned to be designed by accomplishing information processing through integrating knowledge generation, perception, learning, and so forth. By integrating interactive cognition tasks, there is a need to expand signal/data processing (primarily centered on binary computing, coding, manipulation, mining, and other tasks) to information processing. The information theory must be enhanced to explicitly evaluate knowledge generation, perception, and learning by developing an information-theoretic framework of information representation and processing. The information processing at the system and device levels must be evaluated using the cognition measures examining how systems represent and process the information. It is known that information processing depends on the statistical structure of stimuli and data. This statistics may be utilized to attain statistical knowledge generation, learning, adaptation, robustness, and self-awareness. The information-theoretic measures, estimates and limits of cognition, knowledge generation, perception and learning in S must be found and examined to approach fundamental limits and benchmarks. Cognizance has been widely studied from artificial intelligence standpoints. However, limited progress has been achieved in basic theory, design, applications, and technology developments. New theoretical foundations, software, and hardware to support cognitive systems design must be developed. Simple increase of computational power and memory capacity to any level will not result in cognizance and/or intelligence due to entirely distinct functionality, capabilities, measures, and design paradigms. From fundamental, computational, and technology standpoints, problems to be solved are far beyond conventional information theory, signal/data processing, and memory solutions.

Consider a data information set, which is a global knowledge with $\boldsymbol{\Sigma}$ states. By using the observed data $\boldsymbol{D}$, system gains and learns certain knowledge, but not all, $\boldsymbol{\Sigma}$. Before the observations, system possesses some states from distribution $p(\boldsymbol{\Sigma})$ with the information measure $\boldsymbol{M}(\boldsymbol{\Sigma})$. This $\boldsymbol{M}(\boldsymbol{\Sigma})$ must be explicitly defined, and this is an open problem. Once system observes some particular data $\boldsymbol{D}$, the enhanced perception of $\boldsymbol{\Sigma}$ is described by the reciprocal measure estimate $\boldsymbol{M}(\boldsymbol{\Sigma}|\boldsymbol{D})$, and $\boldsymbol{M}(\boldsymbol{\Sigma}|\boldsymbol{D}) \leq \boldsymbol{M}(\boldsymbol{\Sigma})$. The uncertainty about $\boldsymbol{\Sigma}$ reduces through observations, learning, perception, and so forth. We define this process as the information gain that the system acquired about $\boldsymbol{\Sigma}$. Some data $\boldsymbol{D}_{<} \in \boldsymbol{D}$ will increase the uncertainty about $\boldsymbol{\Sigma}$, resulting in the knowledge reduction. For this regret $\boldsymbol{D}_{<}$, one finds $\boldsymbol{M}(\boldsymbol{\Sigma}|\boldsymbol{D}_{<})$, and the information reduction is expressed as $\boldsymbol{I}_{<} = f[\boldsymbol{M}(\boldsymbol{\Sigma}), \boldsymbol{M}(\boldsymbol{\Sigma}|\boldsymbol{D}_{<})]$.

With the goal to achieve cognition and learning by gaining the information (on average) $\boldsymbol{I}_{\boldsymbol{D}\to\boldsymbol{\Sigma}}$, one should derive $\boldsymbol{I}$ using the information measures and estimates. By observing the data, the system cannot learn more about the global knowledge than $\boldsymbol{M}(\boldsymbol{\Sigma})$. In particular, $\boldsymbol{M}(\boldsymbol{\Sigma})$ may represent the number of possible states that the knowledge is mapped, and $\boldsymbol{M}(\boldsymbol{\Sigma})$ indicates the constrained system ability to gain knowledge due to the lack of possibilities in $\boldsymbol{\Sigma}$. The system cannot learn more than the information measure that characterizes the data. In particular, the $\boldsymbol{M}$ of observations limits how much system can learn. In general, $\boldsymbol{M}$ defines the capacity of the data $\boldsymbol{D}$ to provide or to convey information. The information that the system can gain has upper and lower bounds defined by the $\boldsymbol{M}$ limits, while $\boldsymbol{M}$ bounds depend on the statistical properties, structure, and other characteristics of the observable data as well as the system S abilities.

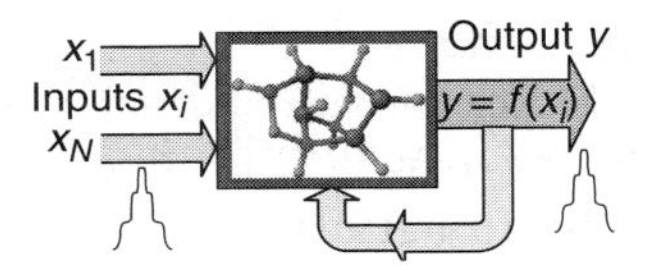

FIGURE 3.15
Cognitive information processing primitive $_PS$.

We consider a cognitive information processing primitive $_P\boldsymbol{S}$ implemented as a multiterminal molecule. Utilizing the continuous information-carrying inputs $\mathbf{x} = [x_1, x_2, x_3, x_4] \in X$, $_P\boldsymbol{S}$ generates a continuous output $y(t)$, $y \in Y$, with distinguished states as shown in Figure 3.15. Hence, the multiple-valued inputs $\mathbf{x}$ are observed and processed by $_P\boldsymbol{S}$ with a transfer function $F(\mathbf{x}, y)$. Cognitive learning can be formulated as utilization and optimization of information measures through the $\boldsymbol{S}$ perception, knowledge generation and reconfiguration. In general, $\boldsymbol{S}$ integrates subsystems $_S\boldsymbol{S}$, modules $_M\boldsymbol{S}$, and primitives (gate/devices level) $_P\boldsymbol{S}$. The primitive $_P\boldsymbol{S}$ statistical model can be described by $\boldsymbol{M}(\mathbf{x}, \mathbf{y})$ as generated through learning and perception using observed $\mathbf{x} = [x_1, x_2, \ldots, x_{n-1}, x_n] \in X$ and $\mathbf{y} = [y_1, y_2, \ldots, y_{m-1}, y_m] \in Y$.

3.8 Reconfigurable 3D Networking-and-Processing and Reconfigurable Vector Neural Networks

Cognitive science is centered on the *neuron doctrine* with a goal to understand a broad spectrum of open fundamental problems of brain functionality. A behaviorist view, with its emphasis on general learning mechanisms, persisted for many years focusing on narrowly conceived issues. Karl Lashley emphasized neural mechanisms and intelligence [22]. Kenneth Craik in *The Nature of Explanation* considered "the conscious working of a highly complex machine, built of parts having dimensions at which the classical laws of mechanics are still very nearly true and space is, to all intents and purposes, Euclidean. This mechanism, I have argued, has the power to represent, or parallel, certain phenomena in the external world as a calculating machine..." [23]. Within this concept, thoughts involve three critical steps (1) external processes were translated into words, numbers, or symbols; (2) these *representations* were manipulated by processes such as reasoning to yield transformed symbols; (3) these transformed symbols were retranslated into external processes to yield a behavior. The idea that the mind creates internal models and then uses these models is of a great interest. The major enterprise is to apply sound biophysics and derive basic fundamentals in the context of the exhibited and utilized biophysical phenomena, effects and mechanisms. *The Computer and the Brain* discusses the fundamental properties of computation in machine and brain, proposing an intriguing "approach toward the understanding of the nervous system from the mathematician's point of view" [24]. The debates have being continued on the cornerstone issues, approaches, and directions. Many recent publications report specific brain correlates of higher cognitive functions (from language to numerical cognition, from decision making to categorization and reasoning, etc.), emphasizing a match between abstract cognitive characterizations and brain structures. The developments of different sensing, recording, and data-acquisition techniques may provide the opportunity to study the brain, neurons, and processing primitives during various cognitive activities. Even assuming that one comprehends the physics of the input/output stimuli (potential, charge, polarization, voltage, current, temperature, intensity, folding, etc.), the use of electroencephalographic methods, event-related potentials, functional magnetic resonance imaging, neuronimaging, magnetoencephalography, transcranial stimulations, chemical labeling, and other techniques have limits on recording and examining spatial and neural activities and do not define their role on cognition or even on signal/data processing.

Reference [25] proposes that the mental activity could be modeled using a network of *on–off neurons*, which can compute and implement logic functions. The basic physics, functionality, and capabilities of a brain neuron and an *artificial neuron* are completely different. Correspondingly, the parallel of *an artificial neuron*, as a brain primitive, is the largely simplified assumption that is not supported by recent basic and experimental findings. Even at the 1942 workshop in New York City (sponsored by the Josiah Macy Jr. Foundation), where Warren McCulloch reported the studies to be published in [25], Arturo Rosenblueth made his presentation. Reference [26] documents that goal-driven behavior emerges in systems with feedback.

The concepts of learning and connectivity was introduced in 1949 [27], while the "artificial intelligence and learning" (training) capabilities of the perceptron model were discussed in [28]. These developments were culminated with a neural networks (NNs) approach proposed by John Hopfield in 1982. In particular, inspired by the *neuronal circuitry*, the NNs were utilized as abstract entities. These NNs, which can be implemented by ICs hardware, are capable to ensure specified by the designer *automatic* training (usually referenced as "learning"), programmed logics, rule extraction, generalization, categorization, extraction, and other functions [29,30]. The Hopfield network is a single-layer recurrent network that allows one to store information as the stable states of a dynamically evolving network configuration. The discrete Hopfield NN can be used as associative memory, while the continuous Hopfield NN can be applied for combinatorial optimization [31]. Utilizing backpropagation and different training algorithms, these *artificial neural networks* (ANNs) can ensure feasibility to solve some relatively complex problems. However, many mathematical issues remain unsolved.

The exploratory research focuses on the NNs addressing some major premises of possible processing tasks of biosystems—logics, reasoning, computing, memory, perception, representation, reconfiguration, decision

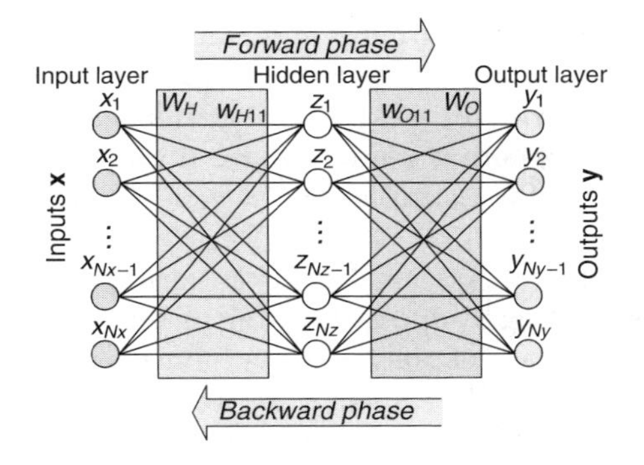

FIGURE 3.16
Multilayer neural network.

making, and so forth. Reference [25] focused on a simple binary threshold-type *artificial neuron* that has both *excitatory*- and *inhibitory*-like inputs. When connected within a network, these *neurons* can accomplish logics and compute Boolean functions. Reference [32] introduced the training algorithm for the multilayer perceptron network, called backpropagation. A great number of network organizations (feedforward and recurrent), training algorithms (supervised, unsupervised, and hybrid), and connectivity (single- and multilayer, competitive, Hopfield, and others) were studied. A two-layer feedforward network is shown in Figure 3.16. This network is fully connected, because there are all-to-all connections between two adjacent neuron layers. The number of neurons in the input, hidden, and output layers is denoted as N_x, N_z, and N_y. In general, the number of hidden layers can be high. The back-propagation training phase for a pattern consists of a *forward phase* followed by a *backward phase*. The main steps are: (1) define the NNs organizations, as well as N_x, N_z, and N_y; (2) initialize the weights; (3) select a training vector pair (input and the corresponding output) from the training set and apply the input vector; (4) *forward phase*—calculate or measure the actual outputs; (5) *backward phase*—according to the difference (error) between actual and desired outputs, adjust the weights matrices W_O (with entities w_{Okj}) and W_H (with entities w_{Hji}) to reduce or eliminate the error (difference); (6) repeat from step 3 for all training vectors; (7) repeat from step 3 until the error is acceptably small. This weight-updating scheme is called *training by pattern*, and it updates the weights after *each* training pattern.

In the *forward phase*, the hidden layer weight matrix W_H is multiplied by the input vector $\mathbf{x} = [x_1, x_2, \ldots, x_{Nx-1}, x_{Nx}]^{\mathrm{T}}$ in order to calculate the hidden layer outputs as $z_j = f(\sum_{i=1}^{N_x} (w_{Hji}x_i - \theta_i))$, where w_{Hji} is the weight (coefficient) between x_i to z_j, and θ_i is the bias; f is the nonlinear continuously differentiable function, for example, the sigmoid function, for example, $f = 1/(1 + \mathrm{e}^{-a})$, $a > 0$.

The output of the hidden layer z_j is used to calculate the NN outputs as $y_k = f\left(\sum_{j=1}^{N_z} (w_{Okj}z_j - \theta_j)\right)$.

In the *backward phase*, the target **t** and output **y** are compared, and the error vector is used to adjust the weights to reduce the error. To ensure the steepest descent in the weight space, the weights change as $\Delta w_{Okj} = \eta_j e_{Ok} z_j$ and $\Delta w_{Hji} = \eta_i e_{Hj} x_i$, where e_{Ok} and e_{Hj} are the errors used to update the weights w_{Okj} and w_{Hji}, for example, $e_{Ok} = y_k(l - y_k)(t_k - y_k)$; η_j and η_i are the coefficients.

A great number of concepts are based on various gradient descent techniques resulting in training by epoch, training by block, and other approaches, which for many problems ensure the convergence, robustness, and stability. The weights are updated with respect to the training patterns using different equations. The number of floating point operations used for weight updating for distinct training approaches are different, but in general very high. The powerful *Neural Network Toolbox* is available in the MATLAB environment, and the reader is directed to a textbook [31] for details.

Various biophysical processes and mechanisms in neurons, as well as associated electrochemomechanical transitions in neurons, their organelles and synapses, are directly or indirectly affect cognition and functionality of brain. Taking into the account that neurons

1. Input and output stimuli by means of biomolecular electrochemomechanical transitions/interactions/events (at pre- and postsynaptic membranes in multiple synapses)
2. Process and store the information utilizing electrochemomechanical transitions/interactions/events in biomolecules housed in the cell body and organelles

we consider a neuron as a multi-input/multi-output processing module that consists of neuronal processing primitives (NPPs). The synapses within axo-dendritic, dendro-axonic and axo-axonic biomolecular hardware establish reconfigurable connectivity, and each synapse is considered to be equivalent to input or output *terminal* to and from NPPs. Numerous neurotransmitters, binding and enzymatic specificity, proteins exclusivity, and other features ensure that the input and output at each terminal are not variables (such as voltage in NNs or spike in neurons) but vectors. One is referred to Figure 3.5, which illustrates and visualizes the reported features. The mircotubules, MAPs, SAPs, and other proteins establish the connectivity between the input, hidden, and output lattices (layers) in NPPs accomplishing the reconfigurable networking. For example, if the neuron has 10,000 input synapses and 1,000 output synapses, one may considers 10,000 NPPs in the input layers and 1,000 NPPs in the output layer. The electrochemomechanically induced transitions result in processing, for example, by varying the weights, number of neurons, connectivity, and so forth. One accomplishes the ^{3D}networking-and-processing utilizing a reconfigurable vector neural network (^{RV}NN), which consists of NPPs. This ^{RV}NN possesses the abilities to perform processing, reconfigurable networking, adaptation, and other tasks (functions) with superior overall performance and enhanced capabilities as compared with conventional NNs. Furthermore, the ^{RV}NN utilizes feedback ensuring dynamic interactive behavior. Referencing to the conventional terminology, a NPP could be considered as an *artificial neuron*, but as was emphasized, a NPP possesses enhanced capabilities due to the utilization of vectors $\mathbf{x}_i$ and $\mathbf{y}_i$. The considered ^{RV}NN consists of massively interconnected NPPs within a 3D hypercube mesh organization. We summarize the reported analysis as follows:

- In ^{RV}NN, each input and output synapse (*terminal*) receives (inputs) and sends (outputs) vectors $\mathbf{x}_i$ and $\mathbf{y}_i$ to and from NPPs.
- ^{RV}NN reconfigurates its organization (utilizing electrochemomechanically induced transitions, NPPs adaptive connectivity, feedback mechanisms, and events) ensuring reconfigurable networking.
- ^{RV}NN performs processing and/or memory storage.

In the conventional NNs, each neuron receives inputs x_i from other neurons, performs a weighted summation, applies an activation function to the weighted sum, and outputs its results to other neurons as y_i. For ^{RV}NN, the input and output vectors are $\mathbf{x} = [\mathbf{x}_1, \mathbf{x}_2, \ldots, \mathbf{x}_{Nx-1}, \mathbf{x}_{Nx}]^T$ and $\mathbf{y} = [\mathbf{y}_1, \mathbf{y}_2, \ldots, \mathbf{y}_{Ny-1}, \mathbf{y}_{Ny}]^T$. Here, $\mathbf{x}_i \in \mathbb{R}^n$ and $\mathbf{y}_i \in \mathbb{R}^m$. The neuron output and hidden layer output are given as

$$y_{ki} = f\left(\sum_{j=1}^{N_z}\sum_{l=1}^{p}(w_{Okjl}z_{jl} - \theta_{jl})\right) \quad \text{and} \quad z_{ji} = f\left(\sum_{i=1}^{N_x}\sum_{h=1}^{n}(w_{Hjih}x_{ih} - \theta_{ih})\right), \quad \mathbf{z}_i \in \mathbb{R}^p.$$

The θ_{jl} and θ_{ih} can be adjusted using feedback as shown in Figure 3.5. The adaptation and training are performed using the error vector.

Though the proposed concept likely has an analogy to bioprocessing, it is primarily developed for ^{M}PPs for which we introduce and apply a reconfigurable ^{3D}networking-and-processing paradigm. The analogies to neurons and possible parallels to neuronal processing are emphasized. However, these results are reported with some cautious understanding that questions may surface. Despite of attractive terminology (neural, learning, adaptive behavior, etc.), used in life science, the real proximity of NNs to biosystems is questionable. Though abstract mathematical formulations can result in the formal equivalence between neural and neuronal networks, no matter how elegant and intellectually satisfactory at an abstract level, serious doubts and concerns remain in the applicability of these models and paralleling NNs to the real brain. For ^{M}PPs, multiterminal solid ^{ME}devices ensure multiple-valued characteristics, ensuring the soundness of the molecular hardware solution in implementation of ^{RV}NN as well as feasibility of reconfigurable ^{3D}networking-and-processing.

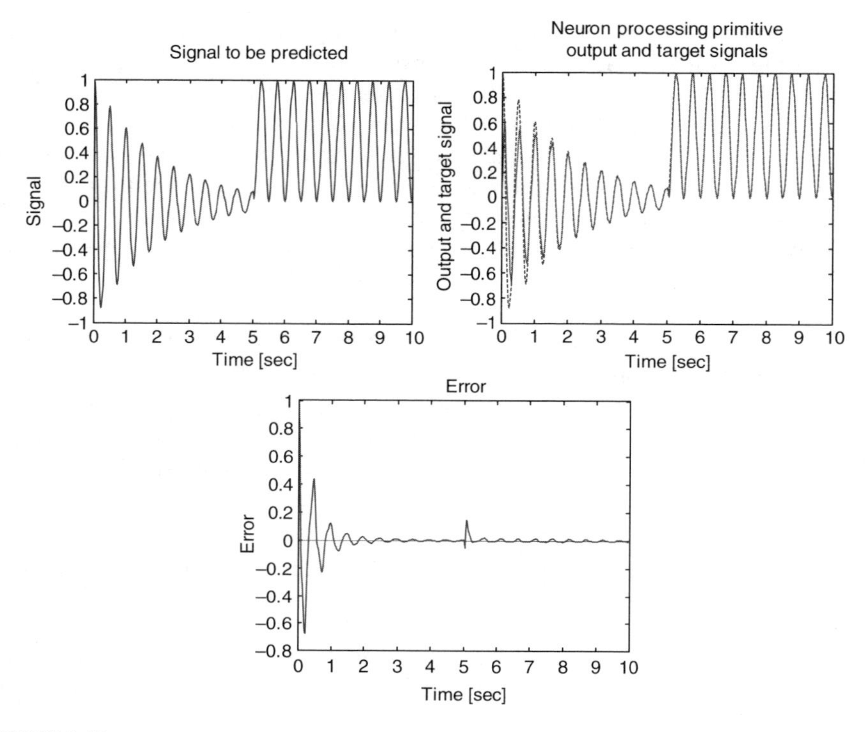

FIGURE 3.17
Training of a NPP.

Example 3.9

Our goal is to map and predict an input (target) signal by training a single NPP. The input signal can be the velocity or position, and let

$$x(t) = \begin{cases} \cos(4\pi t)e^{-0.5t}, & t \in [0 \;\; 5) \text{ sec}, \\ \sin^2(2\pi t), & t \in [5 \;\; 10] \text{ sec}. \end{cases}$$

The input, output and error for a single-layer trained NPP, within a ^{RV}NN, are shown in Figure 3.17. One concludes that the objectives are achieved.

3.9 Biomolecules and Bioinformatics

3.9.1 DNA and Proteins: Introduction

This section dicusses some topics on engineering *bioinformatics*, which complements *biomimetics*. We define engineering bioinformatics as a conceptual paradigm in analysis of system hierarchies, patterns, instructional coding, and synthesis with application to biosystems, engineered bio-inspired systems, novel design, and fabrication strategies.

All ^{BM}PPs are made of biomolecules. One may utilize knowledge of the design and synthesis rules of their nature at the device and system levels. The genome is mapped as a large set of overlapping DNA segments. Bioinformatics can be applied to derive the design and synthesis rules for ^{BM}PPs and ^{M}PPs. From the standpoint of the technological forefront, the overall objective could be formulated as: Understand and utilize the genomic code for synthesis of functional biological entities that exhibit processing capabilities.

The successful completion of this task will result in the understanding and employing of genetic codes to ensure controlled synthesis of molecular assemblies, devices, and systems. Further development is essential characterizing the emergence of biological hierarchies and biotechnology capabilities. The aforementioned baseline features have not been fully understood even for the simplest single-cell bacteria *Escherichia coli* and *Salmonella typhimurium*, which have been studied for many decades.

Amino acid sequences are programmed by a gene. Genes (DNA), and their products (proteins), are the genetic hereditary material. Let us discuss the roles of deoxyribonucleic acid (DNA) and ribonucleic acid (RNA). A DNA molecule is very long (can be several mm) and may consist of thousands genes (thousands or millions of base pairs holding two chains of molecules together) that occupy a specific (exact) position along the single molecule. One DNA molecule represents a large number of genes, each one a particular segment of a helix. The gene can be defined as a DNA sequence coding for a specific polypeptide chain. This definition needs clarification. Most eukaryotic genes contain noncoding regions, called introns, while the coding regions are called exons. Therefore, large portions of genes do not code corresponding segments in polypeptides. Promoters and other regulatory regions of DNA are also included within the gene boundaries. In addition, DNA, which code RNA (rRNA, tRNA, and snRNA), can also be considered. These genes do not have polypeptide products. Correspondingly, a gene can be defined as a region of DNA required for the production of an RNA molecule (distinct definitions for a gene are given for different situations).

The *information-coding* content of DNA is in the form of specific sequences of nucleotides (nucleic acids) along the DNA strand. The DNA inherited by an organism results in specific traits by defining the synthesis of particular proteins. Proteins are the links between genotype and phenotype. Nucleic acids and proteins have specific sequences of monomers that comprise a code, and this *information code* from genes to form proteins is a sequence of nitrogenous bases. Note that the accurate statement is *one gene, one polypeptide*. However, most proteins consist of single polypeptide, and following the common terminology, we will refer proteins as gene products.

In nucleic acids, monomers are four types of nucleotides that differ in their nitrogenous bases. Genes are typically hundreds or thousands nucleotides long, and each gene has a specific sequence of nitrogenous bases. A protein also has monomers arranged in a particular linear order, but its monomers consist of 20 amino acids. Transcription and translation processes (steps) are involved: Transcription is the synthesis of RNA under the direction of DNA. A gene's unique sequence of DNA nucleotides provides a template for assembling a unique sequence of RNA nucleotides. The resulting RNA molecule (called the messenger RNA and denoted as mRNA) is a transcript of the gene's protein-building instructions. Thus, the function of mRNA is to transcript a genetic code from the DNA to the protein-synthesis machinery of the cell. Translation is the synthesis of a polypeptide that occurs under the direction of mRNA. The cell must translate the base sequence of an mRNA molecule into the amino acid sequence of a polypeptide. The sites of translation are ribosomes, with many enzymes and other agents facilitating the orderly linking of amino acids into polypeptide chains. The sequence chain is: DNA → RNA → protein.

The transcription and translation processes are illustrated in Figure 3.18. Transcription results in nucleotide-to-nucleotide transfer of coded information from DNA to RNA. RNA synthesis on a DNA template is catalyzed by RNA polymerase. Promoters (specific nucleotides sequences flanking the start of a gene) signal the initiation of mRNA synthesis. Transcription factors (proteins) help RNA polymerase recognize promoter sequences and bind to the RNA. Transcription continues until the RNA polymerase reaches the termination (stop) sequence of nucleotides on the DNA template. As the mRNA peels away, the DNA double helix re-forms. Translation results in the code transfer from RNA nucleotides to polypeptide amino acids (transfer RNA interprets the genetic code during translation, and each

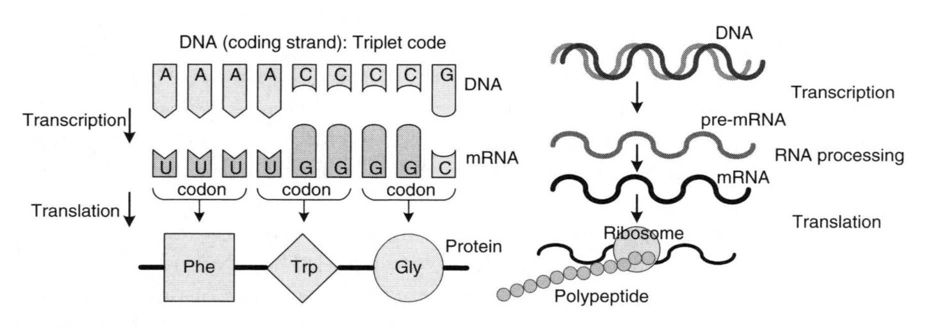

FIGURE 3.18
For each gene, one strand of DNA serves as the coding strand. The sequence of nitrogenous bases specifies the protein (with a certain amino acid sequence). During transcription, the DNA coding strand provides a template for synthesis of mRNA (molecule of complementary sequence with regard to the base-paring rule). During translation, a sequence of base triplets (codon) explicitly specifies the amino acids to be made at the corresponding position along a growing complex protein chain. In a eukaryotic cell, two main steps of protein synthesis (transcription and translation) occur in nucleus (transcription) and cytoplasm (translation)—mRNA is translocated from the nucleus to the cytoplasm through pores in the nuclear envelope (RNA is first synthesized as pre-mRNA that is processed by enzymes before leaving the nucleus as mRNA). In contrast, in a prokaryotic cell (no nucleus), mRNA is produced without this processing.

kind of tRNA brings a specific amino acid to the ribosome). Transfer RNA molecules pick up specific amino acids and line up by means of their *anticodon* triplets at complementary *codon* sites on the mRNA molecule. The ATP process is catalyzed by aminoacryl-tRNA synthetase enzymes. The ribosome controls the coupling of tRNA to mRNA codons. They provide a site for the binding of mRNA, as well as P and A sites (peptidyl-tRNA and aminoacyl-tRNA sites) for holding adjacent tRNA as amino acids are linked in the growing polypeptide chain. There are three major stages—initiation (integrates mRNA with tRNA with the attached first amino acid), elongation (polypeptide chain is completed adding amino acids attached to its tRNA by binding and translocation tRNA and mRNA along the ribosome), and termination (termination codonds cause the protein release freeing the polypeptide chain and dislocation of the ribosome subunits). Several ribosomes can read a single mRNA, forming polyribosome clusters. Complex proteins usually undertake one or several changes during and after translation that affect their 3D structures. This leads to the cell transitional dynamics.

The question of interest is how much information is needed to describe these patterns. Messenger RNA carries information specifying the amino acid sequences of proteins from the DNA to the ribosomes. Transfer RNA (tRNA) translates mRNA nucleotide sequences into protein amino acid sequences. Finally, ribosomal RNA (rRNA) and small nuclear RNA (snRNA) play structural and enzymatic roles. In addition to a binding site for mRNA, each ribosome has two binding sites for tRNA. The P site (peptidyl-tRNA site) holds the tRNA carrying the growing polypeptide chain, while the A site (aminoacyl-tRNA site) holds the tRNA carrying the next amino acid to be added to the chain. The ribosome holds the tRNA and mRNA molecules close together, while proteins catalyze the transfer of an amino acid to the carboxyl end of the growing polypeptide chain. The mRNA is moved through the ribosome in the $5' \rightarrow 3'$ direction only (ribosome and mRNA move relative to each other unidirectionally, codon by codon). RNA polymerase (enzyme) adds nucleotides to the $3'$ end of the growing polymer, and RNA molecule elongates in its $5' \rightarrow 3'$ direction. Specific sequences of nucleotides along the DNA mark the initiation and termination sites where transcription of a gene begins and ends.

The instructions for assembling amino acids into a specific order are encoded by DNA. Four nucleotides specify 20 amino acids. Triplets of nitrogenous bases (codons) code amino acids as the coding instructions. In particular, three consecutive bases specify an amino acid, and there are 64 (4^3) possibilities. As was emphasized, a cell cannot directly translate gene codons into amino acids. The intermediate transcription step (a gene determines the codon sequence of an mRNA molecule) is in place. For each gene, only one of two DNA strands is transcribed, that is, the coding strand of the gene is used. The noncoding strand serves as a template for making a new coding strand when the DNA replicates. The RNA bases are assembled on the template according to the base-pairing rules. Hence, an mRNA molecule is complementary, rather than identical, to its DNA template. For example, if the codon strand of a gene has a codon GCG, the codon at the corresponding position along the mRNA molecule will be CGC. During translation, the sequence of codons along a genetic message (mRNA molecule) is decoded (translated) into a sequence of amino acids resulting in a polypeptide chain. It takes $3N$ nucleotide strands to code N amino acids.

It is important to emphasize that genes can be transcribed and translated after they have been transplanted from one species to another. For example, bacteria can be programmed by the insertion of a human gene to synthesize protein insulin. Proteins are utilized as structures, devices, actuators, sensors, and other units, modules, subsystems, and systems. Having defined a protein sequence, the efforts have been directed to examine protein structure and protein functionality. The protein structure (3D geometry) is due to folding of a peptide chain as well as multiple peptide chains. Proteins are the most structurally complex molecules known. Amino acid bonds determine the folding (α-helix resulting in helix-loop-helix, β-pleated sheet, random conformations, etc.). Most proteins go through several intermediate states to form stable structures and conformation. Figure 3.19 illustrates a protein folding. The folded molecule is the four chains of the ElbB (enhancing lycopene biosynthesis) of *E. coli*.

Nucleic acids are polymers of monomers called nucleotides. Each nucleotide is itself composed of three parts, and a nitrogenous base is joined to a pentose that is bonded to a phosphate group. The DNA molecules consist of two polynucleotide chains (strands) that spiral around forming a double helix, which was discovered by Rosalind Franklin in 1952 through x-ray

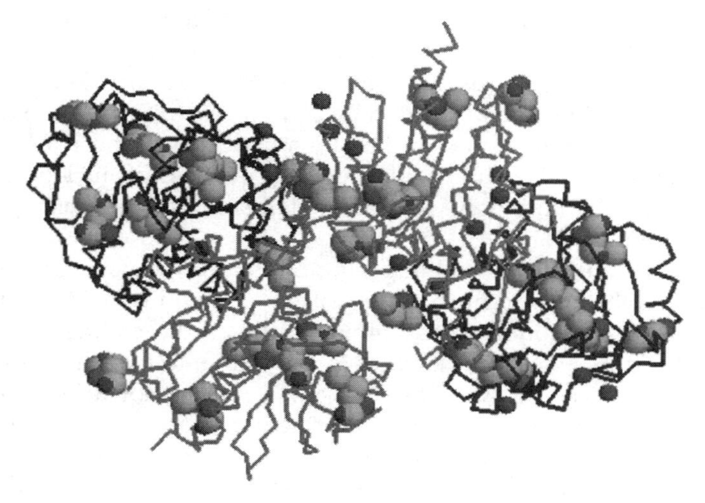

FIGURE 3.19
ElbB geometry comprised of four similar protein sequences.

crystallography. These polynucleotide chains are held together by hydrogen bonds between the paired nitrogenous bases. DNA is a linear double-stranded polymer of the following four nucleotides (bases):

- Deoxyadenosine monophosphate or adenine (A)
- Deoxythymidine monophosphate or thymine (T) in DNA, and uracil (U) in RNA
- Deoxyguanosine monophosphate or guanine (G)
- Deoxycytidine monophosphate or cytosine (C)

Figure 3.20 illustrates that each nucleotide integrates a nitrogenous base joined to a pentose (five-carbon sugar) bonded to a phosphate group. There are two families of nitrogenous bases—pyrimidines (C, T, and U) and purines (A and G). A pyrimidine is characterized by a hexagon ring of carbon and

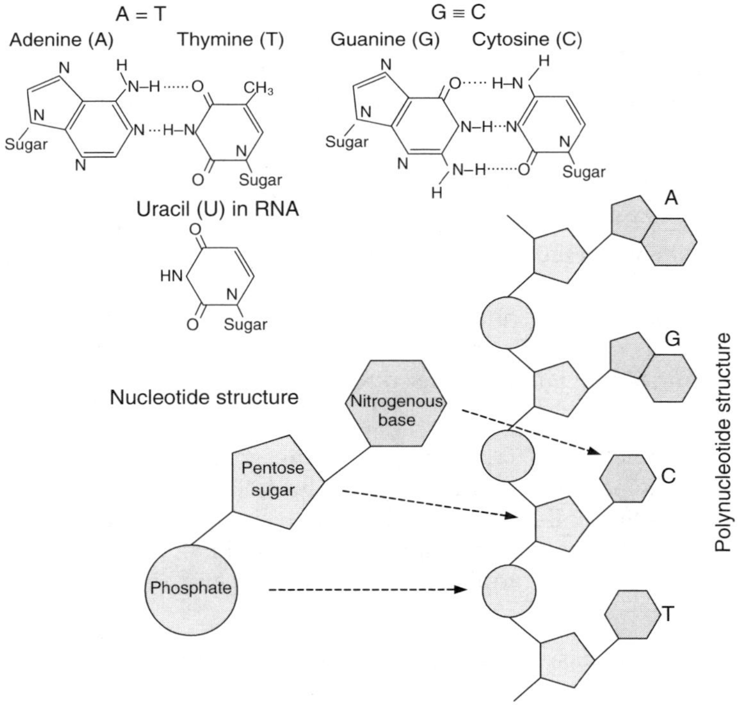

FIGURE 3.20
Nucleotides: Monomers of nucleic acids composed of three molecular building blocks (a purine or pyrimidine nitrogenous base, pentose sugar, and phosphate group) and polynucleotides (a phosphate group bonded to the sugar of the next nucleotide, and the polymer has a sugar–phosphate backbone).

hydrogen atoms. In purines, a pentagon ring is fused to the pyrimidine ring; see Figure 3.20. The monomers are joined by covalent bonds between the phosphate of one nucleotide and the sugar of the next monomer. The backbone, as a repeating pattern of sugar–phosphate–sugar–phosphate, results. The linear order of nitrogenous bases encoded in a gene specifying the amino acid sequence of a protein which exhibits specificity and functionality. The DNA molecule consists of two polynucleotide chains that spiral around forming a double helix. The two sugar–phosphate backbones are on the outside of the helix, and the nitrogenous bases are paired in the interior of the helix. Two polynucleotide chains (strands) are held together by the hydrogen bonds between the paired bases, as shown in Figure 3.20. The Chargaff base-paring rule (A pairs with T, i.e., A≡T, while G pairs with C, i.e., G≡C) specifies that the two strands of the double helix are complementary. If a stretch of a strand has the base sequence AATTGGCC, then the base-paring rule gives the same stretch of the other strand with TTAACCGG sequence. As a cell prepares to divide, two strands of each gene separate, guarantying precise copying (each strand serves as a template to order nucleotides into a new complementary strand).

Figure 3.21 represents the sugar–phosphate backbone of two DNA strands. These two strands are held together by hydrogen bonds between the nitrogenous bases that are paired in the interior of the 2 nm diameter double helix. The base pairs are 0.34 nm apart, and there are 10 pairs per each turn of the helix.

Proteins are large biomolecules. All proteins are chemically similar because they are made from amino acids linked together in long chains. The amino acids can be considered as building blocks that are coded and synthesized by means of nitrogenous base sequences. Amino acids are organic molecules containing both carboxyl and amino groups. Cells build their proteins from 20 amino acids. These 20 amino acids are specified by codon sequences. The following abbreviations are used:

> Ala (alanine), Arg (arginine), Asn (asparagine), Asp (aspartic), Cys (cysteine), Gln (glutamine), Glu (glutamic), Gly (glycine), His (histidine), Ile (isoleucine), Lys (lysine), Leu (leucine), Met (methonine), Phe (phenylalanine), Pro (praline), Ser (serine), Thr (threonine), Trp (tryptophane), Tyr (tyrosine), and Val (valine).

Each amino acid has a hydrogen atom, a carboxyl group, and an amino acid group bonded to the alpha (α) carbon. All amino acids, which make proteins, differ only in what is attached by the fourth bond to the α-carbon. The amino acids are grouped according to the properties of the side chains (R-group). The physical and chemical properties of the side chain define the unique characteristics of the amino acids. Amino acids are classified in the groups based on the chemical and structural properties of their side chains and polarity, for example, nonpolar (hydrophobic), polar (hydrophilic), or electrically

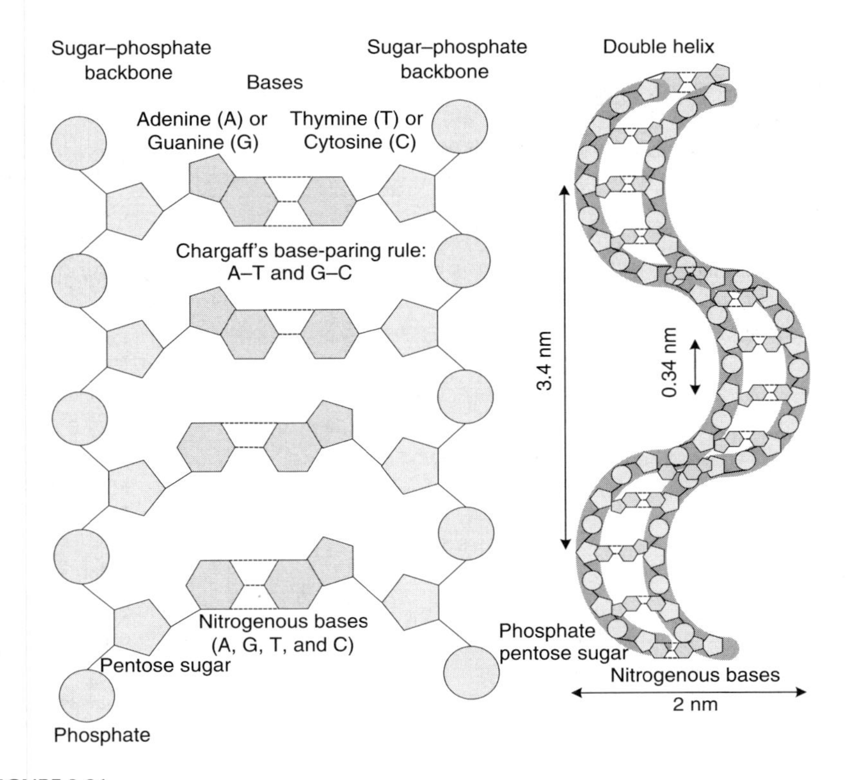

FIGURE 3.21
Two DNA strands are held together by hydrogen bonds between the nitrogenous bases paired in the interior of the double helix (base pairs are 0.34 nm apart and there are 10 pairs per the turn of the helix).

charged. Twenty proteinogenic amino acids, as well as their structures and properties, are reported in Table 3.1.

The amino acid backbone determines the primary sequence of a protein, but the nature of the side chains defines the protein's properties. Amino acid side chains can be polar, nonpolar, or neutral. Polar side chains tend to be present on the surface of a protein, where they can interact with the cell aqueous environment. Nonpolar amino acids tend to reside within the center of the protein, where they can interact with similar nonpolar neighbors (this can create a hydrophobic region within an enzyme where chemical reactions can be conducted in a nonpolar atmosphere). There are exist different methods to classify amino acids by their structure, size, charge, hydrophobicity, and so forth. The charge residue is normally found on the surface of the protein (interacting with water and binding other molecules) and seldom buried in the interior of a folded protein. Arginine, histidine, and lysine are positively charged, while aspartic and glutamic are negatively charged amino acids. Amino acids can be also classified using their distinct optical properties. A tetrahedral carbon atom with four distinct constituents is said to be chiral.

TABLE 3.1

Amino Acids

Amino Acid (Molecular Weight)	Symbol Abbreviation	Molecular Formula	Structure	Properties
Alanine (89.09)	Ala, A	$C_3H_7NO_2$	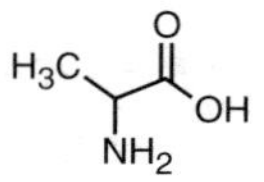	Aliphatic Hydrophobic Neutral
Arginine (174.2)	Arg, R	$C_6H_{14}N_4O_2$	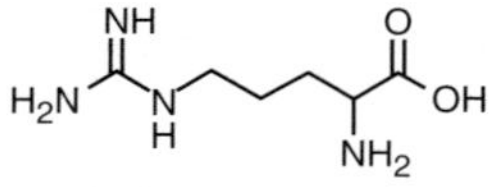	Polar (strongly) Hydrophilic Positively charged (+)
Asparagine (132.12)	Asn, N	$C_4H_8N_2O_3$	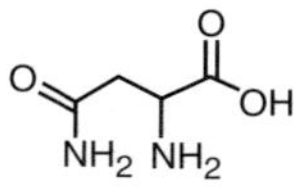	Polar Hydrophilic Neutral
Aspartic acid (aspartate) (133.1)	Asp, D	$C_4H_7NO_4$	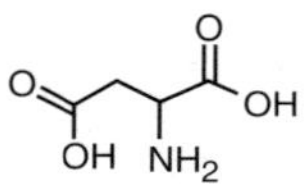	Polar Hydrophilic Negatively charged (−)
Cysteine (121.15)	Cys, C	$C_3H_7NO_2S$	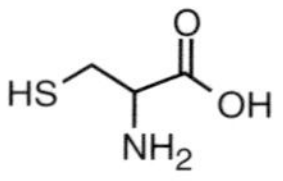	Polar (weakly) Hydrophilic Neutral
Glutamine (146.15)	Gln, Q	$C_5H_9NO_4$	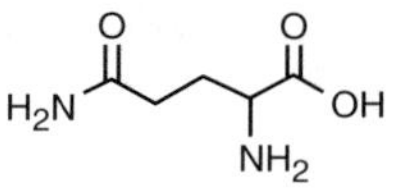	Polar Hydrophilic Neutral
Glutamic acid (glutamate) (147.13)	Glu, E	$C_5H_{10}N_2O_3$	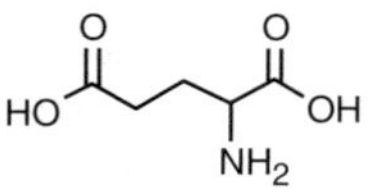	Polar Hydrophilic Negatively charged (−)
Glycine (75.07)	Gly, G	$C_2H_5NO_2$	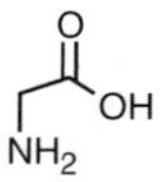	Aliphatic Neutral
Histidine (155.16)	His, H	$C_6H_9N_3O_2$	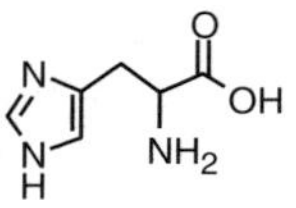	Aromatic Polar (strongly) Hydrophilic Positively charged (+)
Isoleucine (131.17)	Ile, I	$C_6H_{13}NO_2$	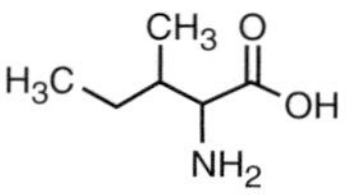	Aliphatic Hydrophobic Neutral

TABLE 3.1

(Continued)

Amino Acid (Molecular Weight)	**Symbol Abbreviation**	**Molecular Formula**	**Structure**	**Properties**
Leucine (131.17)	Leu, L	$C_6H_{13}NO_2$		Aliphatic Hydrophobic Neutral
Lysine (146.21)	Lys, K	$C_6H_{14}N_2O_2$		Polar (strongly) Hydrophilic Positively charged (+)
Methionine (149.21)	Met, M	$C_5H_{11}NO_2S$		Hydrophobic Neutral
Phenylalanine (165.19)	Phe, F	$C_9H_{11}NO_2$		Aromatic Hydrophobic Neutral
Praline (115.13)	Pro, P	$C_5H_9NO_2$		Hydrophobic Neutral
Serine (105.09)	Ser, S	$C_3H_7NO_3$		Polar Hydrophobic Neutral
Threonine (119.12)	Thr, T	$C_4H_9NO_3$		Polar Hydrophobic Neutral
Tryptophan (204.23)	Trp, W	$C_{11}H_{12}N_2O_2$		Aromatic Hydrophobic Neutral
Tyrosine (181.19)	Tyr, Y	$C_9H_{11}NO_3$		Aromatic Polar Hydrophobic
Valine (117.15)	Val, V	$C_5H_{11}NO_2$		Aliphatic Hydrophobic Neutral

The amino acid that does not exhibit chirality is a glycine because its R-group is a hydrogen atom. Chirality describes the handedness of a molecule that is observable by its ability to rotate the plane of polarized light either to the right (dextrorotatory) or to the left (levorotatory). All of the amino acids in proteins are levorotatory α-amino acids (dextrorotatory amino acids have never been found in proteins, although they exist in nature and found in polypetide antibiotics). The aromatic R-groups in amino acids absorb ultraviolet light, and the maximum of absorbance is observed at $\sim$280 nm. The ability of proteins to absorb ultraviolet light is mainly due to the presence of the tryptophan, which strongly absorbs ultraviolet light.

A polymer of many amino acids linked by peptide bonds is a polypeptide chain. At one end of the chain is a free amino group, and at the opposite end there is a free carboxyl group. Hence, the chain has a polarity with an N-terminus (nitrogen of the amino group) and a C-terminus (carbon of the carboxyl group). A protein consists of one or more twisted, wounded, and folded polypeptide chains forming a micromolecule (protein) with a defined 3D shape (geometry). This 3D shape is called conformation. A protein's functionality depends on its geometry that is a function of specific linear sequences of the amino acids that make the polypeptide chain. Some proteins are spherical (globular), while others are fibrous. The structure of a protein is defined by the sequence of amino acids.

Triplets of nitrogenous bases define amino acids. In particular, each arrangement of three consecutive nitrogenous bases specifies an amino acid, and there are 4^3 (64) possible arrangements. The genetic instructions are coded and stored in the DNA as a series of three nucleotides called *codons*. This genetic code is universal and shared by all biosystems (from bacteria to humans). Table 3.2 documents the resulting amino acids as a function of codons (note that some exceptions to this standard table exist). In particular, three bases of an mRNA codon are designated as the first, second, and third bases. The codon AUG stands not only for the methonine amino acid (Met) but also indicates the "Start" mark. Three of the total 64 (4^3) codons are "Stop" marks (termination codon points the end of a genetic message). Hence, four amino acids are "Start" and "Stop," carrying the start and end marks in genetic messages.

The hierarchy, coding, and protein synthesis may allow one to coherently examine and design the synthesis taxonomy. As documented, the genetic instructions from DNA are coded as codons that precisely specify particular amino acids and proteins. The genetic code is shared by all living organisms, from bacteria to humans. This paradigm results in the synthesis of extremely complex functional biosystems. There is a need for further development of engineering bioinformatics to study the synthesis and design taxonomy of complex devices and systems from biomolecules. Both nucleic acids and proteins are informational polymers assembled from linear sequences of nucleotides and amino acids, respectively. Messenger RNA (as the code carrier from the coding strand of a gene) is the intermediate step from DNA to proteins. Can one be able to utilize and apply this paradigm? Likely the

TABLE 3.2

Nucleotides (bases) in Codon and Resulting Amino Acids

Codon First Base	Second Base U	C	A	G	Codon Third Base
U	Phe (UUU)	Ser	Tyr	Cys	U
	Phe (UUC)	Ser	Tyr	Cys	C
	Leu (UUA)	Ser	Stop	Stop	A
	Leu (UUG)	Ser	Stop	Trp	G
C	Leu (CUU)	Pro	His	Arg	U
	Leu (CUC)	Pro	His	Arg	C
	Leu (CUA)	Pro	Gln	Arg	A
	Leu (CUG)	Pro	Gln	Arg	G
A	Ile	Thr	Asn	Ser	U
	Ile	Thr	Asn	Ser	C
	Ile	Thr	Lys	Arg	A
	Met–Start	Thr	Lys	Arg	G
G	Val	Ala	Asp	Gly (GGT)	U
	Val	Ala	Asp	Gly (GGC)	C
	Val	Ala	Glu	Gly (GGA)	A
	Val	Ala	Glu	Gly (GGG)	G

answer is yes, but there are a great number of unsolved problems that may not be resolved in near future.

3.9.2 Overview of Bioinformatics Methods

In biosystems, there are thousands of different proteins, each with specific structure and function, for example, (1) structural proteins, (2) storage (store amino acids) proteins, (3) transport proteins, (4) receptor (sensor) proteins, (5) contractile (actuator) proteins, (6) defensive (protect and combat diseases) proteins, (7) hormonal proteins, (8) enzymatic proteins, and many other functional proteins. The conceptual functional categories for protein can be explicitly defined.

Bioinformatics has become increasingly database driven because of the large-scale functional genomics and proteomics. It is a significant challenge to integrate and utilize databases to coherently examine large-scale data. The single-gene research (functionality, folding, interaction, etc.) has been progressed to a large population of proteins, emphasizing descriptive genomics and functional proteomics. Proteomics includes protein identification as well their characterization and functionality. The corresponding databases have been developed. For example, the SCOP, CATH, and FSSP databases classify proteins based on structural similarity; Pfam and ProtoMap identify families of proteins based on sequence homology, while PartList and GeneCensus examine the occurrence of protein families in various genomes. The large-scale genomics and proteomics are the forefront of biological and genomic research. Genome sequences for different organisms are available. Even human genome sequences are accessible. In particular, (1) GenBank, DDBJ, and EMBL provide nucleic acid sequences; (2) PIR and SWISS-PROT report protein sequences; (3) Protein Data Bank offers 3D protein structures. In addition to sequence and structure databases, efforts have been focused on various functionality aspects. Correspondingly, integrated data-intensive large-scale analysis and heterogeneous intelligent data-mining are essential.

There is a need to develop concepts that will allow one to integrate genomic data. A general problem is to integrate the large-scale diverse genomic information in the viable taxonomies or categories. Currently, majority of the methods are based on the statistical analysis employing unsupervised learning, self-organization, classification, hierarchical clustering, and so forth. For example, a clustering method establishes multitittered partitioning of the data sets. Using the Pearson correlation coefficient $r_{ij} = (1/N - 1)X_i \cdot X_j$, given as a dot product of two *profiles* X_i and X_j, the similarity between genes (or groups of genes) is obtained. Furthermore, the measurement expression ratio profile is found using the average x_{av} and the standard deviation σ_x, that is, $X(k) = (x(k) - x_{\text{av}}/\sigma_x)$. The aggregation of proteomic data from multiple sources must be performed to identify and predict various protein properties and their functionality. However, these are formidable tasks.

The DNA sequences of several human pathogens are known. To achieve reasonable accuracy and high-quality continuous sequences, each base pair was sequenced many times. As a result, 90–93% of the euchromatin sequence has an error rate of less than 1 in 10,000 bases [33]. Different sequencing technologies, mathematical methods, distinct procedures, and measurement techniques have been used. However, it is difficult to estimate the accuracy, and there are many gaps and unknown strings of bases in the large-scale genomic sequence data. There are differences even in the count of genes. For example, the public human genome database reports 31,780 genes (2693 millions of bases sequenced). These include 15,000 known genes and 17,000 predicted genes. Some sources estimate that there can be less than 20,000 actual genes. Some predicted that genes can be "pseudogenes" (noncoding) or fragments of real genes leading to predictions that there could be only 7000 real genes. For example, Celera reported 39,114 genes (2654 million bases sequenced) adding that 12,000 genes are "weak."

E. Coli is one of the most thoroughly studied living organisms. Some *E. coli* are harmless, whereas other distinct genotypes (extraintestinal, enteropathogenic, enterohemorrhagic, enteroinvasive, enterotoxigenic, and enteroaggregative *E. coli*) cause morbidity and mortality as human intestinal pathogens. Distinct *E. coli* genome sequences were reported and compared, for example, MG1655, CFT073, EDL933, and other strains. For different strains (MG1655 and CFT073), the genome length is 4,639,221 and 5,231,428 base pairs (bp), specifying 4293 and 5533 protein-coding genes, and less than 50% of the proteins are known [34]. After sequencing, one may examine genes and proteins. For example, identifying the critical genes involved in virulence mechanisms and host response, one can design effective detectors, vaccines, and treatments. Other genes that provide the *instructional coding* for processing, memory and other systems are of a great interest. Through the analysis, one can (1) examine and characterize changes in genes and proteins of distinct *E. coli* strains; (2) potentially identify protein functionality; (3) compare and examine genes and protein profiles; and (4) integrate the experimental and computational data into the application-specific data-intensive analysis.

Different methods have been applied to attain analysis, comparison, and data recognition. However, conventional approaches may not be well suited, and novel accurate efficient information-theoretic methods must be developed to attain data-intensive robust analysis and heterogeneous data-mining. There is a need to examine genes that are unique and analogous to the same and other species.

Different statistical techniques have been applied to attain global and local sequence comparisons [35–37]. However, under even the simplest random models and scoring systems, the distribution of optimal global alignment scores is unknown. Monte Carlo experiments potentially can provide some promising results for specific scoring systems and sequence compositions, but these results cannot be generalized. Compared with global alignments, statistics for the scores of local alignments, particularly for local alignments lacking gaps, is well posed. A local alignment without gaps consists simply of a pair of equal length segments, one from each of the two sequences being compared. For example, in the BLAST program, the database search can be performed utilizing high-scoring segment pairs (HSPs). To analyze the score probability, a model of random sequences is applied. For proteins, the simplest model chooses the amino acid residues in a sequence independently, with specific background probabilities for the various residues, and the expected score for aligning a random pair of amino acid is required to be negative. For sequence (with lengths m and n), the HSP score statistics is characterized by the scaling parameters K and λ. The expected number of HSPs with score at least S is given as $E = mnKe^{-\lambda S}$. One obtains the E-value for the score S. However, the length of sequence changes E, and sound methods to find the scaling positive parameters K and λ have not been reported. The score is normalized as $S' = (\lambda S - \ln K/\ln 2)$ to obtain the so-called "bit score" S'. The E-value is $E = mn2^{-S'}$. The number of random HSPs with score greater or equal to S is described by a Poisson distribution. For example, the probability of finding exactly A HSPs with score $\geq S$ is given by $\mathrm{e}^{-E}E^{A}/A!$. The probability of finding at least one such HSP is $P = 1 - \mathrm{e}^{-E}$. This is the P-value associated with the score S. In BLAST, the E-value is used to compare two proteins of lengths m and n. To assess the significance of an alignment that arises from the comparison of a protein of length m to a database containing many different proteins of varying lengths, one view is that all proteins in the database are a priori equally likely to be related to the query. This implies that a low E-value for an alignment involving a short database sequence should carry the same weight as a low E-value for an alignment involving a long database sequence. To calculate a "database search" E-value, one multiplies the pairwise-compared

E-value by the number of sequences in the database, for example, the FASTA protein comparison programs.

The approaches applied to date have a sound theoretical foundation only for local alignments that are not permitted to have gaps, short sequences, "estimation" of K and λ. Different amino acid substitution scores $S_{ij} = (1/\lambda)\ln(q_{ij}/p_i p_j)$ are reported. Here, q_{ij} is the target frequency; p_i and p_j are the background frequencies for the various residues. The target frequencies and the corresponding substitution matrix may be calculated for any given evolutionary distance. However, this method has serious deficiencies, and there have been efforts to develop novel methods. For example, utilizing the log-odds matrices, multiple alignments of distantly related protein regions were examined. While we have discussed substitution matrices in the context of protein sequence comparison, the major challenge is to perform the DNA sequence comparison. These DNA sequences contain coding information to be examined. Special attention must be given to all regions (low, medium, and high complexity). The BLAST program filters low-complexity regions from proteins before executing a database search. Owing to the application of vague mathematical methods, low-complexity regions lead to deficiencies in sequence similarity analysis (high-score results for sequences that are not related, existing match cannot be found, etc.). Complete genomes must be analyzed by performing data-intensive analysis coherently utilizing all available data for the sequenced genes to ensure complete topologies and sequence preservation. Therefore, new sound concepts, which are not based on assumed hypotheses and simplifications, must be developed and demonstrated. Recent results expand the statistical methods by developing information-enhanced procedures to perform large-scale analysis. One may intend to overcome the existing formidable challenges to achieve data-intensive analysis and coherent data-mining by developing new methods by utilizing the frequency-domain analysis.

Statistics-based methods test a priori hypotheses against the data with a great number of assumptions and simplifications under which genome–genome comparison can be performed. The information-theoretic methods that ensure robust analysis are developed in the frequency domain utilizing advanced array-based matrix methods attaining systematic analysis. These methods may assist in discovery of new (previously unknown and/or hidden) pattern of the large-scale data sets, relate gene sequence to protein structure, relate protein structure to function, and so forth. One is able to examine (interpret, represent, and map) complex patterns, create and utilize intelligent libraries, and so forth. The key questions are how the instructional coding is accomplished, how to distinguish, detect, and recognize protein functionality under uncertainties, and so forth. The frequency-domain methods, reported in Section 3.9.3, promises to ensure: (1) homology search and genes detection with superior accuracy and robustness under uncertainties; (2) accurate and robust analysis; (3) computational efficiency and soundness; (4) coded information extraction and information retrieval; (5) correlation between large-scale data sets from multiple databases; (6) detecting potential homologues in the databases.

3.9.3 Genome Analysis in the Frequency Domain

Our goal is to examine a sequence similarity for quaternary sequences. One can measure the sum over the length of the sequences of alphabetic similarities at all positions. Alphabetic similarities are symmetrically defined on the Cartesian square of the alphabet. These similarities equal zero whenever the two elements differ, and in contrast to the Hamming similarity, the reported alphabetic similarities take individual values whenever two elements are identical. Hence, lower and upper bounds can be derived.

Let $\mathbf{A} = \{A, C, G, T\}$ is the *symbolic quaternary alphabet*. This *alphabet* can be represented numerically as $\mathbf{A} = \{0\ 1\ 2\ 3\}$ or $\mathbf{A} = \{1+j - 1+j1 - j - 1 - j\}$.

The arbitrary pairs of quaternary N-sequences (words of length N) are

$$x = (x_1, x_2, \ldots, x_{N-1}, x_N), \quad x_i \in \mathbf{A},$$
$$y = (y_1, y_2, \ldots, y_{N-1}, y_N), \quad y_i \in \mathbf{A}.$$

For a pair (x, y) of quaternary words, the similarity is defined as

$$S(x, y) = \sum_{i=1}^{N} s(x_i, y_i).$$

The gene, protein or genome alphabetic similarity can be expressed as

$$s(x_i, y_i) = \begin{cases} 1 & \text{if } x = y = A, \\ 2 & \text{if } x = y = T, \\ 3 & \text{if } x = y = G, \\ 4 & \text{if } x = y = C, \end{cases} \quad \text{or} \quad s(x_i, y_i) = \begin{cases} 1+j & \text{if } x = y = A, \\ 1-j & \text{if } x = y = T, \\ -1+j & \text{if } x = y = G, \\ -1-j & \text{if } x = y = C. \end{cases}$$

Thus, one can consider two *complementary* pairs of symbols. If $x = y$, then $S(x, x)$ corresponds to the *self similarity* of x. If $x \neq y$, then $S(x, y)$ represents the *cross-similarity* of the pair (x, y). We denote by $D(x, y)$ the Hamming distance between x and y, that is, the number of positions in which words x and y are different. The DNA code (quaternary code) or amino acid code similarity can be examined. In fact, genes and proteins are represented by finite sequences (*strands*) of A, C, G, and T nucleotides.

Example 3.10

Let $N = 12$, $x =$ (A A A C G T T A C C T A), and $y =$ (A C G T G A A A T C G G). To perform the analysis, let us, for example, for two quaternary DNA sequences, assign 0 and 3 to C and G, and 1 and 2 to A and T. The cross-similarity of (x, y) is $S(x, y) = 1+0+0+0+3+0+0+1+0+4+0+0 = 9$, and the self-similarity of x S(x, x) can be derived.

One can utilize the symbolic abbreviations using nucleotides (A, U, T, G, and C) and 20 amino acids (from alanine to valine). The symbolic descriptions must be coherent within biophysics and instructional coding. The instructional codes cannot be considered to be analogous to Boolean algebra, binary arithmetic or multiple-valued logics, where the designer without restrictions performs logic and circuits design.

We apply sound mathematical fundamentals to attain analytical, numerical, pattern, visual, and interactive analysis. Consider a sequence of nucleotides A, T, C, and G. We can assign the symbol or number a to the character A, t to the character T, c to the character C, and g to the character G. Here, a, t, c, and g can be complex numbers. There exists a numerical sequence resulting from a character string of length N. In particular,

$$x[n] = au_A[n] + tu_T[n] + cu_C[n] + gu_G[n], \quad n = 0, 1, 2, \ldots, N-1,$$

where $u_A[n]$, $u_T[n]$, $u_C[n]$, and $u_G[n]$ are the binary indicators that take the value of either 1 or 0 at location n depending on whether the corresponding character exists or not at location n; N is the length of the sequence.

For amino acids, we have the following expression for the amino acid sequence:

$$\begin{aligned} x[n] = {} & A_{la}u_{Ala}[n] + A_{rg}u_{Arg}[n] + \cdots \\ & + T_{yr}u_{Tyr}[n] + V_{al}u_{Val}[n], \quad n = 0, 1, 2, \ldots, N-1. \end{aligned}$$

Using the amino acids alphabet (utilizing the common amino acid codes), the *symbolic alphabet* is

$$\mathbf{A} = \{\text{Ala, Arg, \ldots, Tyr, Val}\} \quad \text{or}$$
$$\mathbf{A} = \{\text{A, R, N, D, C, Q, E, G, H, I, L, K, M, F, P, S, T, W, Y, V}\}.$$

Thus, the amino acid sequence is given as

$$x[n] = au_a[n] + ru_r[n] + \cdots + yu_y[n] + vu_v[n], \quad n = 0, 1, 2, \ldots, N-1.$$

We obtain the symbolic strings that map DNA and amino acids. Four-dimensional Fourier transform or DNA and twenty-dimensional Fourier transform for amino acids can be derived. In particular, the discrete Fourier transform of a sequence $x[n]$ of length N is

$$X[k] = \sum_{n=0}^{N-1} x[n]\mathrm{e}^{-j(2\pi/N)kn}, \quad k = 0, 1, 2, \ldots, N-1.$$

This Fourier transform provides a measure of the frequency content at frequency k, which corresponds to a period of N/k samples. The resulting sequences $U_A[k]$, $U_T[k]$, $U_C[k]$, and $U_G[k]$ are the discrete Fourier transforms of the binary indicators $u_A[n]$, $u_T[n]$, $u_C[n]$, and $u_G[n]$. In particular,

$$U_A[k] = \sum_{n=0}^{N-1} u_A[n]e^{-j(2\pi/N)kn}, \quad U_T[k] = \sum_{n=0}^{N-1} u_T[n]e^{-j(2\pi/N)kn},$$
$$U_C[k] = \sum_{n=0}^{N-1} u_C[n]e^{-j(2\pi/N)kn}, \quad U_G[k] = \sum_{n=0}^{N-1} u_G[n]e^{-j(2\pi/N)kn},$$
$$k = 0, 1, 2, \ldots, N-1.$$

Using the numerical values for a, t, c, and g, one obtains

$$X[k] = aU_A[k] + tU_T[k] + cU_C[k] + gU_G[k], \quad k = 0, 1, 2, \ldots, N-1.$$

In general, DNA character strings lead to the sequences $U_A[k]$, $U_T[k]$, $U_C[k]$, and $U_G[k]$ resulting in four-dimensional representation of the frequency spectrum with

$$U_A[k] + U_T[k] + U_C[k] + U_G[k] = \begin{cases} 0, & k \neq 0, \\ N, & k = 0. \end{cases}$$

The total power spectral content of the DNA character string at the frequency k is

$$S[k] = |U_A[k]|^2 + |U_T[k]|^2 + |U_C[k]|^2 + |U_G[k]|^2.$$

For the amino acids, the frequency spectra and power analysis are identical to those reported for DNA. Though the analysis can be accomplished examining multidimensional Fourier transforms, the high dimensionality problem can be resolved by numerically representing the *symbolic alphabet* of nucleotides and amino acids. This concept allows one to obtain well-defined 1D Fourier transform.

Fourier transform may offer superior computational advantages, accuracy, versatility, and coherence in examining complex genomes composed from millions nucleotides. The frequency analysis of DNA and amino acid sequences can find applications in identifying protein-coding genes, defining structural and functional characteristics, identifying patterns in gene sequences, and so forth. The sequences are available with the satisfactory accuracy, and the strings of nucleotides (bases) or amino acids have been found.

E. coli and *S. typhimurium* are among best-studied microorganisms with sequenced genomes [34]. Information for each *E. coli* gene (the EcoGene12 release includes 4,293 genes with 706 predicted or confirmed gene start sites for the MG1655 strain [34]) is organized into separate *gene pages*. The lengths of the genome sequences are different. For example, the *E. coli* MG1655 and CFT073 strains are 4,639,221 and 5,231,428 bp strains. For MG1655, these 4,639,221 bp (A, C, G, and T) specify 4,293 genes. Though there are 717 proteins whose N-terminal amino acids have been verified by sequencing, only 50% of the proteins are known. To locate 4,293 genes (each of which starts with a ribosome binding site) from 4,639,221 possibilities is extremely difficult, and using the information theory, the number of choices is $\log_2(4,639,221/4,293) = 10$ bits.

For the FliG gene, the nucleotide sequence is given as (genomic address of FliG: 2,012,902 bp left end, and 2,013,897 bp right end, 996 length) [34]:

ATGAGTAACCTGACAGGCACCGATAAAAGCGTCATCCTGCTGATGACCATTGGCGAAA
GACCGGGCGGCAGAGGTGTTCAAGCACCTCTCCCAGCGTGAAGTACAAACCCTGAGC
GCTGCAATGGCGAACGTCACGCAGATCTCCAACAAGCAGCTAACCGATGTGCTGGCG
GAGTTTGAGCAAGAAGCTGAACAGTTTGCCGCACTGAATATCAACGCCAACGATTATC
TGCGCTCGGTATTGGTCAAAGCTCTGGGTGAAGAACGTGCCGCCAGCCTGCTGGAAGA
TATTCTCGAAACTCGCGATACCGCCAGCGGTATTGAAACGCTCAACTTTATGGAGCCAC
AGAGCGCCGCCGATCTGATTCGCGATGAGCATCCGCAAATTATCGCCACCATTCTGGTG
CATCTGAAGCGCGCCCAAGCCGCCGATATTCTGGCGTTGTTCGATGAACGTCTGCGCCA
CGACGTGATGTTGCGTATCGCCACCTTTGGCGGCGTGCAGCCAGCCGCGCTGGCGGAG
CTGACCGAAGTACTGAATGGCTTGCTCGACGGTCAGAATCTCAAGCGCAGCAAAATGG
GCGGCGTGAGAACGGCAGCCGAAATTATCAACCTGATGAAAACTCAGCAGGAAGAAG
CCGTTATTACCGCCGTGCGTGAATTCGACGGCGAGCTGGCGCAGAAAATCATCGACGA
GATGTTCCTGTTCGAGAATCTGGTGGATGTCGACGATCGCAGCATTCAGCGTCTGTTGC
AGGAAGTGGATTCCGAATCGCTGTTGATCGCGCTGAAAGGAGCCGAGCAGCCACTGCG
CGAGAAATTCTTGCGCAATATGTCGCAGCGTGCCGCCGATATTCTGCGCGACGATCTCG
CCAACCGTGGTCCGGTGCGTCTGTCGCAGGTGGAAAACGAACAGAAAGCGATTCTGCT
GATTGTGCGCCGCCTTGCCGAAACTGGCGAGATGGTAATTGGCAGCGGCGAGGATACCT
ATGTCTGA

The FliM and FliN sequences are [34]:

- Genomic Address of FliM: 2,018,109 bp left end, and 2,019,113 bp right end, 1,005 length

  ATGGGCGATAGTATTCTTTCTCAAGCTGAAATTGATGCGCTGTTGAATGGTGACA
  GCGAAGTCAAAGACGAACCGACAGCCAGTGTTAGCGGCGAAAGTGACATTCG
  TCCGTACGATCCGAATACCCAACGACGGGTTGTGCGCGAACGTTTGCAGGCGCT
  GGAAATCATTAATGAGCGCTTTGCCCGCCATTTTCGTATGGGGCTGTTCAACCTG
  CTGCGTCGTAGCCCGGATATAACCGTCGGGGCCATCCGCATTCAGCCGTACCAT
  GAATTTGCCCGCAACCTGCCGGTGCACCAACCTGAACCTTATCCATCTGAAACC
  GCTGCGCGGCACTGGGCTGGTGGTGTTCTCACCGAGTCTGGTGTTTATCGCCGTG
  GATAACCTGTTTGGCGGCGATGGACGCTTCCCGACCAAAGTGGAAGGTCGCGAG
  TTTACCCATACCGAACAGCGCGTCATCAACCGCATGTTGAAACTGGCGCTTGAA
  GGCTATAGCGACGCCTGGAAGGCGATTAATCCGCTGGAAGTTGAGTACGTGCGT
  TCGGAAATGCAGGTGAAATTTACCAATATCACCACCTCGCCGAACGACATTGTG

GTTAACACGCCGTTCCATGTGGAGATTGGCAACCTGACCGGCGAATTTAATATCT
GCCTGCCATTCAGCATGATCGAGCCGCTACGGGAATTGTTGGTTAACCCGCCGCT
GGAAAACTCGCGTAATGAAGATCAGAACTGGCGCGATAACCTGGTGCGCCAGGT
GCAGCATTCACAGCTGGAGCTGGTCGCCAACTTTGCCGATATCTCGCTACGCCT
GTCGCAGATTTTAAAACTGAACCCCGGCGACGTCCTGCCGATAGAAAAACCCGA
TCGCATCATCGCCCATGTTGACGGCGTCCCGGTGCTGACCAGTCAGTATGGCACC
CTCAACGGTCAGTATGCGTTACGGATAGAACATTTGATTAACCCGATTTTAAATTC
TCTGAACGAGGAACAGCCCAAATGA

- Genomic Address of FliN: 2,019,110 bp left end, and 2,019,523 bp right end, 414 length

ATGAGTGACATGAATAATCCGGCCGATGACAACAACGGCGCAATGGACGATCT
GTGGGCTGAAGCGTTGAGCGAACAAAAATCAACCAGCAGCAAAAGCGCTGCC
GAGACGGTGTTCCAGCAATTTGGCGGTGGTGATGTCAGCGGAACGTTGCAGGA
TATCGACCTGATTATGGATATTCCGGTCAAGCTGACCGTCGAGCTGGGCCGTAC
GCGGATGACCATCAAAGAGCTGTTGCGTCTGACGCAAGGGTCCGTCGTGGCGC
TGGACGGTCTGGCGGGCGAACCACTGGATATTCTGATCAACGGTTATTTAATCG
CCCAGGGCGAAGTGGTGGTCGTTGCCGATAAATATGGCGTGCGGATCACCGATA
TCATTACTCCGTCTGAGCGAATGCGCCGCCTGAGCCGTTAG

A high-performance interactive software has been developed in the MATLAB environment to support the frequency-domain analysis. We utilize the power spectral density (PSD) analysis applying different methods of PSD estimation (covariance, multiplier, periodogram, etc.). For example, the Welch method is based on dividing the sequence of data into (possibly overlapping) segments, computing a modified periodogram of each segment and averaging the PSD estimates. That is, we consider $x_m[n] = x[(N/M)m-(L/2)+n], n = 0, 1, \ldots, L-1$ to be the mth segment of the sequence $x \in C^N$ divided into M segments of length L. The Welch PSD estimate is given as $R_x = \{|X_m[k]|^2\}_m$, where $\{\cdot\}_m$ denotes averaging across the data segments.

Figure 3.22 reports the frequency analysis of amino acids for FliG.

Figure 3.23 illustrates the power spectra of the DNA sequences for FliG, FliM, and FliN.

The PSD estimation and analysis can be utilized to distinguish genomic sequences versus nongenomic sequences. Figure 3.24 shows four PSDs for: (1) stand-alone *E. coli* FliG gene; (2) FliG gene surrounded by the FliM and FliN genes; (3) FliG surrounded by random nucleotides. The documented results demonstrate very distinct PSDs for the studied sequences. Thus, the proposed concept allows one to distinguish genomic and nongenomic sequences.

Using distinct methods, the results of the developed interactive software for bacterial (*E. coli* and *S. typhimurium*) complete genome sequences, genes, and distinct human genes are reported in [34,38–40]. Depending on their functionality, proteins (defined by DNA) are different. Potentially, biomolecular devices and ^{BM}PPs can be devised, designed, and synthesized using the

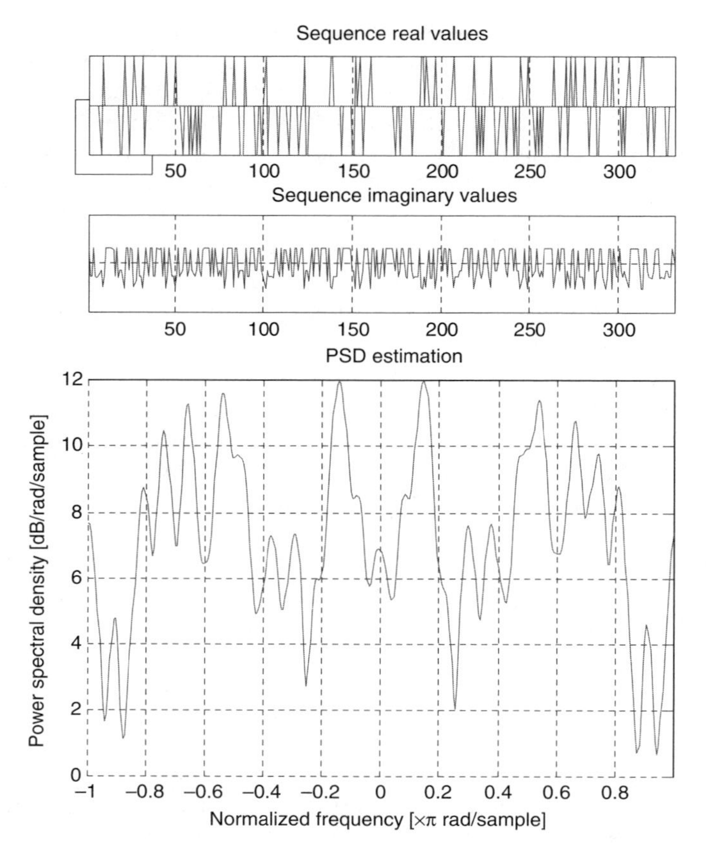

FIGURE 3.22
Power spectral density for FliG amino acids.

DNA instructional coding to synthesize functional proteins as modules, subsystems, and systems. We showed that genes have distinctive PSDs. The correlation between the similar genes, which may result in proteins that exhibit similar functionality, is researched. The autocorrelation analysis is performed. The deterministic autocorrelation sequence $r_{xx}[n]$ of a sequence $x[n]$ is given as

$$r_{xx}[n] = \sum_{k=-\infty}^{\infty} x[k]x[n+k], \quad n = 0, 1, 2, \ldots, N-1,$$

where $x[n]$ is a sequence of either nucleotides or amino acids.

The autocorrelation sequence measures the dependence of values of the sequence at different positions in the sequence. A finite random sequence has an autocorrelation sequence of all zeros with the exception of a single large value at zero. We examine the "randomness" of the studied protein sequence

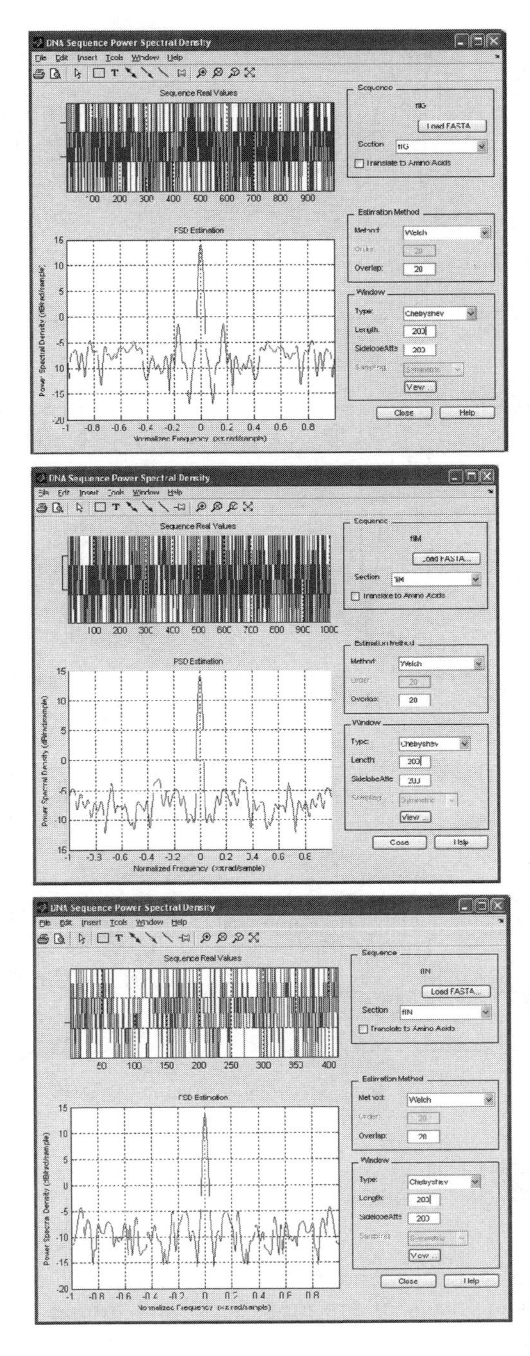

FIGURE 3.23
Power spectral density for FliG, FliM, and FliN.

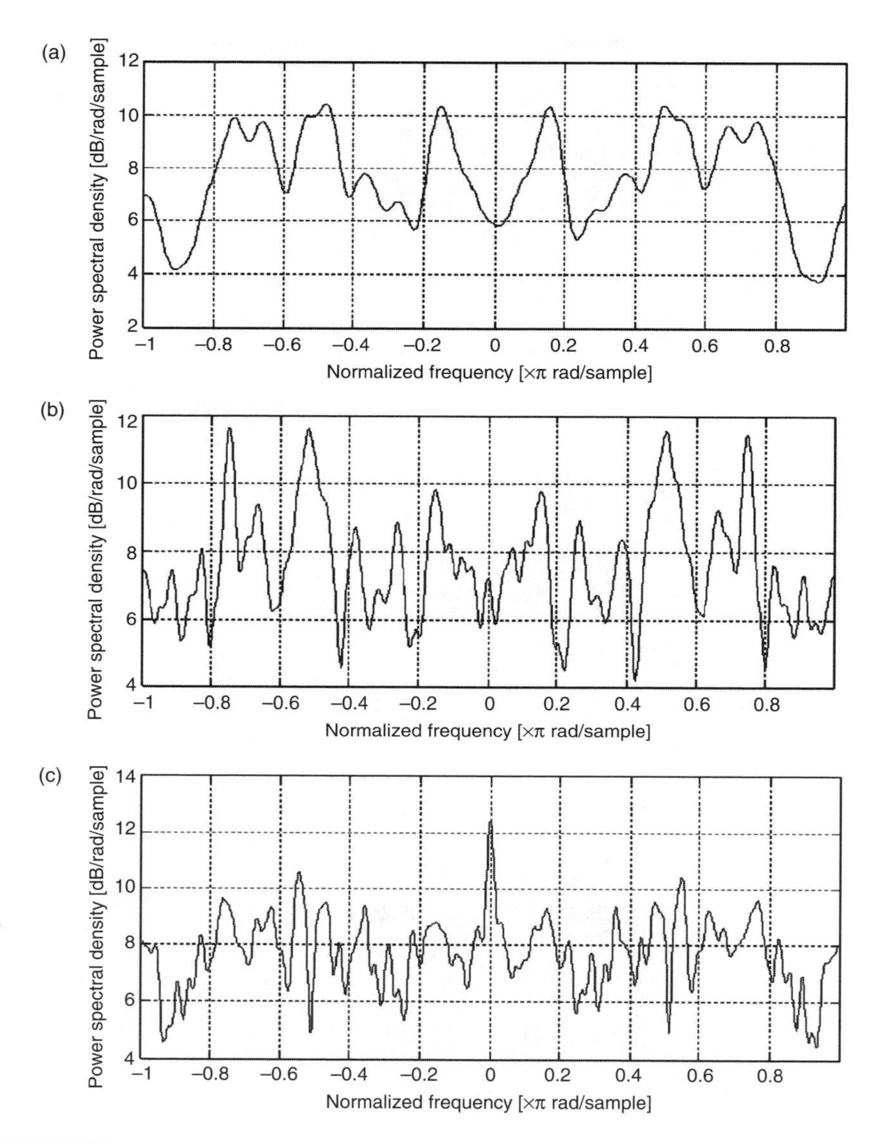

FIGURE 3.24
Power spectral density for the sequences of (a) stand-alone FliG; (b) FliM–FliG–FliN genes; (c) random nucleotides–FliG–random nucleotides.

applying the autocorrelation analysis [41]. Figures 3.25 and 3.26 show the autocorrelation sequences of the nucleotide and amino acid sequences of the ElbB and FliG genes.

The results, shown in Figures 3.25 and 3.26, indicate that the DNA and amino acid sequences are examined utilizing the frequency-domain analysis. The results for ElbB gene and FliG gene (the so-called nanobiomotor *switch*),

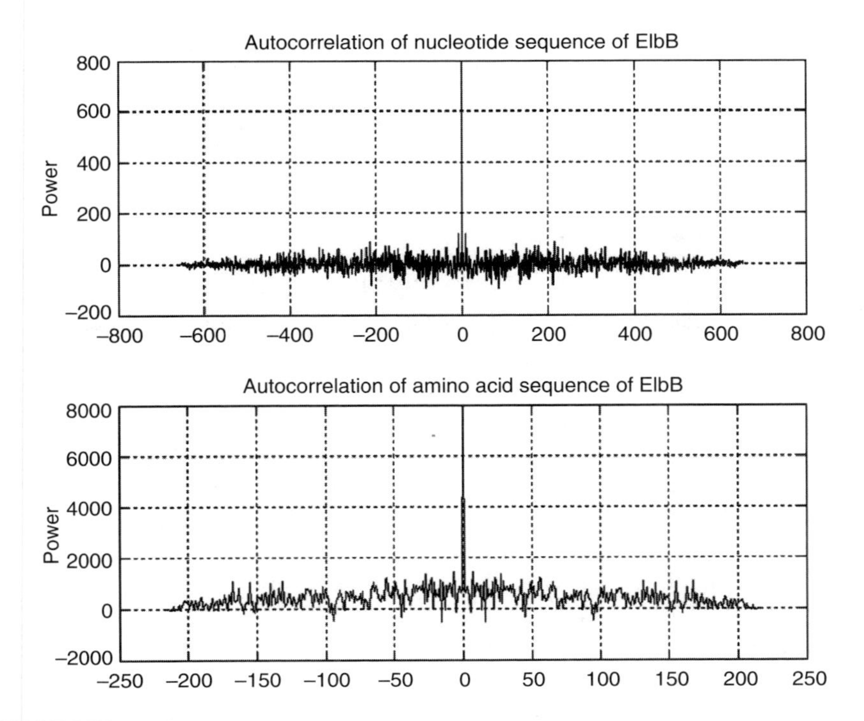

FIGURE 3.25
Autocorrelation of the nucleotide and amino acid sequence for ElbB gene.

as well as for other genes, indicate that there is a coding, and there are specific templates that must be examined. Large DNA and amino acid sequences can be examined using the proposed concept, while statistical methods have the well-known deficiencies.

Example 3.11

FliG and FliN proteins have attracted attention because mutant phenotypes suggest that these *switch* proteins are needed not only for *E. coli* biomolecular motor assembly but also for control. We perform the correlation analysis to analyze the genes patterns. The cross-correlation analysis is achieved by using the following equation

$$r_{xy}[n] = \sum_{k=-\infty}^{\infty} x[k]y[n+k].$$

Figure 3.27 shows the cross-correlation of FliG and FliN. The correlation coefficients $\rho_{xy}[n] = (r_{xy}[n]/\sqrt{r_x[0]r_y[0]})$ provide a normalized measure of correlation. If $\rho_{xy} > 0$, then the sequences are positively correlated, while if $\rho_x = 0$, the sequences are uncorrelated. For the studied case, $\rho_{xy} = 0$, which validates the results.

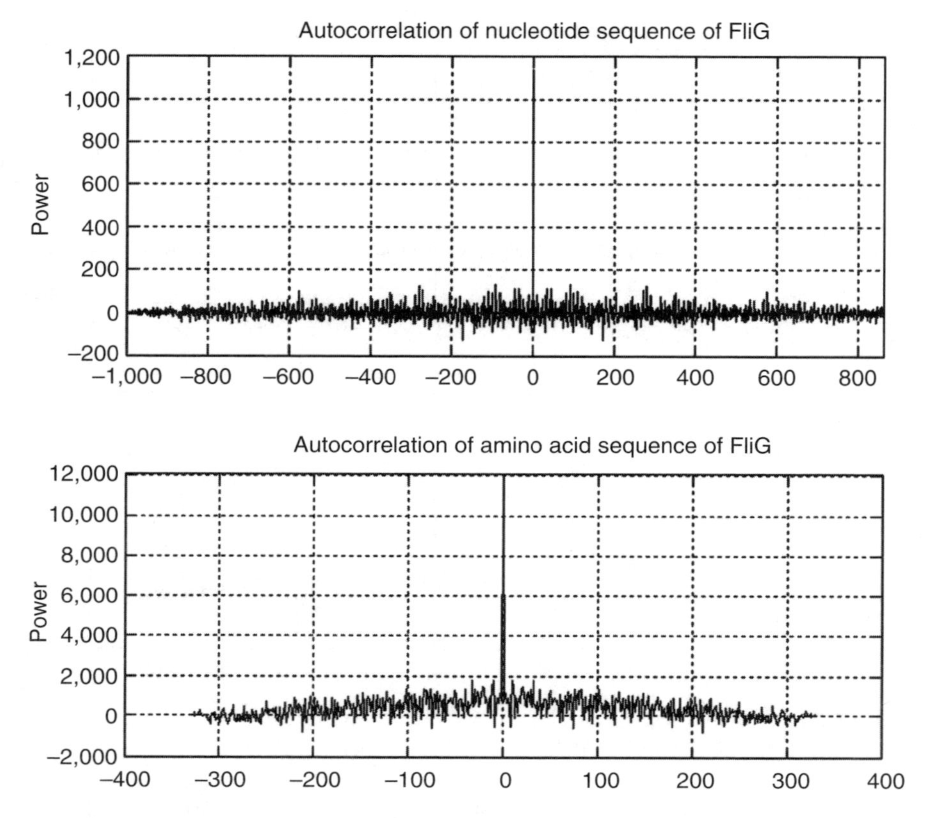

FIGURE 3.26
Autocorrelation of the nucleotide and amino acid sequence for FliG gene.

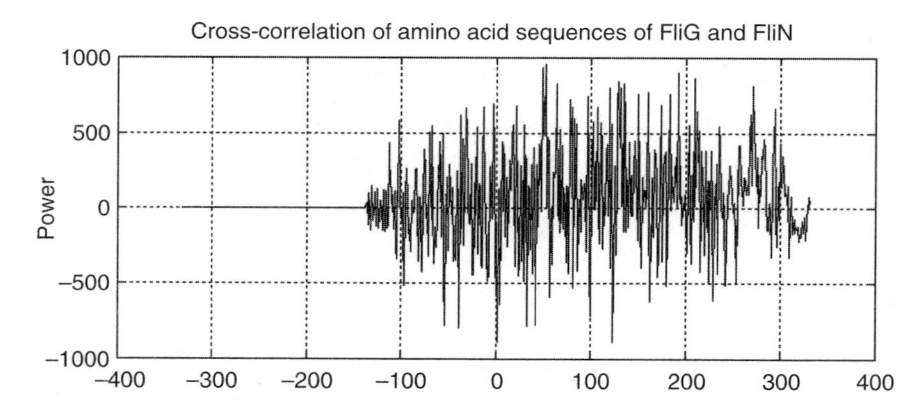

FIGURE 3.27
Cross-correlation of the FliG and FliN.

Entropy Analysis: Entropy, which is the Shannon quantity of information, measures the complexity of the set. The uncertainty after binding for each site (Shannon entropy of position l) is

$$H(l) = -\sum_{b\in \mathbf{A}} f(b,l)\log_2 f(b,l),$$

where $\mathbf{A}$ is the cardinality of the four-letter DNA *symbolic alphabet*, $\mathbf{A} = \{\text{A, C, G, T}\}$; $f(b,l)$ is the frequency of base b at position l.

For DNA, the maximum uncertainty at any given position is $\log_2 \mathbf{A} = 2$ bits. For amino acids, the alphabet is $\mathbf{A} = \{\text{Ala, Arg, \ldots, Tyr, Val}\}$. Therefore, for amino acids, the maximum entropy at any given position is $\log_2 \mathbf{A} = 4.32$ bits.

Using the entropy $H(l)$, one derives the information at every position in the site as

$$R(l) = \log_2 \mathbf{A} - \left(-\sum_{b\in \mathbf{A}} f(b,l)\log_2 f(b,l)\right).$$

The total amount of pattern in ribosome binding sites is found by adding the information from each position, that is, $R_\Sigma(l) = \sum_l R(l)$ bits per site. For *E. coli* and *S. typhimurium* one finds 11.2 and 11.1 bits per site. We apply probability methods to study *E. coli* and *S. typhimurium* genomes. One can identify interesting sections of a genome including the *low-* and *high-complexity* regions. Examining DNA as a coding system, it is fond that different DNA segments have different entropies. In general, entropy depends on the probability model attributed to the source and other assumptions. Figure 3.28 shows the entropy of complete gene sequences for *E. coli* EDL933 with 5476 genes [39]. The entropy analysis is also performed for the *S. typhimurium* genome with 4596 genes as illustrated in Figure 3.28.

3.9.4 DNA- and Protein-Centered Electronics: Feasibility Assessment

Alternatively to the synthesis of natural and *synthetic* ^{BM}PPs by using a biomimetics paradigm (reported in Section 3.5), attempts have been made to utilize DNA, DNA derivatives, and proteins as electronic devices. Though the idea of DNA electronics, memories, and computing is intriguing, there are fundamental questions on their overall fundamental soundness as well as on technological feasibility. This ultimately results in questioned practicality. From the electronics and processing perspective, it is very unlikely that DNA (located in nucleus) can be utilized or entailed to perform any processing, computing, memory, and logics tasks. The role of DNA was emphasized in Section 3.9.1. As will be reported, it is highly unlikely that DNA and protein can be utilized as electronic devices or circuits. Therefore, the terminology DNA- and protein-centered electronics likely is not well funded, and the use of a word electronics may be questioned. In this subsection we attempt to use this not-coherent terminology because DNA derivatives potentially can be

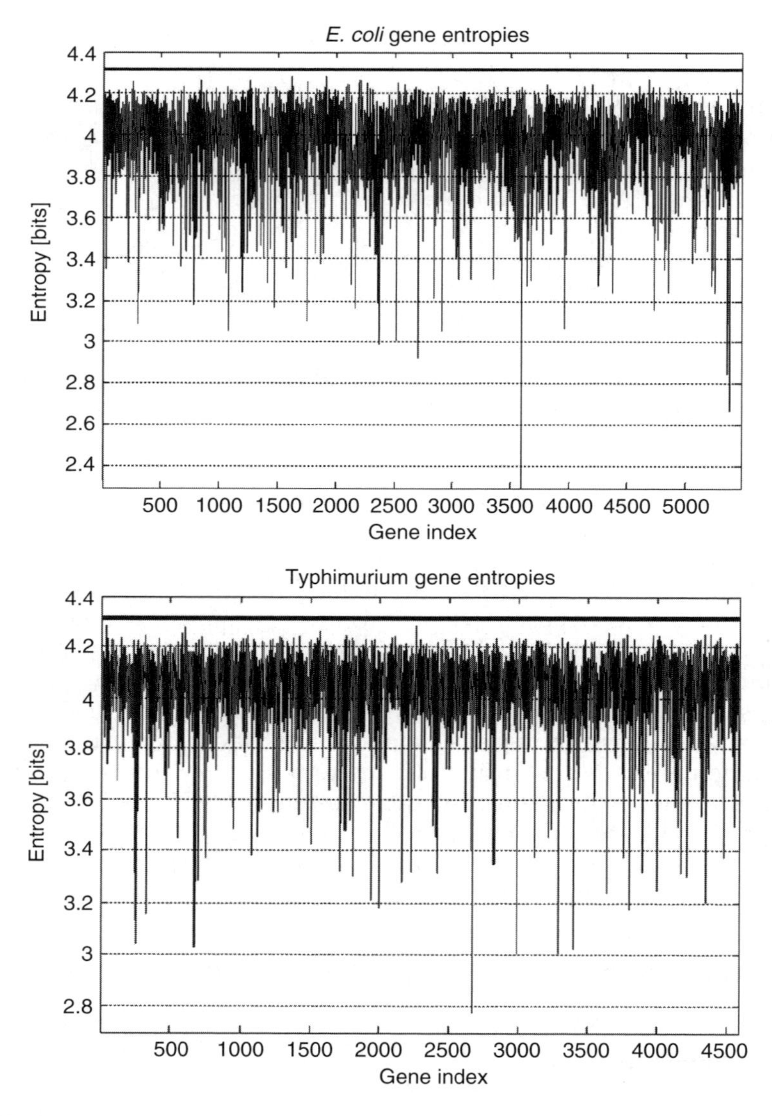

FIGURE 3.28
Entropies for *E. coli* and *Salmonella typhimurium*.

applicable to form structures in electronic devices, though not providing any advantages to the conventional CMOS technology solutions.

Consider some most successful examples. Microelectronic proof-of-concept devices (transistors and sensors) with polymeric guanine (polyG) and other various DNA derivatives have been designed, analyzed, tested, characterized, and evaluated [2,42–45]. The polyG FETs have been fabricated slightly modifying the existing CMOS processes. However, the employed biocentered materials to form the channel do not result in alternative device physics or departure from the CMOS technology. Though the DNA derivatives have been studied in the CMOS-centered devices, the baseline performance characteristics are found to be impractical. Furthermore, the deposition of DNA and DNA derivatives significantly complicates the overall fabrication and packaging. These DNA FETs can be viewed only as a proof-of-concept merging the silicon technology with potential biomolecular materials.

The electron transport through ~30–200 nm in length, 2 nm in diameter polyG ropes, which form the channel on the insulator (silicon oxide, silicon nitride, etc.) between the source and drain in FETs, are of interest. The experimental I–V and conductance (dI/dV)–voltage (G–V) characteristics were obtained. The fabrication of these FETs was performed utilizing the CMOS technology and depositing A-DNA on insulator with the adsorption-based source–DNA–drain attachment. It is reported that an A-DNA bundle of poly(dA)–poly(dT) is an n-type semiconductor, while poly(dG)–poly(dC) is a p-type [42]. DNA derivatives may ensure ordered bundles, layers, and ropes, which are organized and interconnected due to specific reactivity of molecules' functional groups with affinity to distinct surfaces and molecules. Guanine exhibits a low oxidation potential that results in electron transport. An FET with polyG $(dG(C_{10})_2)_n$, which is deposited (dropped with subsequent evaporation) on the silicon oxide (or silicon nitride) between source and drain, is shown in Figure 3.29 [46]. For these polyG nFETs, though the weakly-controlled current–voltage (I–V) characteristics have been found in the linear region, the saturation region and control features are not adequate. Overall, this solution is found to be not viable.

The channels are straightforwardly formed using well-established CMOS processes. Other options have been introduced to form channels in FETs. For example, in p- and n-type carbon nanotube FETs, single- and multiwall carbon nanotubes form the channel. Carbon nanotube FETs have been fabricated, tested, and characterized [47–49]. These carbon nanotubes FETs utilize the same device physics as conventional FETs, and the fabrication technology remains CMOS with significant additional challenges due to the deposition of carbon nanotubes. This solution may not ensure the expected advantages and benefits. Significant technological difficulties of utilizing carbon nanotubes in microelectronic devices remain unsolved.

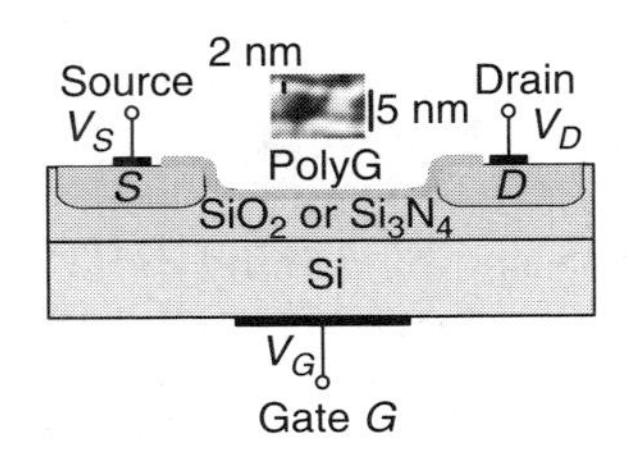

FIGURE 3.29
PolyG nFET with ~100 nm channel length.

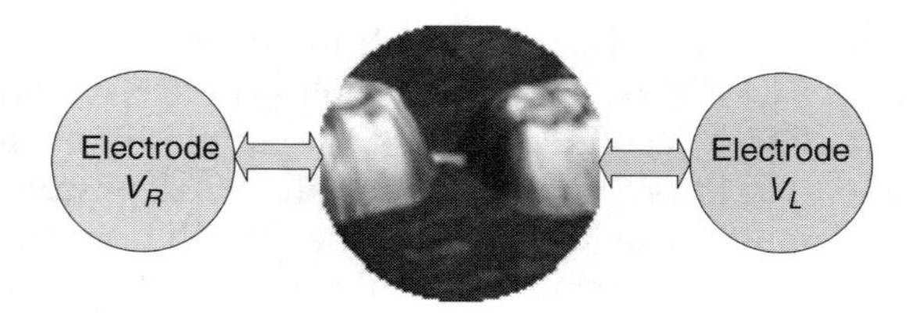

FIGURE 3.30
Illustrative representation of a double-stranded DNA with two electrodes.

Sequence-dependent self-assembled DNA and templated protein synthesis can be used to build patterned 2D and 3D structures with desired geometrical topologies. Integrated circuits are designed utilizing functional electronic devices that must exhibit desired electronic phenomena possessing required characteristics. Insulating, conducting, and semiconducting behaviors have been reported for the contact–biomolecule–contact complexes. The test beds and proof-of-concept prototypes of two-terminal biomolecular and organic *rectifiers* were fabricated utilizing CMOS technology [42–45,50–55]. Figure 3.30 illustrates DNA functionalized to electrodes. There are debates regarding the electronic properties of biomolecules and DNA owing to the fact that some experiments have been questioned and several experiments are found to be not very convincing. For two-terminal *rectifiers*, the experimental I–V characteristics leave much to be desired [43–45,50–55].

In devising new fabrication technologies, one may focus on robust *bottom-up* biomolecular assembling, which is consistently performed by biosystems. Progress has been made in the synthesis of DNA with the specified sequences. By utilizing a motif-based DNA self-assembly, complex 3D structures were synthesized [53]. There are precise binding rules. For example, adenine A with complementary guanine G, and thymine T with complementary cytosine C. However, thermodynamic and other designs to minimize sequence mismatches are still complex tasks. The desired assemblies frequently cannot be synthesized owing to geometric, thermodynamic, and other limits of DNA hybridization [56]. Most importantly, the experimental results [43–45,53] provide evidence that DNA and DNA derivatives do not exhibit the suitable characteristics even for two-terminal *rectifiers*, and it is unlikely that suitable device functionality may be achieved. Though the 3D "multi-interconnected" DNA structures can be synthesized, it seems that it is virtually impossible to expect sound DNA-centered electronic devices.

The resistor–diode logic will be covered in Section 4.1, and Figure 4.3 illustrates the implementation of ^{M}AND gate. Though the simple resistor–diode circuits can be theoretically assembled using the DNA motif, this solution may be impractical. The metallization of 3D DNA lattices, which potentially can modify the electron transport in the metalized segments, cannot be viewed as a sound solution. DNA with the appropriate sequence potentially can be used only as biomolecular *wire*, if needed. The qualitative achievable I–V characteristics for the *admissible* applied voltage for DNA are shown in Figure 3.31.

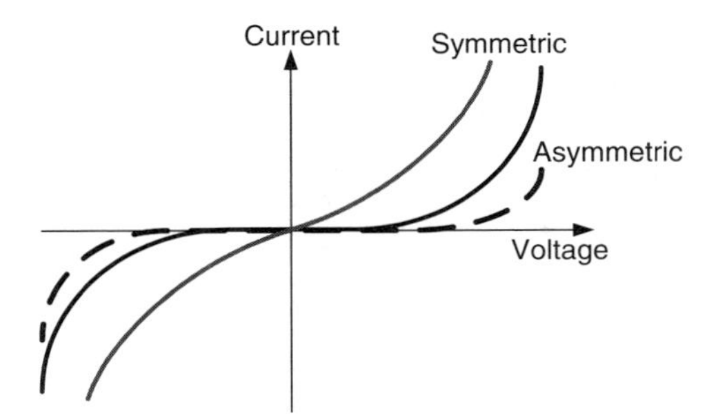

FIGURE 3.31
Symmetric and asymmetric I–V characteristics.

The magnitude of the applied voltage is bounded due to the thermal stability of the molecule, that is, $|V| \leq V_{\max}$, and $V_{\max}$ is ~1 V depending upon sequence, single versus double stranded, number of base pairs, temperature, functionalization, end group, and so forth. For different DNA sequences, the I–V characteristics vary [53], and for the conducting DNA, the characteristics are similar to some organic molecules [51,54,55]. The quantitative I–V characteristics of a single DNA strain, functionalized to the nitrogenous bases are shown in Figure 3.31. The asymmetry of the I–V characteristics could be due to the effects related to the asymmetric DNA fictionalization to the electrodes, contact nonuniformity, and so forth. It is difficult to expect that, with the reported characteristics, the DNA exhibits a significant potential for electronics.

The sequence-dependent assembled DNA and templated protein synthesis can be used to build patterned 3D structures with the desired geometrical topologies. As reported in Section 3.9.1, the protein 3D geometry is due to folding of a peptide chain as well as multiple peptide chains. Amino acid bonds determine the folding (α-helix resulting in helix-loop-helix, β-pleated sheet, random conformations, etc.). Most proteins evolve through several intermediate states to form stable structures. The conformation of proteins can be reinforced by strong covalent bonds called disulfide bridges. Disulfide bridges form where two cysteine monomers (amino acids with sulfhydryl groups on their side chains) are positioned close by the folding of the protein. Figure 3.32 illustrates the schematics of the folded protein with hydrogen bonds, ionic bonds, and hydrophobic interactions between side chains. These weak bonds and strong covalent (disulfide bridges) bonds can be considered to be similar to 3D biomolecular *circuits*. However, the practicality and feasibility of biomolecular *circuits*, where proteins function as the ^{ME}device or ^{M}gates, remain to be examined.

Overall Assessment on the Feasibility of DNA- and Protein-Centered Electronics: The existence and superiority of ^{BM}PPs are undisputable facts. However, ^{BM}PPs are profoundly different as compared with synthetic (organic and inorganic) molecular electronics, solid and fluidic ^{M}ICs, and

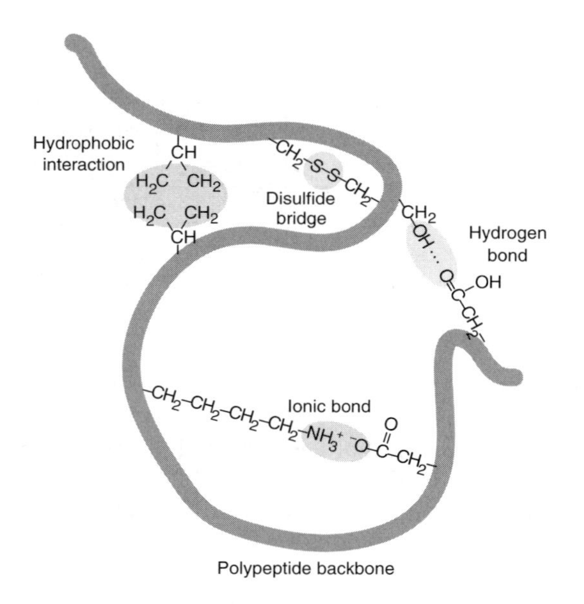

FIGURE 3.32
Protein tertiary structure with weak and strong bonds.

MPPs. It is highly unlikely that biomolecules are utilized (within BMPPs) or can be utilized (within MPPs) as *electronic* elements, components, devices, or primitives. The challenges in the possible application of biomolecules as MEdevices were discussed in this section. It seems that the overall feasibility of using DNA motifs (including the metalized, modified, and other solutions) and proteins as electronic devices or circuits does not look promising. In fact, electronic devices ultimately depend on the electron transport. Suitable device (or circuit) characteristics and performance are virtually unachievable by DNA or proteins. There is no evidence that DNA and proteins can or could be utilized as MEdevices. Departing from the unfounded notion of using DNA and protein as *circuits*, Sections 3.1 and 3.2 introduced *biomolecular processing hardware*, where electrochemomechanically induced transitions, interactions, and events lead to a sound biocentered premise.

Alternative Solution: Solid Molecular Electronics: In biosystems, the information coding (accomplished by the DNA) and the transcription–translation mechanism (DNA → RNA → protein) result in the synthesis of *biomolecular processing hardware* with embedded *software*. Utilizing *biomolecular processing hardware* and *software*, BMPPs accomplish information processing by making use of specific biophysical phenomena, effects, and mechanisms. These phenomena and transitions are not electron transport (flow) centered. For example, the controlled electron flow, which is a defining phenomenon in solid-state electronic devices and MEdevices, is not a key factor in neuronal processing-and-memory primitives and BMPPs, where other biophysics ultimately results in processing and memory. Biomolecules provide unique capabilities to synthesize complex 3D structures and lattices. With the attempt to utilize these synthesis abilities, natural and modified nitrogenous bases have been studied departing from the DNA-centered theme towards exploring the ^{ME}device-centered solution. Different nitrogenous bases, along with bioconjugation and immobilization/quantification/quencher reagents, have been used attempting to ensure synthesis soundness and attain the desired electronic characteristics. In devising novel device physics and discovering novel ^{ME}devices one may not entirely focus on merely natural biomolecules motif. Different biomolecules and organic mono- and polycyclic molecules, which have a strong biocentered premise, are proposed and covered in Chapter 5. These multiterminal molecules significantly enlarge the class of molecules to be examined and used in order to ensure the desired electronic characteristics suitable for molecular electronics. In particular, functional ^{ME}devices are synthesized utilizing *input*, *control*, and *output* terminals as reported in Chapter 5, and the *design rules* are reported in Section 4.6.2. Chapter 6 documents that organic molecules exhibit quantum effects and electronic properties that can be utilized in new generation of electronic devices.

References

1. S. R. Hameroff and J. Tuszynski, Search for quantum and classical modes of information processing in microtubules: implications for The Living Atate. In *Handbook on Bioenergetic Organization in Living Systems*, Eds. F. Musumeci and M.-W. Ho, World Scientific, Singapore, 2003.
2. S. E. Lyshevski, Three-dimensional molecular electronics and integrated circuits for signal and information processing platforms. In *Handbook on Nano and Molecular Electronics*, Ed. S. E. Lyshevski, CRC Press, Boca Raton, FL, pp. 6-1–6-104, 2007.
3. D. D. Georgiev and J. F. Glazebrook, Subneuronal processing of information by solitary waves and stochastic processes. In *Handbook on Nano and Molecular Electronics*, Ed. S. E. Lyshevski, CRC Press, Boca Raton, FL, pp. 17-1–17-41, 2007.
4. P. Dayan and L. F. Abbott, *Theoretical Neuroscience: Computational and Mathematical Modeling of Neural Systems*, MIT Press, Cambridge, MA, 2001.
5. W. Freeman, Tutorial on neurobiology from single neurons to brain chaos, *Int. J. Biforcation Chaos*, vol. 2, no. 3, pp. 451–482, 1992.
6. S. Grossberg, On the production and release of chemical transmitters and related topics in cellular control, *J. Theor. Biol.*, vol. 22, pp. 325–364, 1969.
7. S. Grossberg, *Studies of Mind and Brain*, Reidel, Amsterdam, The Netherlands, 1982.
8. L. F. Abbott and W. G. Regehr, Synaptic computation, *Nature*, vol. 431, pp. 796–803, 2004.

9. H. Markram, J. Lubke, M. Frotscher and B. Sakmann, Regulation of synaptic efficacy by coincidence of postsynaptic APs and EPSPs, *Science*, vol. 275, pp. 213–215, 1997.
10. C. C. Rumsey and L. F. Abbott, Equalization of synaptic efficacy by activity- and timing-dependent synaptic plasticity, *J. Neurophysiol.*, vol. 91, no. 5, pp. 2273–2280, 2004.
11. S. Grossberg, Birth of a learning law, *Neural Networks*, vol. 11, no. 1, 1968.
12. F. Frantherz, Neuroelectronics interfacing: semiconductor chips with ion channels, nerve cells, and brain. In *Handbook of Nanoelectronics and Information Technology*, Ed. R. Waser, Wiley-VCH, Darmstadt, Germany, pp. 781–810, 2005.
13. A. Renyi, On measure of entropy and information, *Proc. Berkeley Symp. Math. Stat. Prob.*, vol. 1 , pp. 547–561, 1961.
14. Yu. M. Kabanov, The capacity of a channel of the Poisson type, *Theory Prob. Appl.*, vol. 23, pp. 143–147, 1978.
15. D. Johnson, Point process models of single-neuron discharges, *J. Comp. Neurosci.*, vol. 3, pp. 275–299, 1996.
16. S. P. Strong, R. Koberle, R. R. de Ruyter van Steveninck and William Bialek, Entropy and information in neuronal spike trains, *Phys. Rev. Lett.*, vol. 80, no. 1, 1998.
17. F. Rieke, D. Warland, R. R. de Ruyter van Steveninck and W. Bialek, *Spikes: Exploring the Neural Code*, MIT Press, Cambridge, MA, 1997.
18. K. C. Smith, Multiple-valued logic: A tutorial and appreciation, *Computer*, vol. 21, 4, pp. 17–27, 1998.
19. J. R. Buitenweg, W. L. C. Rutten and E. Marani, Modeled channel distributions explain extracellular recordings from cultured neurons sealed to microelectrodes, *IEEE Trans. Biomed. Engi.*, vol. 49, no. 11, pp. 1580–1590, 2002.
20. F. J. Sigworth and K. G. Klemic, Microchip technology in ion-channel research, *IEEE Trans. Nanobiosci.*, vol. 4, no. 1, pp. 121–127, 2005.
21. H. Suzuki, Y. Kato-Yamada, H. Noji and S. Takeuchi, Planar lipid membrane array for membrane protein chip, *Proc. Conf. MEMS*, pp. 272–275, 2004.
22. K. S. Lashley, *Brain Mechanisms and Intelligence*, University of Chicago Press, Chicago, 1929.
23. K. Craik, *The Nature of Explanation*, Cambridge University Press, Cambridge, UK, 1943.
24. J. von Neumann, *The Computer and the Brain*, Yale University Press, New Haven, CT, 1957.
25. W. S. McCulloch and W. H. Pitts, A logical calculus of the ideas immanent in nervous activity, *Bulletin of Mathematical Biology*, vol. 5, pp. 115–133, 1943.
26. A. Rosenblueth, A., N. Wiener and J. Bigelow, Behavior, purpose and teleology, *Philos. Sci.*, vol. 10, pp. 18–24, 1943.
27. D. O. Hebb, *The Organization of Behavior: A Neuropsychological Theory*, Wiley-Interscience, New York, 1949.
28. M. Minsky, Steps toward artificial intelligence, *Proc. IRE*, vol. 49, no. 1, pp. 8–30, 1961.
29. J. J. Hopfield, Neural networks and physical systems with emergent collective computational abilities, *Proc. Natl. Acad. Sci. USA*, vol. 79, pp. 2554–2558, 1982.
30. J. J. Hopfield, Neurons with graded response have collective computations properties like those of two-state neurons, *Proc. Natl. Acad. Sci. USA*, vol. 81, pp. 3088–3092, 1984.

31. S. Haykin, *Neural Networks: A Comprehensive Foundation*, MacMillan College Publishing Co, New York, 1994.
32. D. E. Rumelhart, G. E. Hinton and R. J. Williams, Learning internal representations by error propagation. In *Parallel Distributed Processing: Explorations in Microstructure of Cognition*, Eds. D. E. Rumelhart and J. L. McClelland, vol. 1, MIT Press, Cambridge, MA, pp. 318–362, 1986.
33. S. F. Altschul, T. L. Madden, A. A. Schäffer, J. Zhang, Z. Zhang, W. Miller and D. J. Lipman, Gapped BLAST and PSI-BLAST: A new generation of protein database search programs, *Nucl. Acids Res.*, vol. 25, pp. 3389–3402, 1997.
34. E. Martz, *Protein Explorer*, 2003. http://www.proteinexplorer.org
35. P. Bertone and M. Gerstein, Integrative data mining: The new direction in bioinformatics, *IEEE Eng. Med. Biol.*, vol. 20, pp. 33–40, 2001.
36. G. G. Lusman and D. Lancet, Visualizing large-scale genomic sequences, *IEEE Eng. Med. Biol.*, no. 4, pp. 49–54, 2001.
37. E. W. Myers and W. Miller, Optimal alignments in linear space, *Comput. Appl. Biosci.*, vol. 4, no. 1, pp. 11–17, 1988.
38. *Protein Data Bank*, 2003. http://www.rcsb.org/pdb/
39. K. E. Rudd, EcoGene: A genome sequence database for *Escherichia coli* K-12, *Nucl. Acids Res.*, vol. 28, pp. 60–64, 2000.
40. R. A. Welch, V. Burland, G. Plunkett, P. Redford, P. Roesch, D. Rasko, E. L. Buckels, et al., Extensive mosaic structure revealed by the complete genome sequence of uropathogenic *Escherichia coli*, *Microbiology*, vol. 99, no. 26, pp. 17020–17024, 2002.
41. S. E. Lyshevski, *NEMS and MEMS: Fundamentals of Nano- and Microengineering*, 2nd edn., CRC Press, Boca Raton, FL, 2005.
42. K. H. Yoo, D. H. Ha, J. O. Lee, J. W. Park, J. Kim, J. J. Kim, H. Y. Lee, T. Kawai and H. Y. Choi, Electrical conduction through poly(dA)-poly(dT) and poly(dG)-poly(dC) DNA molecules, *Phys. Rev. Lett.*, vol. 87, no. 19, 2001.
43. D. Porath, Direct measurement of electrical transport through DNA molecules, *Nature*, vol. 403, pp. 635–638, 2000.
44. D. Porath, G. Cuniberti and R. Di Felice, Charge transport in DNA-based devices, *Top. Curr. Chem.*, vol. 237, pp. 183–227, 2004.
45. A. Rakitin, P. Aich, C. Papadopoulos, Y. Kobzar, A. S. Vedeneev, J. S. Lee and J. M. Xu, Metallic conduction through engineered DNA: DNA nanoelectronic building blocks, *Phy. Rev. Let.*, vol. 86, no. 16, pp. 3670–3673, 2001.
46. M. A. Lyshevski, Multi-valued DNA-based electronic nanodevices, *IEEE Conf. Multiple-Valued Logic Design*, Calgary, Canada, pp. 39–42, 2005.
47. J. Appenzeller, R. Martel, P. Solomon, K. Chan, P. Avouris, J. Knoch, J. Benedict, et al., A 10 nm MOSFET concept, *Microelect. Eng.*, vol. 56, no. 1–2, pp. 213–219, 2001.
48. V. Derycke, R. Martel, J. Appenzeller and P. Avouris, Carbon nanotube inter- and intramolecular logic gates, *Nano Lett.*, 2001.
49. R. Martel, H. S. P. Wong, K. Chan and P. Avouris, Carbon nanotube field effect transistors for logic applications, *Proc. Electron Devices Meeting, IEDM Technical Digest*, pp. 7.5.1–7.5.4, 2001.
50. J. C. Ellenbogen and J. C. Love, Architectures for molecular electronic computers: Logic structures and an adder designed from molecular electronic diodes, *Proc. IEEE*, vol. 88, no. 3, pp. 386–426, 2000.

51. K. Lee, J. Choi and D. B. Janes, Measurement of *I–V* characteristic of organic molecules using step junction, *Proc. IEEE Conf. on Nanotechnology*, Munich, Germany, pp. 125–127, 2004.
52. A. K. Mahapatro, S. Ghosh and D. B. Janes, Nanometer scale electrode separation (nanogap) using electromigration at room temperature, *Proc. IEEE Trans. Nanotechnol.*, vol. 5, no. 3, pp. 232–236, 2006.
53. A. K. Mahapatro, D. B. Janes, K. J. Jeong and G. U. Lee, Electrical behavior of nano-scale junctions with well engineered double stranded DNA molecules, *Proc. IEEE Conference on Nanotechnology*, Cincinnati, OH, 2006.
54. R. M. Metzger, Unimolecular electronics: results and prospects. In *Handbook on Molecular and NanoElectronics*, Ed. S. E. Lyshevski, CRC Press, Boca Raton, FL, pp. 3-1–3-25, 2007.
55. J. Reichert, R. Ochs, D. Beckmann, H. B. Weber, M. Mayor and H. V. Lohneysen, Driving current through single organic molecules, *Phys. Rev. Lett.*, vol. 88, no. 17, 2002.
56. A. Carbone and N. C. Seeman, Circuits and programmable self-assembling DNA structures, *Proc. Natl. Acad. Sci.* USA, vol. 99, no. 20, pp. 12577–12582, 2002.

4

Design of Molecular Integrated Circuits

4.1 Molecular Electronics and Gates: Device and Circuits Prospective

This chapter introduces the reader to the analysis and design of analog and digital ^{M}ICs. There are the electronic hardware and software aspects of design. Distinct ^{M}gates and $^{\aleph}$hypercells can be used to perform logic functions and memory storage. To store the data, memory cells are used. A systematic arrangement of memory cells and peripheral ^{M}ICs (to address and write data into the cells as well as to delete stored data from the cells) constitute the memory. The ^{M}devices can be used to implement static and dynamic random access memory (RAM) as well as programmable and alterable read-only memory (ROM). Here, RAM is the read–write memory in which each individual memory primitive can be addressed at any time, while ROM is commonly used to store instructions of an operating system. Static RAM (SRAM) may consist of a basic flip-flop ^{M}device with stable states (e.g., binary 0 and 1, or multiple values). In contrast, dynamic RAM, which can be implemented using a ^{M}device and a storage capacitor, stores one bit of information by charging the capacitor—see, as an example, the dynamic RAM cell in Figure 4.1. The binary information is stored as the charge on the molecular storage capacitor $^{M}C_s$ (logic 0 or 1). This RAM cell is addressed by switching *on* the access ^{ME}device via the worldline signal, resulting in the charge transfer into and out of $^{M}C_s$ on the dataline. The capacitor $^{M}C_s$ is isolated from the rest of the circuitry when the ^{ME}device is *off*. However, the leakage current through the ^{ME}device may require RAM cell refreshment to restore the original signal. Dynamic shift registers can be implemented using transmission ^{M}gates and ^{M}inverters, flip-flops can be synthesized by cross-coupling NOR ^{M}gates, while delay flip-flops can be built using transmission ^{M}gates and feedback ^{M}inverters. In particular, the SRAM and memory elements designed using ^{M}gates, without $^{M}C_s$, which presents a challenge, are shown in Figure 4.7.

Among the specific characteristics under consideration are the read–write speed, memory density, power dissipation, volatility (data should be maintained in the memory array when the power is off), and so forth. The address,

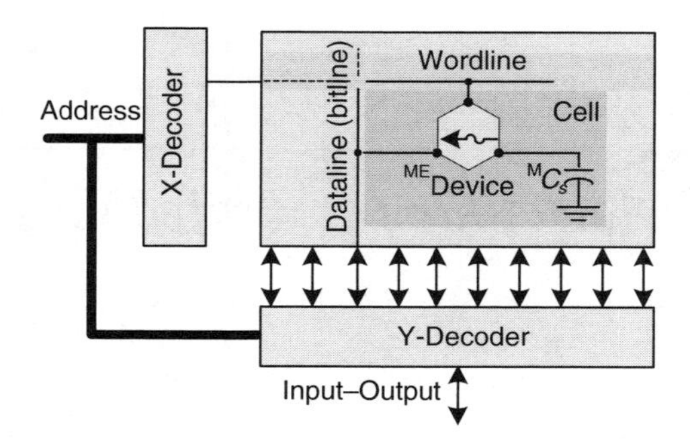

FIGURE 4.1
Dynamic RAM cell with ^{ME}device and storage molecular capacitor $^{M}C_S$.

data-, and control lines are connected to the memory array. The control lines define the function to be performed or the status of the memory system. The address lines and datalines ensure data manipulation and provide address into or out of the memory array. The address lines are connected to an address row decoder, which selects a row of cells from an array of memory cells. A RAM organization, as shown in Figure 4.1, consists of an array of storage cells arranged in an array of 2^n columns (bitlines) and 2^m rows (wordlines). To read the data stored in the array, a row address is supplied to the row decoder, which selects a specific wordline. All cells along this wordline are activated, and the content of each of these cells is placed onto each of their corresponding bitlines. The storage cells can store one (or more) bit of information. The signal available on the bitlines is directed to a decoder. As shown in Figure 4.1, a binary (or high-radix) cell stores binary (or ternary, quaternary, etc.) information utilizing a ^{ME}device at the intersection of the wordline and bitline. The ROM cell can be implemented as: (1) a parallel molecular NOR (^{M}NOR) array of cells, or, (2) a series molecular NAND (^{M}NAND) array of cells. The ROM cell is programmed by either connecting or disconnecting the ^{M}device output (a familiar example is the drain in an FET) from the bitline. Though a parallel ^{M}NOR array is faster, a series ^{M}NAND array is a sound alternative.

In Figure 4.1, the schematic representation of the multiterminal ^{ME}device is shown as . There is a need to design ^{M}devices for which robustly controllable dynamics results in a sequence of state transitions (electron- or photon-induced transport, vibrations, etc.). These transitions and events correspond to a sequence of computational, logic, or memory states. This is guaranteed for quantum ^{M}devices because quantum dynamics can be considered to be deterministic. Nondeterminism of quantum mechanics arises when a device interacts with an uncontrolled outside environment, leaks information to an environment or disturbance. In ^{M}devices, the *global* state evolutions (state transitions) should be deterministic, predictable, and controllable. It should be emphasized that stochastic computing effectively utilizes nondeterminism and uncertainties [1].

From the device physics perspective, the bounds posed by the Heisenberg uncertainty principle restrict the experiment-centered features (measurement, accuracy, data acquisition, etc.), and do not impose limits on device physics, performance, and capabilities. For logic primitives, the device physics defines the mechanism of physical encoding of the logical states in the device. Quantum computing concepts emerged proposing, for example, to utilize quantum spins of electrons or atoms to store information. In fact, a spin is a discrete two-state composition theoretically allowing encoding of a bit. One can encode information using electromagnetic waves and cavity oscillations in optical devices. The feasibility of different state-encoding concepts depends on the ability to maintain the logical state for a required period. The stored information must be reliable. That is, the probability of the stored logical state spontaneously changing to another value should be small. One can utilize energy barriers and wells in the controllable energy space for a set of physical states encoding a given logical state. For example, the ^{ME}device may change the logical state as the electron passes the energy barrier. To prevent this, quantum tunneling can be suppressed by using high and wide potential barriers, minimizing excitation, noise, and so forth. To change the logical state, one varies the energy barrier as illustrated in Figure 4.2. Examining the logical transition processes, the logical states can be retained reliably by potential energy barriers that may separate the physical states. The logical state is changed by varying the energy surface barriers as schematically depicted in Figure 4.2 for a one-dimensional case. Adiabatic transitions between logical states that are located at stable or metastable local energy minima result.

We have discussed conventional and emerging memories and logic devices. Some details to implement memory cells will be reported later. We focus our attention on logics and combinational circuits. In VLSI design, resistor–transistor logic (RTL), diode–transistor logic (DTL), transistor–transistor logic (TTL), emitter-coupled logic (ECL), integrated-injection logic (IIL), merged-transistor logic (MTL), and other logic families have been used. All logic families and subfamilies (within TTL, there are Schottky, low-power Schottky, advanced Schottky, etc.) have advantages and drawbacks. Molecular electronics offers unprecedented capabilities compared with microelectronics. Correspondingly, some logic families that ensure marginal performance using

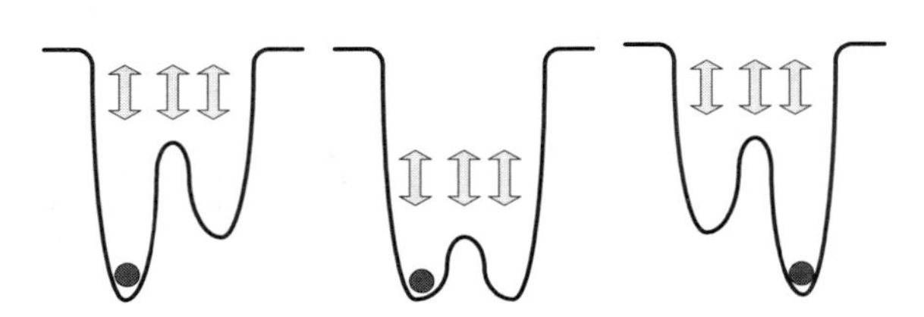

FIGURE 4.2
Logical states and energy barriers.

solid-state devices may provide exceptional performance when ^{M}devices are utilized.

A molecular AND (^{M}AND) gate can be realized utilizing a molecular resistor (schematics representation is -ꟷꟷ-, while the symbol is ^{M}r) and molecular diodes (representation and symbol are -▶|- and ^{M}d). The three-input ^{M}AND gate, constructed within the diode logic family, is shown in Figure 4.3. The important feature is the use of two-terminal ^{ME}devices within the molecular resistor–diode logic (^{M}RDL).

The considered ^{M}RDL, although could lead to the relatively straightforward synthesis and verification of proof-of-concept ^{M}gates (see Chapter 5), may not ensure acceptable performance. Other solutions are considered. The ^{M}NOR gate, realized using the molecular resistor–transistor logic (^{M}RTL), is shown in Figure 4.4a. In electronics, NAND is one of the most important gates. The ^{M}NAND gate, designed by applying the molecular diode–transistor logic (^{M}DTL), is shown in Figure 4.4b. In Figures 4.4a and b, we use different symbols to designate molecular resistors (-ꟷꟷ-; ^{M}r), molecular diodes (-▶|-; ^{M}d), and molecular transistors (⊢; ^{M}T). It will be discussed subsequently that the definition ^{M}T may be used with a great caution because of distinct device physics of molecular and semiconductor devices. In order to introduce the subject, we use, for the moment, this incoherent terminology applying the argument that ^{M}T may ensure characteristics similar to FETs and BJTs.

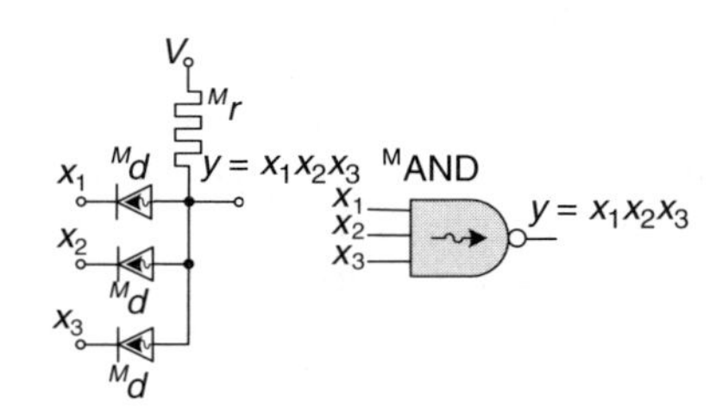

FIGURE 4.3
Implementation of ^{M}AND gate using ^{M}RDL.

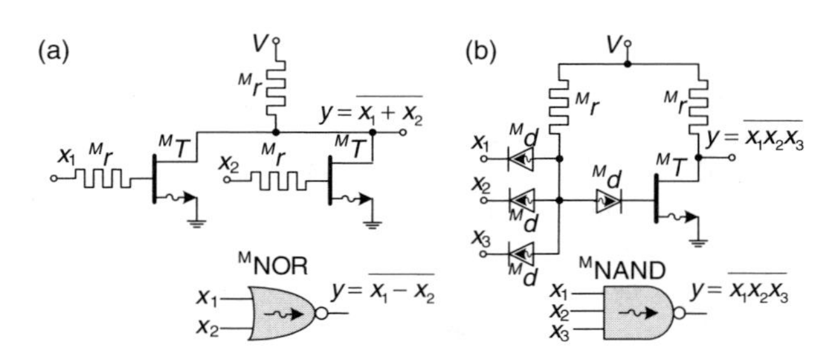

FIGURE 4.4
Circuit schematics: (a) Two-input ^{M}NOR gate; (b) Three-input ^{M}NAND gates.

The device physics of solid ^{ME}devices is entirely different compared with conventional three-, four-, and many-terminal FETs and BJTs. Therefore, we depart from a conventional solid-state terminology, microelectronics-centered definitions and symbols. Even a three-terminal solid ^{ME}device with controlled *I–V* characteristics may not be referenced as a transistor. New terminology can be developed in the observable future, reflecting the device physics of ^{M}device. We define functional solid ^{M}devices in which electrons are utilized as the charge carriers ensuring controllable *I–V* characteristics by means of controlled electron flow to be ^{ME}devices. The device physics of these devices is covered and examined in Chapter 6.

The ^{M}NAND gate, implemented within a ^{M}DTL logic family using the $^{\aleph}$hypercell primitive realization, is illustrated in Figure 4.5a. We emphasized the need of developing a new terminology, definitions, and symbols for ^{ME}devices. Quantum effects (emission, interference, resonance, tunneling, etc.), electron- and photon-induced transitions, electron–photon assisted interactions, as well as other phenomena can be uniquely utilized. In Figure 4.5b, a multiterminal ^{ME}device (^{ME}D) is represented as ⬡, which was already used in Figure 4.1. Using the proposed ^{ME}D symbol, the illustrated ^{ME}D may have six (or more) *input*, *control*, and *output* terminals (ports) with the corresponding molecular bonds accomplishing the interconnect; see Sections 5.1 and 6.9. This interconnect can be *chemical-bond fabrics* (by means of atomic bonds) and/or *energy-based* (energy exchange/conversion/transmission by means of, for example, radiation and absorption of electromagnetic, thermal, and vibrational energy). A six-terminal monocyclic ^{ME}device with a carbon interconnecting framework, which establishes the chemical bonds (*chemical-bond fabrics* interconnect), can be depicted as ⌬. As an illustration, a 3D $^{\aleph}$hypercell primitive to implement a logic function $y = f(x_1, x_2, x_3)$ is shown in Figure 4.5b. Two-terminal molecular devices (^{M}d and ^{M}r) are shown using symbols ▷|

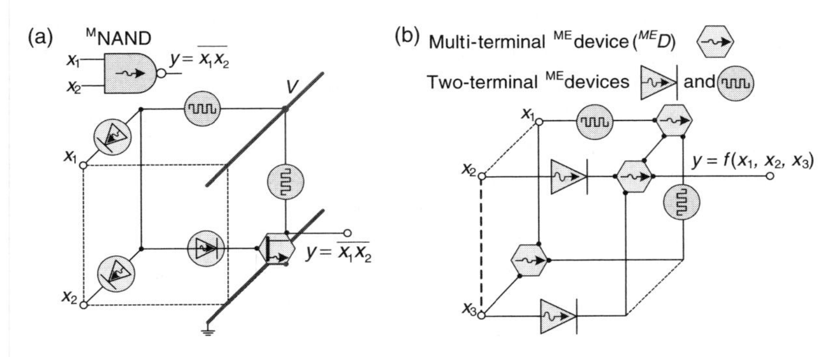

FIGURE 4.5
(a) Implementation of a ^{M}NAND gate realized by a $^{\aleph}$hypercell primitive; (b) $^{\aleph}$Hypercell primitive with two- and multiterminal ^{ME}devices.

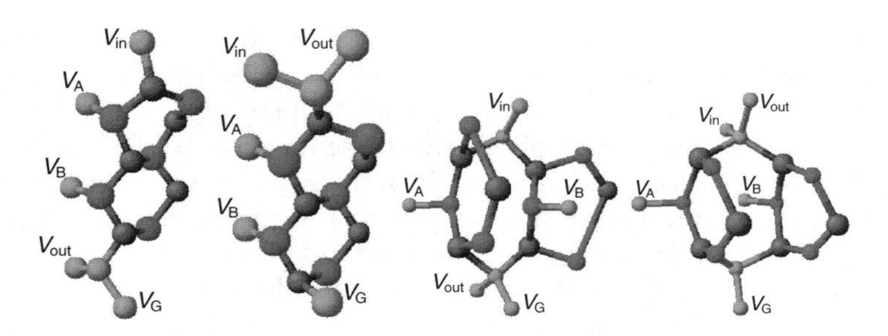

FIGURE 4.6
^{M}AND, ^{M}NAND, ^{M}OR, and ^{M}NOR gates designed within the molecular ^{ME}D–^{ME}D logic family.

and , respectively. The input signals (x_1, x_2, and x_3) and output switching function f are documented in Figure 4.5b.

Molecular gates (^{M}AND, ^{M}NAND, ^{M}OR, and ^{M}NOR), designed within the molecular multiterminal ^{ME}D–^{ME}D logic family, are shown in Figure 4.6. Here, as covered in Section 5.1, three-terminal cyclic molecules are utilized as ^{ME}devices. The device physics of the multiterminal ^{ME}device is based on quantum interaction and controlled electron transport. The inputs signals V_A and V_B are supplied to the *input* terminals, while the output signals is V_{out}. These ^{M}AND and ^{M}NAND gates are designed using cyclic molecules within the carbon interconnecting framework. The details of synthesis, device physics, and phenomena utilized are reported in Chapters 5 and 6. A coherent design should be performed in order to ensure the desired performance, functionality, characteristics, aggregability, topology, and other features. Complex ^{M}gates can be synthesized utilizing a $^{\aleph}$hypercell-centered hardware solution forming combinational and memory ^{M}ICs. The *design rules* are reported in Section 4.6.2.

The memories can be implemented by utilizing combinational circuits avoiding the use of the capacitor. For example, memory cells using transmission gates and ^{M}gates are depicted in Figure 4.7. A memory storage cell with the feedback path formed by two ^{M}NOT gates is shown in Figure 4.7a. To store data in the cell or to read data, the *Select* input is set to 1 or to 0. The stored data, which are accessible, remain indefinitely in the feedback loop. Figure 4.1 depicts the address input to the decoder, and one implements $2^m \times n$ SRAM cells with $[a_1, a_2, \ldots, a_{m-1}, a_m]$ addresses resulting in 2^m *Select* inputs, which are used to read or write the contents of the cells. A 2×2 array of SRAM cells is shown in Figure 4.7b. The memories can be implemented using ^{M}NOR and ^{M}NAND gates. The well-known basic latch is a cross-coupled connection of two ^{M}NOR gates, as illustrated in Figure 4.7c. Two inputs (*Set* and *Reset*) provide the means for changing the *states* Q_1 and Q_2. When *Set* $= 0$ and *Reset* $= 0$, the latch maintains its existing *state*, which can be either $Q_1 = 0$ and $Q_2 = 1$, or $Q_1 = 1$ and $Q_2 = 0$. As *Set* $= 1$ and *Reset* $= 0$, the latch is set in the *state* $Q_1 = 1$ and $Q_2 = 0$. When *Set* $= 0$ and *Reset* $= 1$, the latch rests into a *state* $Q_1 = 0$ and $Q_2 = 1$. If *Set* $= 1$ and *Reset* $= 1$, we have $Q_1 = 0$ and $Q_2 = 0$.

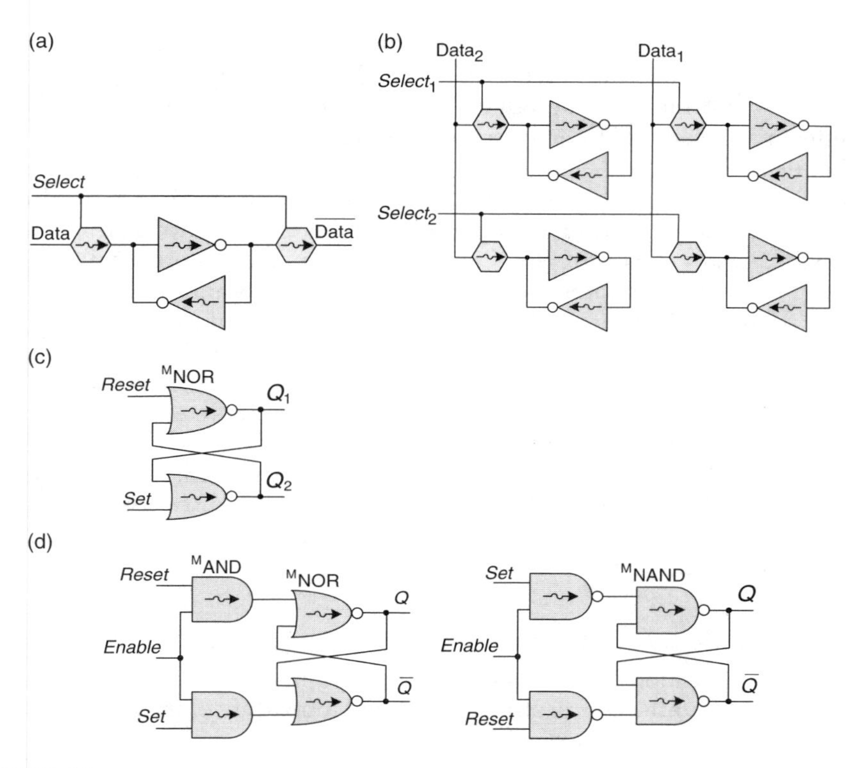

FIGURE 4.7
Molecular memory elements and cells: (a) SRAM cell; (b) 2 × 2 array of SRAM cells; (c) basic latch as a memory; (d) gated ^{M}NOR and ^{M}NAND latches.

The gated ^{M}NOR and ^{M}NAND latches with the control input (*Enable*), as memory cells, are shown in Figure 4.7d. To analyze delays, timing diagrams, switching, and other important features, which define the overall performance, one must examine the dynamics of ^{ME}devices and ^{M}gates that constitute combinational and memory circuits. This device-level analysis is covered in Chapter 6.

4.2 Decision Diagrams and Logic Design of Molecular Integrated Circuits

Innovative solutions to perform the system-level logic deign for ^{M}ICs needs to be examined. One needs to depart from 2D-centered logic design (VLSI, ULSI, and post-ULSI) as well as from planar IC topologies and organizations. We proposed ^{M}ICs which resemble hierarchical ^{BM}PPs, mimicking topologies and organizations observed in living biosystems. The SLSI design, which is under development, should comply with the envisioned device-level outlook and fabrication technologies. The use of $^{\aleph}$hypercells as baseline hardware ^{M}primitives in design of ^{M}ICs and processing/memory platforms results in a feasible technology-centric solution.

Let us focus on the design of combinational and memory circuits. For 2D CMOS ICs, decision diagram (unique canonical structure) is derived as a reduced decision tree by using topological operators. In contrast, a 3D realization of ^{M}ICs by means of $^{\aleph}$hypercells as well as other features emphasized result in the need for a new class of decision diagrams and design methods. These decision diagrams and design concepts must be developed to handle the complexity, 3D features, *bottom-up* fabrication, etc. of ^{M}ICs. A concept of design of linear decision diagram, mapped by hypercubes, was proposed in [1,2]. In general, the hypercube (cube, pyramid, hexagon, or other 3D topological aggregates) is a unique canonical structure that is a reduced decision tree. Hypercubes are synthesized by using topological operators (deleting and splitting nodes). Optimal and suboptimal technology-centric topology mappings of complex switching functions can be accomplished and analyzed. The major optimization criteria are

1. Minimization of decision diagram nodes and circuit realization
2. Simplification of topological structures—linear arithmetic leads to simple synthesis and straightforward embedding of linear decision diagrams into 3D topologies
3. Minimization of path length in decision diagrams
4. Routing simplification
5. Verification and evaluation

Optimal topology mapping results in evaluation simplicity, testability enhancement, and other important features including overall hardware performance enhancements. For example, switching power is not only a function of devices/gates/switches, but also a function of circuit topology, organization, design methods, routing, dynamics, switching activities, and other factors that can be optimized. In general, a novel CAD-supported SLSI should be developed for optimal technology-centric designs to obtain high-performance molecular platforms. Through a concurrent design, the designer should be able to perform the following major tasks:

1. Logic design of ^{M}ICs utilizing novel representations of data structures
2. Design and aggregation of hypercubes
3. Design of binary and multiple-valued decision diagrams
4. CAD developments to concurrently support technology-centric design tasks

SLSI utilizes a coherent top-down/bottom-up synthesis taxonomy as an important part of a ^{M}*architectronics* paradigm. The design complexity should be emphasized. Current CAD-supported postULSI design does not allow one to design ICs with a number of gates more than 1,000,000. For ^{M}ICs, the design complexity significantly increases and novel methods are sought [1,2]. The binary decision diagram (BDD) for representing Boolean functions is the advanced approach in high-level logic design [1]. The reduced-order and optimized BDDs ensure large-scale data manipulations, which are used to perform the logic design and circuitry mapping utilizing hardware description languages. The design scheme is

$$\text{Function (Circuit)} \leftrightarrow \text{BDD model} \leftrightarrow \text{Optimization}$$
$$\leftrightarrow \text{Mapping} \leftrightarrow \text{Realization.}$$

The dimension of a decision diagram (number of nodes) is a function of the number of variables and their ordering. In general, the design complexity is $O(n^3)$. This enormous complexity significantly limits the abilities to design complex ICs without partitioning and decomposition. The commonly used word-level decision diagrams further increase the complexity due to processing of data in word-level format. Therefore, novel sound software-supported design approaches are needed. Innovative methods in data structure representation and data structure manipulation are developed and applied to ensure design specifications and objectives. We design ^{M}ICs utilizing the linear word-level decision diagrams (LWDDs) that allow one to perform the compact representation of logic circuits using linear arithmetic polynomials (LP) [1,3]. The design complexity becomes $O(n)$. The proposed concept ensures compact representation of circuits compared with other formats and methods. The following design algorithm guarantees a compact circuit representation:

$$\text{Function (Circuit)} \leftrightarrow \text{BDD Model} \leftrightarrow \text{LWDD Model} \leftrightarrow \text{Realization.}$$

The LWDD is embedded in hypercubes that represent circuits in a 3D space. The polynomial representation of logical functions ensure the description of multioutput functions in a word-level format. The expression of a Boolean function f of n variables $(x_1, x_2, \ldots, x_{n-1}, x_n)$ is

$$LP = a_0 + a_1x_1 + a_2x_2 + \ldots + a_{n-1}x_{n-1} + a_nx_n = a_0 + \sum_{j=1}^{n} a_jx_j.$$

To perform a design, the mapping LWDD $(a_0, a_1, a_2, \ldots, a_{n-1}, a_n) \leftrightarrow$ LP is used. The nodes of LP correspond to a Davio expansion. The LWDD is used to represent any m-level circuit with levels L_i, $i = 1, 2, \ldots, m-1, m$ with elements of the molecular primitive library. Two data structures are defined in the algebraic form by a set of LPs as

$$L = \begin{cases} L_1: & \text{inputs } x_j; \text{outputs } y_{1k} \\ L_2: & \text{inputs } y_{1k}; \text{outputs } y_{2l} \\ & \cdots \\ L_{m-1}: & \text{inputs } y_{m-2,t}; \text{outputs } y_{m-1,w} \\ L_m: & \text{inputs } y_{m-1,w}; \text{outputs } y_{m,n} \end{cases}$$

that corresponds to

$$\text{LP}_1 = a_0^1 + \sum_{j=1}^{n_1} a_j^1 x_j, \ \ldots \ \text{LP}_m = a_0^n + \sum_{j=1}^{n_m} a_j^n y_{m-1,j},$$

or in the graphic form by a set of LWDDs as

$$\text{LWDD}_1(a_0^1, \ldots, a_{n_1}^1) \leftrightarrow \text{LP}_1, \ldots, \text{LWDD}_m(a_0^n, \ldots, a_{n_m}^n) \leftrightarrow \text{LP}_m.$$

The use of LWDDs is a departure from the existing logic design tools. This concept is compatible with the existing software, algorithms, and circuit representation formats. Circuit transformation, format transformation, modular organization, library functions over primitives, and other features can be accomplished. All combinational circuits can be represented by LWDDs. The format transformation can be performed for circuits defined in the Electronic Data Interchange Format (EDIF), Berkeley Logic Interchange Format (BLIF), International Symposium on Circuits and Systems Format (ISCAS), Verilog, etc. The library functions may have a library of LWDDs for multiinput gates, as well as libraries of ^{M}devices, ^{M}gates, and $^{\aleph}$hypercells. The important feature is that these primitives are realized (through logic design) and synthesized as primitive aggregates. The reported LWDD simplifies analysis, verification, evaluation, and other tasks.

Arithmetic expressions underlying the design of LWDDs are canonical representations of logic functions. They are alternatives of the sum-of-product, product-of-sum, Reed–Muller, and other forms of representation of Boolean functions. Linear word-level decision diagrams are obtained by mapping LPs, where the nodes correspond to the Davio expansion and functionalizing vertices to the coefficients of the LPs. The design algorithm is given as

$$\text{Function (Circuit)} \leftrightarrow \text{LP Model} \leftrightarrow \text{LWDD Model} \leftrightarrow \text{Realization.}$$

Any m-level logic circuit with fixed order of elements is uniquely represented by a system of m LWDDs. The proposed concept is verified by designing ^{M}ICs representing Boolean functions by hypercubes. The CAD toolbox for logic design is based on the principles of 3D realization of logic functions with a library of primitives. LWDDs are extended by embedding the decision tree into the hypercube structure. For two graphs $G = (V, E)$ and $H = (W, F)$, we embed the graph G into the graph H. The results are partitioned according to the new structural properties of the cell and the type of the embedded tree. The embedding of a guest graph G into a host graph H is a one-to-one mapping M_{GV}:$V(G) \rightarrow V(H)$, along with the mapping M that maps an edge $(u;v) \in E(G)$ to a path between $M_{GV}(u)$ and $M_{GV}(v)$ in H. Thus, the embedding of G into H is a one-to-one mapping of the nodes in G to the nodes in H.

The design execution performance estimates can be evaluated [1]. Decision trees are designed using the Shannon and Davio expansions. There is a need to find the best variable and expansion for any node of the decision tree in terms of information estimates in order to optimize the design synthesizing optimal ^{M}ICs. The optimization algorithm should generate the *optimal paths* in a decision tree with respect to the design criteria. The decision tree is designed by arbitrarily chosen variables using either Shannon (S), positive Davio (pD) or negative Davio (nD) expansions for each node. The decision tree design process is a recursive decomposition of a switching function. This recursive decomposition corresponds to the expansion of switching function f with respect to the variable x. The variable x carries information that influences f. The initial and final state of the expansion $\sigma \in \{S, pD, nD\}$ can be characterized by the performance estimates. The information-centered optimization of ^{M}ICs design is performed in order to derive optimal decision diagrams. A path in the decision tree starts from a node and finishes at a terminal node. Each path corresponds to a term in the final expression for f. For the c17 circuit, implemented using ^{M}NAND gates, as shown in Figure 4.8a, Davio expansions ensure optimal design as compared with the Shannon expansion [1]. We have emphasized the fundamental differences between molecular fabrication and solid-state CMOS technology. For c17 circuit, the CMOS layout is depicted in Figure 4.8b, and to underline the difference, we recall that the 3D-topology ^{M}NAND gate was documented in Figure 4.6.

The CAD-supported logic design of proof-of-concept ^{M}ICs is successfully accomplished for complex benchmarking ICs in order to verify and examine the concept proposed. The size of LWDDs is compared with the best results

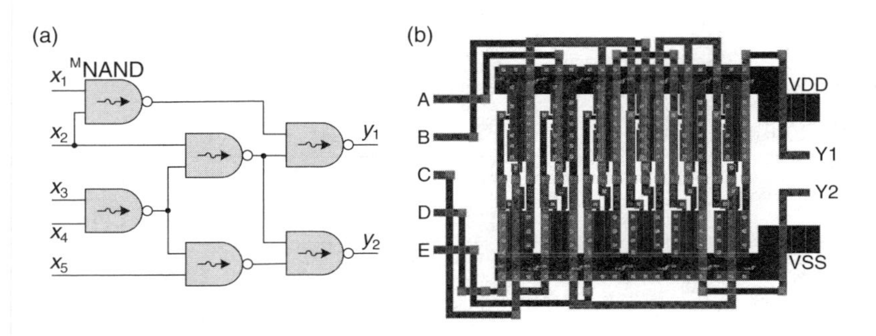

FIGURE 4.8
(a) C17 circuit consists of 6 ^{M}NAND gates; (b) CMOS layout.

received by other Decision Diagram Packages developed for 2D VLSI design. The method reported and software algorithms were tested and validated [1]. The number of nodes, number of levels, and CPU time (in seconds) required to design decision diagrams for ^{M}ICs are examined. In addition, volumetric size, topological parameters, and other design performance variables are analyzed. We assume: (1) feedforward neural network with no feedback; (2) threshold ^{M}gates as the processing primitives; (3) aggregated $^{\aleph}$hypercells comprised from ^{M}gates; (4) multilevel combinational circuits over the library of NAND, NOR, and EXOR ^{M}gates implemented using three-terminal ^{ME}devices.

Experiments were conducted for various ICs, and representative results and data are reported in Table 4.1. The space size is given by X, Y, and Z. The volumetric quantity is $V = X \times Y \times Z$. The topological characteristics are analyzed using the total number of terminal (N_T) and intermediate (N_I) nodes. For example, c880 is an 8-bit arithmetic logic unit (ALU). ALU is a combinational circuit that performs arithmetic and logical operations on a pair of a-bit operands. The operations to be performed are specified by a set of function-selected inputs. The core of this c880 circuit is the 8-bit adder that has 60 inputs and 26 outputs. A planar design leads to 383 gates. In contrast, 3D design results in 294 ^{M}gates. A 9-bit ALU (c5315) with 178 inputs and 123 outputs is implemented using 1413 ^{M}gates, while a c6288 multiplier (32 inputs and 32 outputs) has 2327 ^{M}gates. Molecular gates are aggregated, networked, and grouped in 3D. The number of incompletely specified hypercubes was minimized. The hypercubes in the ith layer were connected to the corresponding hypercubes in $(i - 1)$th and $(i + 1)$th layers. The number of terminal nodes and intermediate nodes are 3750 and 2813 for a 9-bit ALU. For a multiplier, we have $N_T = 9248$ and $N_I = 6916$. To combine all layers, more than 10,000 connections were generated. The design in 3D was performed within 0.344 sec for 9-bit ALU. The studied 9-bit ALU performs arithmetic and logic operations simultaneously on two 9-bit input data words and computes the parity of the results. The conventional 2D logic design for c5315 with 178 inputs and 123 outputs results in 2406 gates. In contrast, the proposed design, as performed using a proof-of-concept SLSI software, leads to 1413 ^{M}gates that

TABLE 4.1
Design Results for ^{M}ICs

Circuit	I/O	Space Size				Nodes and Connections		
		#G	#X	#Y	#Z	#N_T	#N_I	CPU Time (sec)
c432	36/7	126	66	64	66	2022	1896	<0.03
8-bit ALU c880	60/26	294	70	72	70	612	482	<0.04
9-bit ALU c5315	178/123	1413	138	132	126	3750	2813	<0.35
16 × 16 Multiplier c6288	32/32	2384	248	248	244	9246	6916	<0.46

are networked and aggregated within a 3D topology. In addition to conventional parameters (diameter, dilation cost, expansion, load, etc.), we use the number of variables in the logic function described by hypercubes, number of links, *fan-out* of the intermediate nodes, statistics, and others to perform the evaluation. For the sound comparison reasons, to ensure the similarity to 2D design, binary three-terminal ^{ME}devices were used. The use of multiple-valued multiterminal ^{ME}devices results in superior performance.

The representative proof-of-concept CAD SLSI toolbox and software solutions were developed in order to demonstrate the design soundness, feasibility, and advantages for various combinational ^{M}ICs. The compatibility with hardware description languages is important. Three netlist formats (EDIF, ISCAS, and BLIF) are used in a proof-of-concept SLSI software that features:

1. New design concept uniquely suitable for ^{M}ICs
2. Synthesis and partitioning linear decision diagrams for given functions or circuits
3. Spectral representation of logic functions
4. Circuit testability, verification, and evaluation
5. Compact format ensuring robustness and rapid prototyping
6. Compressed optimal representation of complex ^{M}ICs

For ^{M}ICs, the design outcomes are shown in Figure 4.9 displaying the results in the Command Window. In particular, the design of c17 circuit, 8-bit ALU (c880), 9-bit ALU (c5315), and 16×16 multiplier (c6288) are displayed.

4.3 Hypercube Design

The binary tree is a networked description that carries information about dual connections of each node. The binary tree also carries information about functionality of the logic circuit and its topology. The nodes of the binary tree are associated with the Shannon and Davio expansions with respect to each variable and coordinate in 3D. A node in the binary decision tree realizes the Shannon decomposition

$$f = x_i f_0 \oplus x_i f_1,$$

where $f_0 = f|_{x_i=0}$ and $f_1 = f|_{x_i=1}$ for all variables in f.

Thus, each node realizes the Shannon expansion, and the nodes are distributed over levels. The classical hypercube contains 2^n nodes, while

FIGURE 4.9
Design of ^{M}ICs using a proof-of-concept SLSI software.

the proposed hypercube has $2^n + \sum_{i=0}^{n} 2^{n-1} C_i^m$ nodes in order to guarantee design soundness ensuring a technology-centric design of ^{M}ICs. The hypercube consists of terminal nodes, intermediate nodes, and roots. This ensures a straightforward implementation, for example, by using the multiplexer with inputs (D, EN, and SEL) and *b* data outputs Y. The inputs tuple (D, EN, and SEL) represents *n* data sources D with *b* bits leading to $b \times n$ data inputs to

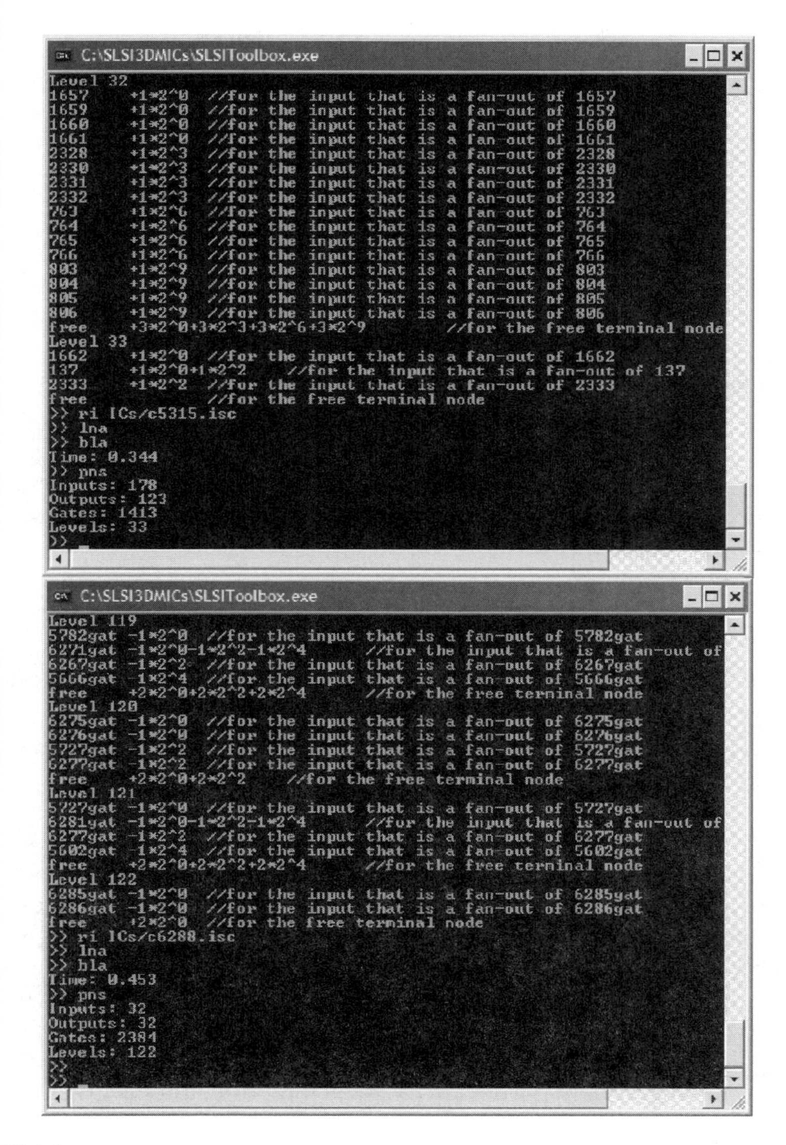

FIGURE 4.9
Continued.

be switched to the b outputs, b enable commands EN, and s select commands SEL. The n-input b-bit multiplexer output is expressed as a logical sum of product terms:

$$\mathrm{iY} = \sum_{j=0}^{n-1} \mathrm{EN} \cdot \mathrm{M}_j \cdot \mathrm{iDj},$$

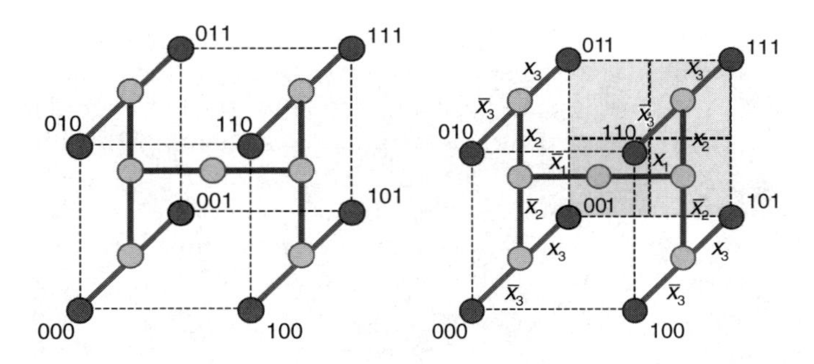

FIGURE 4.10
Multiplexer-based hypercube and implementation of a switching function $f = \bar{x}_1 x_2 \vee x_1 \bar{x}_2 \vee x_1 x_2 x_3$.

where iY is the particular output bit ($1 \leq i \leq b$); EN is the enable input, and when EN $= 0$, all outputs are 0; M_j is the minterm j of the l-select inputs; iDj is the input bit i of source j.

The design steps are

Step 1: Connect the terminal node with the intermediate nodes.

Step 2: Connect the root with two intermediate nodes located symmetrically on the opposite faces.

Step 3: Pattern the terminal and intermediate nodes on the opposite faces and connect them through the root.

Figure 4.10 reports a hypercube implemented using a two-to-one multiplexer.

There are several methods for representing logic functions. We utilize a hypercube solution. In general, a hypercube is a homogeneous aggregated assembly for massive super-high-performance parallel computing. We apply the enhanced switching theory integrated with a novel logic design concept. In the design, graph-based data structures and 3D topology are utilized. The hypercube is a topological representation of a switching function by n-dimensional graph. In particular, the switching function f is given as

$$\underset{f}{\text{Switching function}} \Rightarrow \underset{\underset{\text{Operation}}{\Uparrow}}{\mathop{\mathbf{L}}_{i=0}^{2^n-1}} \; \overset{\overset{\text{Coefficient}}{\Downarrow}}{\mathbf{K}_i}(x_1^{i_1} \ldots x_n^{i_n}) \Rightarrow \underset{f_F}{\text{Form of switching function}}.$$

The data structure is described in matrix form using the truth vector $\boldsymbol{F}$ of a given switching function f as well as the vector of coefficients $\mathbf{K}$. The logic operations are represented by $\mathbf{L}$. Hypercubes compute f. Figure 4.10 reports a hypercube to implement $f = \bar{x}_1 x_2 \vee x_1 \bar{x}_2 \vee x_1 x_2 x_3$. From the technology-centric viewpoints, we propose a concept that employs ^{M}gates and $^{\aleph}$hypercells coherently mapping the device/system molecular hardware and data structure solutions. Aggregated hypercubes, implemented as $^{\aleph}$hypercell lattices, can implement switching functions f of arbitrary complexity. The logic design in spatial dimensions is based on advanced methods and enhanced data structures to satisfy the requirements and specifications imposed by molecular hardware. The appropriate data structure of logic functions and methods of embedding this structure into hypercubes and its implementation by $^{\aleph}$hypercells are developed. The algorithm in a logic functions manipulation in order to change the carrier of information from the algebraic form (logic equation) to the hypercube structure consists of three steps:

Step 1: The logic function is transformed to the appropriate algebraic form (Reed–Muller, arithmetic, or word-level in a matrix or algebraic representation).

Step 2: The derived algebraic form is converted to the graphical form (decision tree or decision diagram).

Step 3: The obtained graphical form is embedded in technology-implementable hypercube. This results to the hypercube–$^{\aleph}$hypercell technology-centric mapping for complex ^{M}ICs.

The design is expressed as

$$\underset{\text{Step 1}}{\text{Logic Function}} \Leftrightarrow \underset{\text{Step 2}}{\text{Graph}} \Leftrightarrow \underset{\text{Step 3}}{\text{Hypercube–Hypercell Mapping}} / {}^{M}\text{ICs}.$$

The proposed procedure results in

- Algebraic representations and robust manipulations of complex switching logic functions
- Matrix representations and manipulations providing consistency of logic relationships for variables and functions from the spectral theory viewpoint
- Graph-based representations using decision trees
- Direct mapping of decision diagrams into logical networks, as demonstrated for multiplexer-centered hypercubes
- Robust embedding of data structures into hypercube with the following hypercube–$^{\aleph}$hypercells mapping

From the synthesis viewpoint, the complexity of the molecular interconnect corresponds to the complexity of ^{ME}devices. We introduce a 3D *directly interconnected molecular electronics* (^{3D}DIME) concept in order to reduce the synthesis complexity, minimize delays, ensure robustness, enhance reliability, and so forth. This solution minimizes the interconnect by utilizing a direct atomic bonding of *input, control,* and *output* terminals by means of a direct device-to-device aggregation. Chapter 5 documents that ^{ME}devices and ^{M}gates are engineered and implemented using cyclic molecules within a carbon framework. The *output* terminal of the ^{ME}device can be directly connected to or netted with the *input* terminal of other ^{ME}device; for an illustrative example, see Figure 4.6. The ^{3D}DIME concept promises to ensure synthesis feasibility, compact implementation of $^{\aleph}$hypercells, modularity, applicability of ^{M}primitives from the primitive library, and so forth. Section 4.6.2 provides the *design rules*.

4.4 Molecular Signal/Data Processing and Memory Platforms

Advanced computer architectures (beyond von Neumann architecture) can be devised and implemented to guarantee superior processing, reconfigurability, robustness, networking, and so forth. In the von Neumann computer architecture, the CPU executes sequences of instructions and operands, which are fetched by the program control unit (PCU), executed by the data processing unit (DPU), and then placed in the memory. Caches (high-speed memory in which data is copied when it is retrieved from the RAM, improving the overall performance by reducing the average memory access time) are used. The CPU may have more than one processors and coprocessors with various execution units and multilevel instruction and data caches. These processors can share or have their own caches. The *datapath* contains ICs to perform arithmetic and logical operations on words such as fixed- or floating-point numbers. The CPU design involves the trade-off between the hardware/software requirements, performance, and affordability. The CPU is usually partitioned on the control and *datapath* units. The control unit selects and sequences the data processing operations. The core interface unit is a switch that can be implemented as autonomous cache controllers operating concurrently and feeding the specified number (64 or 128) of bytes of data per cycle. This core interface unit connects all controllers to the data or instruction caches of processors. Additionally, the core interface unit accepts and sequences information from the processors. A control unit is responsible for controlling data flow between controllers that regulate the *in* and *out* information flows. The interface is accomplished by means of input/output devices and units. On-chip debuging, error detection, sequencing logic, self-test, monitoring, and other units must be integrated to control a pipelined computer. The computer performance depends on the architecture, organization, and hardware components. Figure 4.11 illustrates the conventional computer architecture.

Consider signal/data and information processing between nerve cells. The key to understand processing, memory, learning, intelligence, adaptation, control, hierarchy, and other system-level basics lies in the ability to comprehend biophysical phenomena exhibited, organization utilized, and architecture possessed by the central nervous system, the neurons, and their

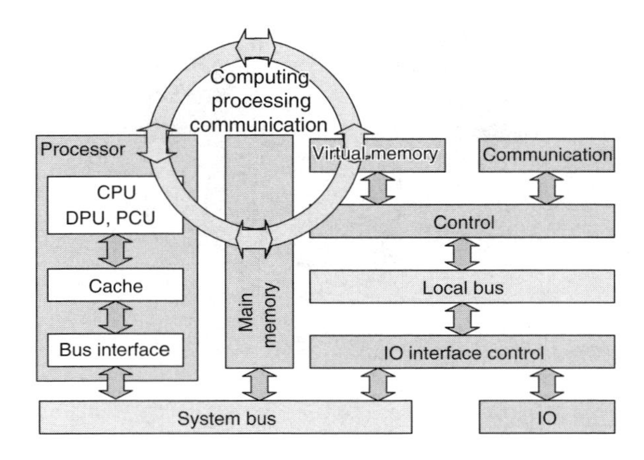

FIGURE 4.11
Computer architecture.

organelles. Unfortunately, many problems have not been resolved. A single neuron can perform processing, memory storage, and other tasks. A neuron has thousands of synapses with binding sites, membrane channels, microtubules, microtubule–associated proteins (MAPs), synapse–associated proteins (SAPs), and other proteins, and so forth. As reported in Chapter 3, the electrochemomechanically induced transitions lead to various processing, memory, communication, networking, and other tasks. The processing and memories are reconfigurable and constantly evolve and adapt. Neurons function within

- 3D-centered hierarchically distributed, robust, adaptive, parallel, and networked organization
- Unknown architecture

Figure 4.12 shows a schematic of a possible integrated processor-and-memory ^{M}PP architecture. Processor executes sequences of instructions and operands, which are fetched (by the control unit) and placed in memory. The instructions and data form *instruction* and *data streams* that flow to and from the processor. The core interface unit concurrently controls operations and data retrieval. This interface unit interfaces all controllers to the data or processor instruction caches. The interface unit accepts and sequences information from the processors. A control unit is responsible for controlling data flow regulating the information in and outflows.

There is a need to study and comprehend hierarchical distributed computing in nervous systems, which process, store, code, compress, manipulate, route, and network information in the optimal manner. Figure 4.13 shows the principle of organization of a nervous system that has similarity to the ^{M}PP shown in Figure 4.12. The distributed central nervous system adaptively reconfigures based on information processing, memory storage,

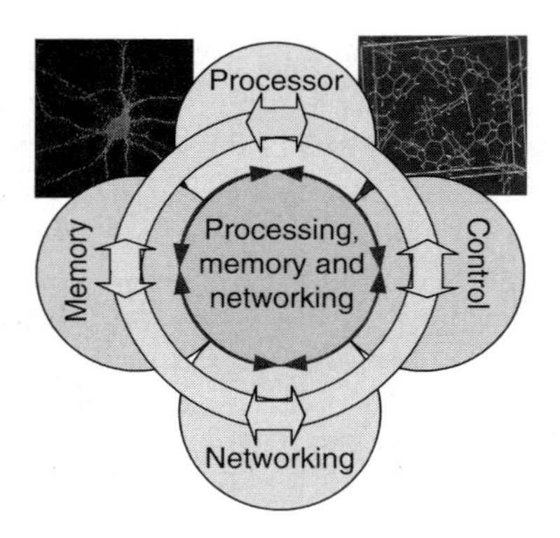

FIGURE 4.12
Processing-and-memory ^{M}PP architecture.

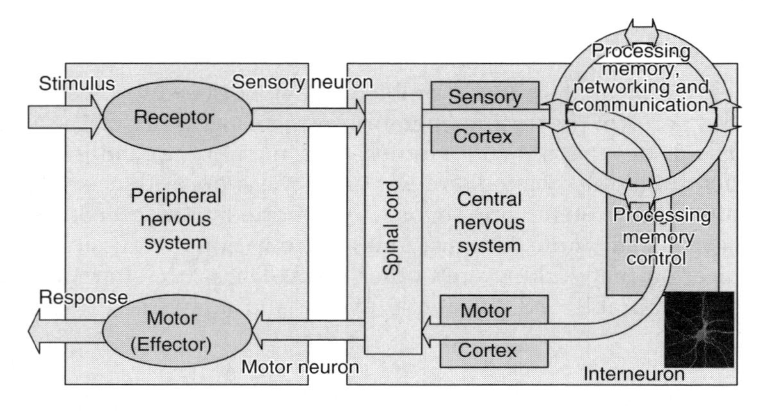

FIGURE 4.13
The vertebrate nervous system.

communication, and control (instruction) parallelisms. This principle can be effectively used in design of various processing and memory platforms within 3D organization and enabling architectures.

As there are difficulties to comprehend, utilize, and implement ^{BM}PPs, the overall goal is to design ^{M}PPs. The implementation of envisioned ^{M}PPs primarily depends on the progress in device physics, system organization/architecture, CAD-supported SLSI design, and molecular fabrication technologies.

The critical problems in the design are the development, optimization, and utilization of hardware and software. The current status of fundamental, applied, and technological developments suggests that the ^{M}PPs will be likely designed utilizing a digital paradigm. Numbers in binary digital processors and memories are represented as a string of zeros and ones, and circuits perform Boolean operations. Arithmetic operations are performed based on a hierarchy of operations that are built upon simple operations. The methods to compute and algorithms used are different. Therefore, speed, robustness, accuracy, and other performance characteristics vary. Information is represented as a string of bits. The number of bits depends on the length of the word (quantity of bits on which hardware is capable to operate). The operations are performed over the string of bits. There are rules that associate a numerical value $X$with the corresponding bit string $x = \{x_0, x_1, \ldots, x_{n-2}, x_{n-1}\}, x_i \in 0, 1$. The associated word (string of bits) is n bits long. If for every value X there exists one, and only one, corresponding bit string x, the number system is nonredundant. If there can exist more than one x that represents the same value X, the number system is redundant. A *weighted* number system is used, and a numerical value is associated with the bit string x as $x = \sum_{i=0}^{n-1} x_i w_i$, $w_0 = 1, \ldots, w_i = (w_i - 1)(r_i - 1)$, where r_i is the *radix* integer.

By making use of the multiplicity of instructions and data streams, the following classification can be applied:

1. Single instruction stream–single data stream: conventional word-sequential architecture including pipelined computing platforms with parallel arithmetic logic unit (ALU)
2. Single instruction stream–multiple data stream: multiple ALU architectures, for example, parallel-array processor (ALU can be either bit-serial or bit-parallel)
3. Multiple instruction stream–single data stream
4. Multiple instruction stream–multiple data stream: the multiprocessor system with multiple control units

In biosystems, multiple instruction stream–multiple data stream is observed. There is no evidence that technology will provide the ability to synthesize even simple biomolecular processors, not to mention biocomputers, in near future. Therefore, efforts are concentrated on ^{M}PPs designed using solid molecular electronics that ensures soundness and technological feasibility.

Three-dimensional topologies and organizations significantly improve the performance of processing platforms guarantying—for example, massive parallelism and optimal utilization. Using the number of instructions executed (N), number of cycles per instruction (C_{PI}), and clock frequency (f_{clock}), the program execution time is

$$T_{\mathrm{ex}} = NC_{\mathrm{PI}}/f_{\mathrm{clock}}.$$

In general, the circuit hardware determines the clock frequency f_{clock}, while the software affects the number of instructions executed N. The architecture defines the number of cycles per instruction C_{PI}. Processing platforms integrate functional controlled hardware units and systems that perform processing, memory storage, execution, and so forth. The ^{M}PP accepts digital or analog input information, processes and manipulates it according to a list of internally stored machine instructions, stores the information, and

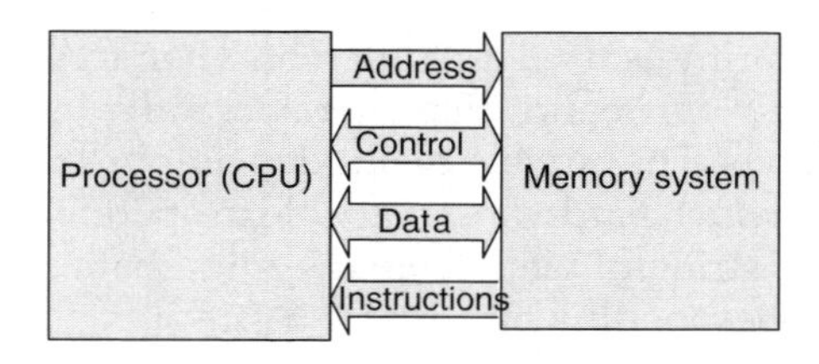

FIGURE 4.14
Memory–processor interface.

produces the resulting output. The list of instructions is called a program, and internal storage is called memory. A memory unit integrates different memories. The processor accesses (reads or loads) the data from the memory systems, performs computations, and stores (writes) the data back to memory. The memory system is a collection of storage locations. Each storage location (memory word) has an address. A collection of storage locations forms an address space. Figure 4.14 documents the data flow and its control, representing how a processor is connected to a memory system via address, control, and data interfaces. High-performance memory systems should be capable to serve multiple requests simultaneously.

When a processor attempts to load or read the data from the memory location, the request is issued, and the processor stalls while the request returns. While ^{M}PPs can operate with overlapping memory requests, data cannot be optimally manipulated if there are long memory delays. Therefore, the key performance parameter is the effective memory speed. The following limitations are imposed on any memory systems: (1) cannot be infinitely large; (2) cannot contain an arbitrarily large amount of information; (3) cannot operate infinitely fast. Hence, the major characteristics are speed and capacity. The memory system performance is characterized by the latency τ_l and bandwidth B_w. Memory latency is the delay from when the processor first requests a word from the memory until that word arrives and is available for use by the processor. Bandwidth is the rate at which information can be transferred from the memory system. Using the number of requests that the memory can service concurrently (N_{request}), we have $B_w = N_{\text{request}}/\tau_l$. Using ^{M}ICs, it become feasible to design and build superior memory systems with exceptional capacity, low latency, and high bandwidth, approaching physical and technological limits of a molecular hardware solution. Furthermore, using molecular electronics, it becomes possible to match the memory and processor performance characteristics and capabilities.

Memory hierarchies ensure decreased latency and reduced bandwidth requirements, whereas parallel memories provide higher bandwidth. In ^{M}PP one can utilize an organization with a fast memory located in front of a large but relatively slow memory. This increases speed and enhances memory capacity. However, this solution results in the application of registers in the processor unit, and most commonly accessed variables should be allocated at registers. A variety of techniques employing either hardware, software, or a combination of hardware and software must be employed to ensure that most references to memory are fed by the faster memory.

The locality principle is based on the fact that some memory locations are referenced more often than others. The implementation of spatial locality, due to the sequential access, provides one with the property that an access to a given memory location increases the probability that neighboring locations will soon be accessed. Making use of the frequency of program looping behavior, temporal locality ensures the access to a given memory location, increasing the probability that the same location will be accessed again soon. If a variable was not referenced for a while, it is unlikely that this variable will be needed soon. The performance parameter, which can be used to quantitatively examine different memory systems, is the effective latency τ_{ef}. We have $\tau_{ef} = \tau_{hit}R_{hit} + \tau_{miss}(1 - R_{hit})$, where τ_{hit} and τ_{miss} are the hit and miss latencies; R_{hit} is the hit ratio, $R_{hit} < 1$. If the needed word is found in a level of the hierarchy, it is called a hit. Correspondingly, if a request must be sent to the next lower level, the request is said to be a miss. The miss ratio is given as $R_{miss} = (1 - R_{hit})$. Both R_{hit} and R_{miss} are affected by the program being executed and influenced by the high- or low-level memory capacity ratio. The access efficiency E_{ef} of multiple-level memory ($i - 1$ and i) is found using the access time and hit and miss ratios. In particular, $E_{ef} = ((t_{\text{access time } i-1}/t_{\text{access time } i})R_{miss} + R_{hit})^{-1}$.

The hardware can dynamically allocate parts of the cache memory to addresses likely to be accessed soon. The cache contains only redundant copies of the address space. The cache memory can be associative or content-addressable. In an associative memory, the address of a memory location is stored along with its content. Rather than reading data directly from a memory location, the cache is given an address and responds by providing data that might or might not be the data requested. When a cache miss occurs, memory access is then performed from the main memory and the cache is updated to include the new data. The cache should hold the most active portions of the memory, and the hardware dynamically selects portions of main memory to store in the cache. When the cache is full, some data must be transferred to the main memory or deleted. A strategy for cache memory management is needed. These cache management strategies are based on the locality principle. In particular, spatial (selection of what is brought into the cache) and temporal (selection of what must be removed) localities are embedded. When a cache miss occurs, that hardware copies a contiguous block of memory into the cache, which includes the word requested. This fixed-size memory block can be small, medium, or large. Caches can require all fixed-size memory blocks to be aligned. When a fixed-size memory block is brought into the cache, it is likely that another fixed-size memory block must be removed. The selection of the removed fixed-size memory block is based on effort to capture temporal locality.

The cache can integrate the data memory and the tag memory. The address of each cache line contained in the data memory is stored in the tag memory. The state can also track which cache line is modified. Each line contained in the data memory is allocated by a corresponding entry in the tag memory to indicate the full address of the cache line. The requirement that the cache memory be associative (content-addressable) complicates the design because addressing data by content is more complex than by its address (all tags must be compared concurrently). The cache can be simplified by embedding a mapping of memory locations to cache cells. The mapping limits the number of possible cells in which a particular line may reside. Each memory location can be mapped to a single location in the cache through direct mapping. There is no choice of where the line resides and which line must be replaced; however, poor utilization results. In contrast, a two-way set-associative cache maps each memory location into either of two locations in the cache. Hence, this mapping can be viewed as two identical, directly mapped caches. In fact, both caches must be searched at each memory access, and the appropriate data selected and multiplexed on a tag match hit and on a miss. Then, a choice must be made between two possible cache lines as to which is to be replaced. A single, least recently used bit can be saved for each such pair of lines to remember which line has been accessed more recently. This bit must be toggled to the current state each time. To this end, an M-way associative cache maps each memory location into M memory locations in the cache. Therefore, this cache map can be constructed from M identical directly mapped caches. The problem of maintaining the least recently used ordering of M cache lines is primarily due to the fact that there are $M!$ possible orderings. In fact, it takes at least $\log_2 M!$ bits to store the ordering.

Multiple memory banks, formed by ^{M}ICs, can be integrated together to form a parallel main memory system. Because each bank can service a request, a parallel main memory system with N_{mb} *banks* can service N_{mb} requests simultaneously, increasing the bandwidth of the memory system by N_{mb} times the bandwidth of a single bank. The number of banks is a power of two, that is, $N_{mb} = 2^p$. An n-bit memory word address is partitioned into two parts: a p-bit bank number and an m-bit address of a word within a bank. The p bits, used to select a bank number, could be any p bits of the n-bit word address. Let us use the low-order p address bits to select the bank number. The higher-order $m = (n - p)$ bits of the word address is used to access a word in the selected bank. Multiple memory banks can be connected using *simple paralleling* and *complex paralleling*. Figure 4.15 shows the structure of a simple parallel memory system in which m address bits are simultaneously supplied to all memory banks. All banks are connected to the same read–write control line. For a read operation, the banks perform the read operation and accumulate the data in the latches. Data can then be read from the latches one by one by setting the switch appropriately. The banks can be accessed again to carry out another read or write operation. For a write operation, the latches are loaded one by one. When all latches have been written, their contents can be written into the memory banks by supplying m bits of address. In a simple parallel memory, all banks are cycled at the same time. Each bank starts and completes its individual operations at the same time as every other bank, and a new memory cycle starts for all banks once the previous cycle is

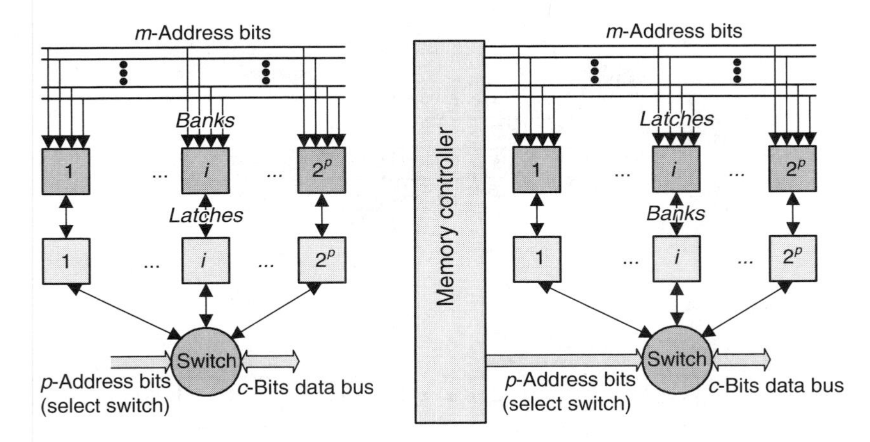

FIGURE 4.15
Simple and complex parallel main memory systems.

complete. A complex parallel memory system is shown in Figure 4.15. Each bank is set to operate on its own, independent of the operation of the other banks. For example, the ith bank performs a read operation on a particular memory address, while the $(i + 1)$th bank performs a write operation on a different and unrelated memory address. Complex paralleling is achieved using the address latch and a read–write command line for each bank. The *memory controller* handles the operation of the complex parallel memory. The processing unit submits the memory request to the memory controller, which determines which bank needs to be accessed. The controller then determines if the bank is busy by monitoring a busy line for each bank. The controller holds the request if the bank is busy, submitting it when the bank becomes available to accept the request. When the bank responds to a read request, the switch is set by the controller to accept the request from the bank and forward it to the processing unit. It can be foreseen that complex parallel main memory systems will be implemented ensuring vector processing. If consecutive elements of a vector are present in different memory banks, then the memory system can sustain a bandwidth of one element per clock cycle. Memory systems in ^{M}PPs can have thousands of banks with multiple memory controllers that allow multiple independent memory requests at every clock cycle.

Pipelining is a technique used to increase the processor throughput with limited hardware in order to implement complex *datapath* (data processing) units (multipliers, floating-point adders, etc.). A pipeline processor should integrate a sequence of i data processing ^{M}primitives that cooperatively perform a single operation on a stream of data operands passing through them. Design of pipelining ^{M}ICs involves deriving multistage balanced sequential algorithms to perform the given function. Fast buffer registers are placed between the ^{M}primitives to ensure the transfer of data between them without interfering with one another. These buffers should be clocked at the maximum

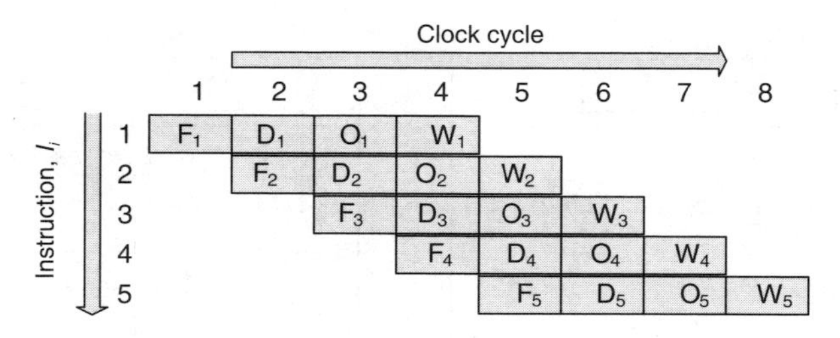

FIGURE 4.16
Pipelining of instruction execution.

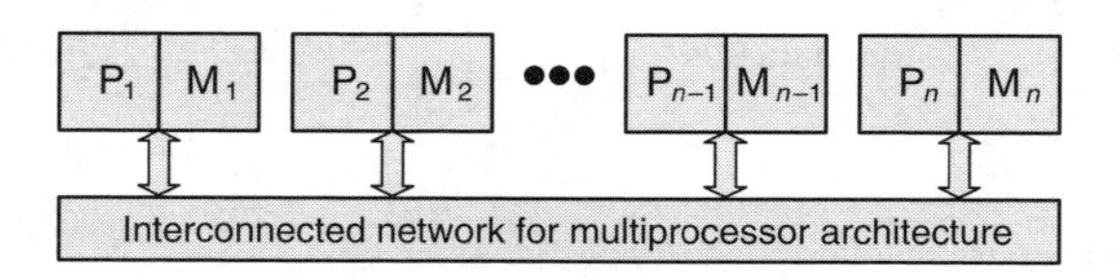

FIGURE 4.17
Multiprocessor architecture.

rate that guarantees the reliable data transfer between ^{M}primitives. As illustrated in Figure 4.16, ^{M}PPs must be designed to guarantee the robust execution of overlapped instructions using pipelining. Specific hardware units are needed to achieve these four basic steps—fetch F_i, decode D_i, operate O_i, and write W_i. The execution of instruction can be overlapped. When the execution of some instruction I_i depends on the results of a previous instruction I_{i-1} that is not yet completed, instruction I_i must be delayed. The pipeline is said to be stalled, waiting for the execution of instruction I_{i-1} to be completed. While it is not possible to eliminate such situations, it is important to minimize the probability of their occurrence. This is a key consideration in the design of the instruction set and the design of the compilers that translate high-level language programs into machine language.

The parallel execution capability, called superscalar processing, when added to pipelining of the individual instructions, means that more than one instruction can be executed per basic step. Thus, the execution rate can be increased. The rate R_{T} of performing basic steps in the processor depends on the processor clock rate. The use of multiprocessors speeds up the execution of large programs by executing subtasks in parallel. The main difficulty in achieving this is decomposition of given task into its parallel subtasks and ordering these subtasks to the individual processors in such a way that communication among the subtasks is performed efficiently and robustly. Figure 4.17 documents a block diagram of a multiprocessor system with the interconnection network needed for data sharing among the processors P_i. Parallel paths are needed in this network to ensure parallel activity in the processors as they access the global memory space as represented by the multiple memory units M_i. This is performed utilizing 3D-centered organization.

In general, multiassociative caches, multiple memories, pipelining, multiprocessing, and other designs are envisioned to be implemented resembling possible ^{BM}PPs solutions, that is, hierarchy, parallelism, redundancy, locality, mapping, and so forth. This does not imply that ^{BM}PPs utilize digital paradigm, clocking, binary solution, and other concepts commonly utilized in conventional signal/data processing platforms.

4.5 Finite-State Machines and Their Use in Hardware and Software Design

Simple register-level subsystems perform a single data-processing operations, for example, summation $X := x_1 + x_2$, subtraction $X := x_1 - x_2$, and so forth. To do complex data processing operations, multifunctional register-level subsystems should be designed and utilized. These register-level subsystems are partitioned as a data processing unit (*datapath*) and a controlling unit (control unit). The control unit is responsible for collecting and controlling the data processing operations (actions) of the *datapath*. To design the register-level subsystems, one studies a set of operations to be executed and then designs ^{M}ICs using a set of register-level components that implement the desired functions. The ultimate goal is to ensure optimal achievable performance under various constraints and limits. It is difficult to impose meaningful mathematical structures on register-level behavior using Boolean algebra and conventional gate-level design. Owing to these difficulties, heuristic synthesis is commonly accomplished within a sequential algorithm as follows:

1. Define the desired behavior as a set of sequences of register-transfer operations (each operation can be implemented using the available components) comprising the algorithm to be executed.
2. Examine the algorithm to determine the types of components and their number to ensure the required *datapath*.
3. Design a complete block diagram for the *datapath* using the components chosen.
4. Examine the algorithm and *datapath* in order to derive the control signals with ultimate goal to synthesize the control unit for the found *datapath* that meets the algorithm's requirements.
5. Accomplish test, verification, and evaluations tasks performing analysis and design.

We perform the design of virtual control units that ensure extensibility, flexibility, adaptability, robustness, and reusability. The design is performed using the hierarchic graphs (HGs). The most important problem is to develop straightforward algorithms that ensure implementation (nonrecursive and recursive calls) and utilize hierarchical specifications. We will examine the behavior, perform the logic design of , and implement reusable control units modeled as hierarchical finite-state machines with virtual states. The goal is to attain the top-down sequential well-defined decomposition to develop complex robust control algorithm step by step.

Consider *datapath* and control units. The *datapath* unit consists of memory and combinational units. A control unit performs a set of instructions by generating the appropriate sequence of microinstructions that depend on intermediate logic conditions or on intermediate states of the *datapath* unit. To describe the evolution of a control unit, behavioral models are developed. We use the direct-connected HGs containing nodes. Each HG has an entry (*Begin*) and an output (*End*). Rectangular nodes contain microinstructions, macroinstructions, or both.

A microinstruction set U_i includes a subset of micro-operations from the set $U = \{u_1, u_2, \ldots, u_{u-1}, u_u\}$. Micro-operations $\{u_1, u_2, \ldots, u_{u-1}, u_u\}$ control the specific actions in the *datapath* as shown in Figure 4.18. For example, one can specify that u_1 sends the data in the local stack, u_2 sends the data in the output stack, u_3 forms the address, u_4 calculates the address, u_5 forwards the data from the local stack, u_6 stores the data from the local stack in the register, u_7 forwards the data from the output stack to external output, and so forth. A micro-operation is the output causing an action

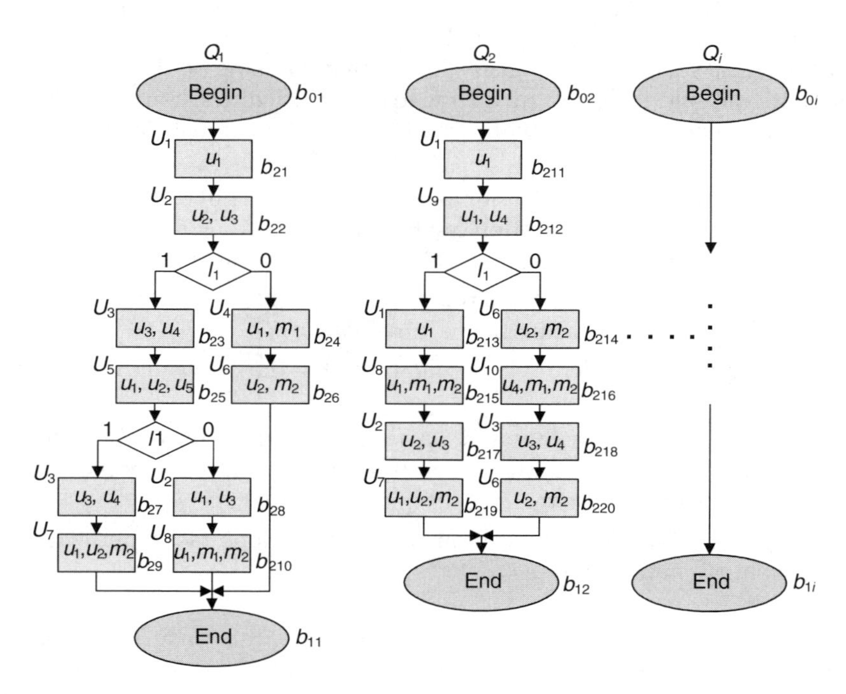

FIGURE 4.18
Control algorithm represented by HGs $Q_1, Q_2, \ldots, Q_{i-1}, Q_i$.

in the *datapath*. Any macroinstruction incorporates macro-operations from the set $M = \{m_1, m_2, \ldots, m_{m-1}, m_m\}$. Each macro-operation is described by another lower-level HG. Assume that each macroinstruction includes one macro-operation. Each rhomboidal node contains one element from the set $L \cup G$, where $L = \{l_1, l_2, \ldots, l_{l-1}, l_l\}$ is the set of logic conditions; $G = \{g_1, g_2, \ldots, g_{g-1}, g_g\}$ is the set of logic functions. Using logic conditions as inputs, logic functions are derived by examining predefined set of sequential steps that are described by a lower-level HG. Directed lines connect the inputs and outputs of the nodes. Consider a set $E = M \cup G$, $E = \{e_1, e_2, \ldots, e_{e-1}, e_e\}$. All elements $e_i \in E$ have HGs, and each e_i has the corresponding HG Q_i, which specifies either an algorithm for performing e_i (if $e_i \in M$) or for calculating e_i (if $e_i \in G$). Assume that $M(Q_i)$ is the subset of macro-operations and $G(Q_i)$ is the subset of logic functions that belong to the HG Q_i. If $M(Q_i) \cup G(Q_i) = \emptyset$, then the well-known scheme results [4]. The application of HGs enables one to gradually (linguistically) and sequentially synthesize complex control algorithm, concentrating the efforts at each stage on a specified level of abstraction because specific elements of the set E are used. Each component of the set E is simple and can be checked and debugged independently. Figure 4.18 shows the HGs $Q_1, Q_2, \ldots, Q_{i-1}, Q_i$, which describe the control algorithm.

The execution of HGs is examined studying complex operations $e_i = m_j \in M$ and $e_i = g_j \in G$. Each complex operation e_i that is described by a HG Q_i must be replaced with a new subsequence of operators that produces the result executing Q_i. In the illustrative example shown in Figure 4.19, Q_1 is the first HG at the first level Q^1, the second level Q^2 is formed by Q_2, Q_3, and Q_4, and so forth. We consider the following hierarchical sequence of HGs

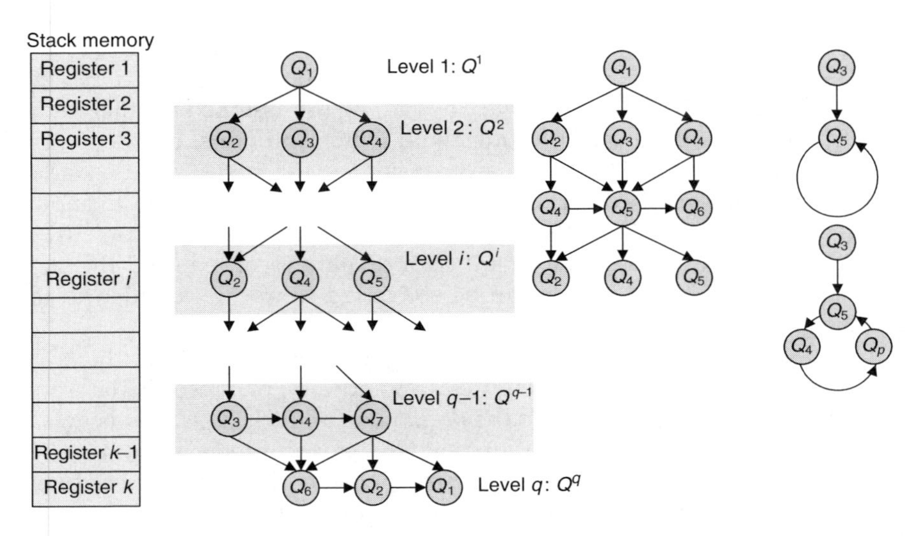

FIGURE 4.19
Stack memory with multiple-level sequential HGs with an illustration of recursive call.

$Q_{1(\text{level 1})} \Rightarrow Q^2_{(\text{level 2})} \Rightarrow \cdots \Rightarrow Q^{q-1}_{(\text{level } q-1)} \Rightarrow Q^q_{(\text{level } q)}$. All $Q_{i(\text{level } i)}$ have the corresponding HGs. For example, Q^2 is a subset of the HGs that are used to describe elements from the set $M(Q_1) \cup G(Q_1) = \emptyset$, while Q^3 is a subset of the HGs that are used to map elements from the sets $\cup_{q \in Q^2} M(q)$ and $\cup_{q \in Q^2} G(q)$. In Figure 4.19, $Q^1 = \{Q_1\}$, $Q^2 = \{Q_2, Q_3, Q_4\}$, $Q^i = \{Q_2, Q_4, Q_5\}$, and so forth.

Micro-operations u^+ and u^- are used to increment and decrement the stack pointer. The problem of switching to various levels can be solved using a stack memory; see Figure 4.19. Consider an algorithm for $e_i \in M(Q_1) \cup G(Q_1) = \emptyset$. The stack pointer is incremented by the micro-operation u^+, and a new register of the stack memory is set as the current register. The previous register stores the state when it was interrupted. The new Q_i becomes responsible for the control until terminated. After termination of Q_i, the micro-operation u^- is generated to return to the interrupted state. As a result, control is passed to the state in which Q_f is called. The design algorithm is: for a given control algorithm A, described by the set of HGs, construct the finite-state machine that implements A. The design includes the following steps:

1. Transformation of the HGs to the state transition table
2. State encoding
3. Combinational logic optimization and verification
4. Final design, analysis, and evaluation

The first step is divided into three tasks: (1) mark the HGs with labels b (see Figure 4.19); (2) record transitions between the labels in the extended state transition table; (3) convert the extended table to ordinary form. The labels b_{01} and b_{11} are assigned to the nodes *Begin* and *End* of the Q_1. The labels $b_{02}, \ldots, b_{0i}$ and $b_{12}, \ldots, b_{1i}$ are assigned to nodes *Begin* and *End* for $Q_2, \ldots, Q_i$, respectively. The labels $b_{21}, b_{22}, \ldots, b_{2j}$ are assigned to other nodes of HGs, inputs and outputs of nodes with logic conditions, etc. Repeating labels is not allowed. The labels are considered the states. The extended state transition table is designed using the state evolutions due to inputs (logic conditions) and logic functions that cause the transitions from $\mathbf{x}(t)$ to $\mathbf{x}(t+1)$. All evolutions of the state vector $\mathbf{x}(t)$ are recorded, and the state $x_k(t)$ has the label k. The table can be converted from the extended to the ordinary form.

To program the Code converter, using the algorithm flow-charts as illustrative examples shown in Figures 4.18 and 4.19, one records the transition from the state x_1 assigned to the *Begin* node of the HG Q_1, that is, $x_{01} \Rightarrow x_{21}(Q_1)$. The transitions between different HGs are recorded as $x_{ij} \Rightarrow x_{nm}(Q_j)$. For all transitions, the data-transfer instructions are derived. The hardware is illustrated in Figure 4.20. Robust control algorithms are derived using the HGs by employing the hierarchical behavior specifications and top-down decomposition. The reported method guarantees exceptional adaptation and

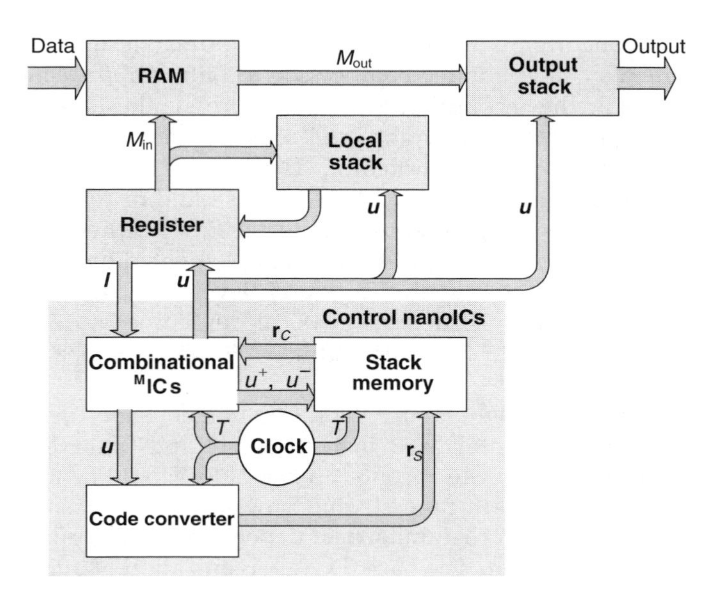

FIGURE 4.20
Hardware schematics.

reusability features through reconfigurable hardware and reprogrammable software for complex ^{M}ICs.

4.6 Adaptive Defect-Tolerant Molecular Processing Platforms

4.6.1 Programmable Gate Arrays

Some molecular fabrication processes, such as organic synthesis, self-assembly, and so forth, have been shown to be quite promising [2,5,6]. However, it is unlikely that near-future technologies will guarantee the reasonable repeatable characteristics, affordable high-quality high yield, satisfactory uniformity, desired failure tolerance, needed testability, and other important specifications and features imposed on ^{M}devices and ^{M}ICs. Therefore, design of robust defect-tolerant adaptive (reconfigurable) hardware and software to accommodate failures, inconsistence, variations, nonuniformity, and defects is critical.

For conventional ICs, programmable gate arrays (PGAs) have been developed and utilized. These PGAs lead one to on-chip reconfigurable circuits. The reconfigurable logics can be utilized as a functional unit in the *datapath* of the processor having access to the processor register file and to on-chip memory ports. Another approach is to integrate the reconfigurable part of the processor as a coprocessor. For this solution, the reconfigurable logic operates concurrently with the processor. Optimal design and memory port assignments can guarantee coprocessor reconfigurability and concurrency. In general, the reconfigurable architecture synthesis emphasizes a high-level design, rapid prototyping, and reconfigurability so as to reduce time and cost, improving performance. The goal is to design affordable high-performance high-yield ^{M}ICs. These ^{M}ICs should be testable to detect defects and faults. Design of the application-specific ^{M}ICs involves mapping application requirements into specifications implemented by the hardware. The specifications are represented at every level of abstraction including the system, behavior, structure, physical, and process domains. The designer should be able to differently utilize ^{M}ICs to meet the application requirements.

Reconfigurable ^{M}PPs should use reprogrammable logic units, such as PGAs, to implement a specialized instruction set and arithmetic units to optimize performance. Ideally, reconfigurable ^{M}PPs should be reconfigured in real time (runtime), enabling the existing hardware to be reused depending on its interaction with external units, data dependencies, algorithm requirements, faults, and so forth. The basic PGAs organization is built using the programmable logic blocks (PLBs) and programmable interconnect blocks (PIBs); see Figure 4.21. The PLBs and PIBs will hold the current configuration setting until adaptation will be accomplished. The PGA is programmed by downloading the information in the file through a serial or parallel logic connection. The time required to configure a PGA is called the configuration time, and PGAs can be configured in series or in parallel. Figure 4.21 illustrates the basic organizations. For example, pipelined interfaced PGAs organization is suitable for functions that have streaming data at specific intervals, while arrayed PGAs organization is appropriate for functions that require a systolic array. A hierarchy of configurability is different for

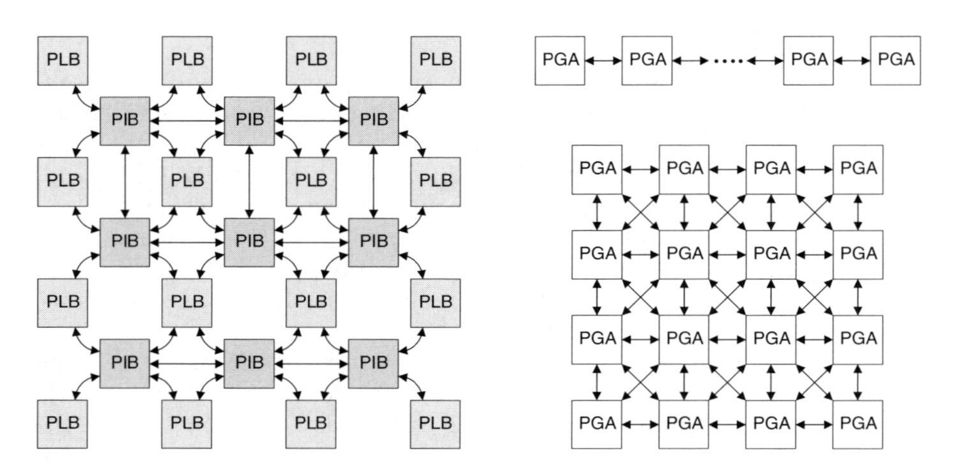

FIGURE 4.21
Programmable gate arrays and multiple PGAs organization.

different PGAs organizations, and the ^{M}ICs specifics impose constraints on the technology-centric SLSI.

4.6.2 Reconfigurable Molecular Integrated Circuits and Design Rules

Molecular ICs can be synthesized through hierarchical synthesis motifs utilizing $^{\aleph}$hypercells as molecular hardware primitives [2]. Section 4.3 introduced a 3D *directly interconnected molecular electronics* (^{3D}DIME) concept utilizing a direct device-to-device aggregation. The ^{ME}device–^{ME}device interconnect can be *chemical-bonding fabrics* (atomic bonding), *energy based* (for example, utilizing the exchange/conversion/transmission of radiated and absorbed electromagnetic, thermal, and vibrational energy), etc.

One needs to design reconfigurable ^{M}IC and ^{M}PPs, and develop complimentary software tools to cope with imperfect (partially defective and faulty) $^{\aleph}$hypercells and circuits in arithmetic, control, input–output, memory, and other units. Molecular electronics will result in ^{M}ICs with a significant number of entirely or partially defective and faulty devices and interconnect. The redundancy concept may not be effectively applied, while reconfiguration ensures the soundness. The circuit reconfigurability capability is defined by the yield, complexity, software abilities (to detect, identify, and tolerate the hardware deficiencies), and so forth. Adaptability and reconfigurability can be achieved through hardware diagnostics, testing, and analysis with the following mapping, matching, switching, controlling, rerouting, and networking tasks performed by software. Thus, one designs, optimizes, builds, tests/evaluates, and reconfigures ^{M}ICs. We develop the following *design rules*:

1. Design and optimize ^{M}ICs or ^{M}PPs
2. Apply the *target* ^{M}ICs realization using the *modular* $^{\aleph}$hypercell primitives
3. Design a *specific* $^{\aleph}$hypercell *template* assessing the expected yield and error rates
4. Analyze and perform the *bottom-up* synthesis developing and specifying the technology, processes, sequence order, and other tasks to synthesize ^{M}ICs as an assembly of $^{\aleph}$hypercell aggregates
5. Utilizing hierarchical
   - a. Random assembly with random sequence
     b. Near-random assembly with near-random sequences
     c. Ordered assembly with deterministic sequences
   - Specificity (terminal/interconnect-recognition, terminal/interconnect site recognition, self-binding, paring, and complimentary compliance of $^{\aleph}$hypercell primitives within node *lattices*)

- Nearest-neighboring $^{\aleph}$hypercell placement motifs
  Synthesize $^{\aleph}$hypercells aggregates forming node *lattices* that should realize ^{M}ICs

6. Perform diagnostics, verification, and testing
7. Reconfigurate, characterize, evaluate, and validate ^{M}ICs

These *design rules* define the random, near-random, or ordered (directed) ordering of $^{\aleph}$hypercells and their aggregates. Using this hierarchical strategy, we ensure the soundness of the integrated design-synthesis-networking-and-reconfiguration tasks. The proposed concept results in suboptimal solution- and design complexity as random or near-random ordering are utilized. However,

- Affordability and high yield with tolerable error rate
- Selectivity and specificity of $^{\aleph}$hypercells as processing and memory primitives
- Controllable self-assembling and robust binding/paring by utilizing $^{\aleph}$hypercell–$^{\aleph}$hypercell and $^{\aleph}$hypercell–interconnect uniformity, complimentary compliance, and recognition
- Overall aggregability, reconfigurability, and functionality may be ensured

As an illustrative example, Figure 4.22 documents a $4 \times 3 \times 1$ *lattice* with 12 interconnected nodes N_{ijk}. Each node comprises $2 \times 2 \times 2$ *modular* $^{\aleph}$hypercells D_{ijk} (given as $\boxed{D_{ijk}}$) engineered from ^{M}gates. Each ^{M}gate can comprise of two or more ^{M}devices depending on its device-level implementation, functionality, application, etc. For example, memory $^{\aleph}$hypercells can be designed using the ^{M}NAND gates, while multiplexer, adder, and multiplier $^{\aleph}$hypercells have tens of ^{M}gates. The nodes and $^{\aleph}$hypercells are connected through the exterior (peripheral) and interior interconnects. A ***single hypercell core*** represents a fixed motif and specific $^{\aleph}$hypercells from the primitive library, while a ***split hypercell core is applied to*** assemble and network $^{\aleph}$hypercells aggregates. As illustrated in Figure 4.22, the synthesis will result in the defective or faulty hypercells (D_{ijk}), hypercells *miss* D_{ijk}, and interconnect/link *miss*, which should be detected and handled.

As mentioned, node N_{12}, $^{\aleph}$hypercells (D_{123}, D_{124}, D_{125}, D_{126}, D_{128}, D_{223}, D_{238}, D_{312}, D_{322}, D_{323}, D_{338}, and D_{342}), interconnects between (D_{121}–D_{122}, D_{127}–D_{128}, etc.), and a link (from D_{347}) are defective or faulty. Furthermore, D_{345} is missing. One needs to identify and isolate defective $^{\aleph}$hypercells and nodes through reconfiguration. While partially functional nodes can be utilized, it is very challenging to employ partially functional $^{\aleph}$hypercells. The missing $^{\aleph}$hypercells, faulty interconnect, and links must be identified. In microelectronics, various diagnostics and verification algorithms using different test signals, vectors, patterns, protocols, and routing schemes are available. Those concepts, to some extent, potentially can be utilized in

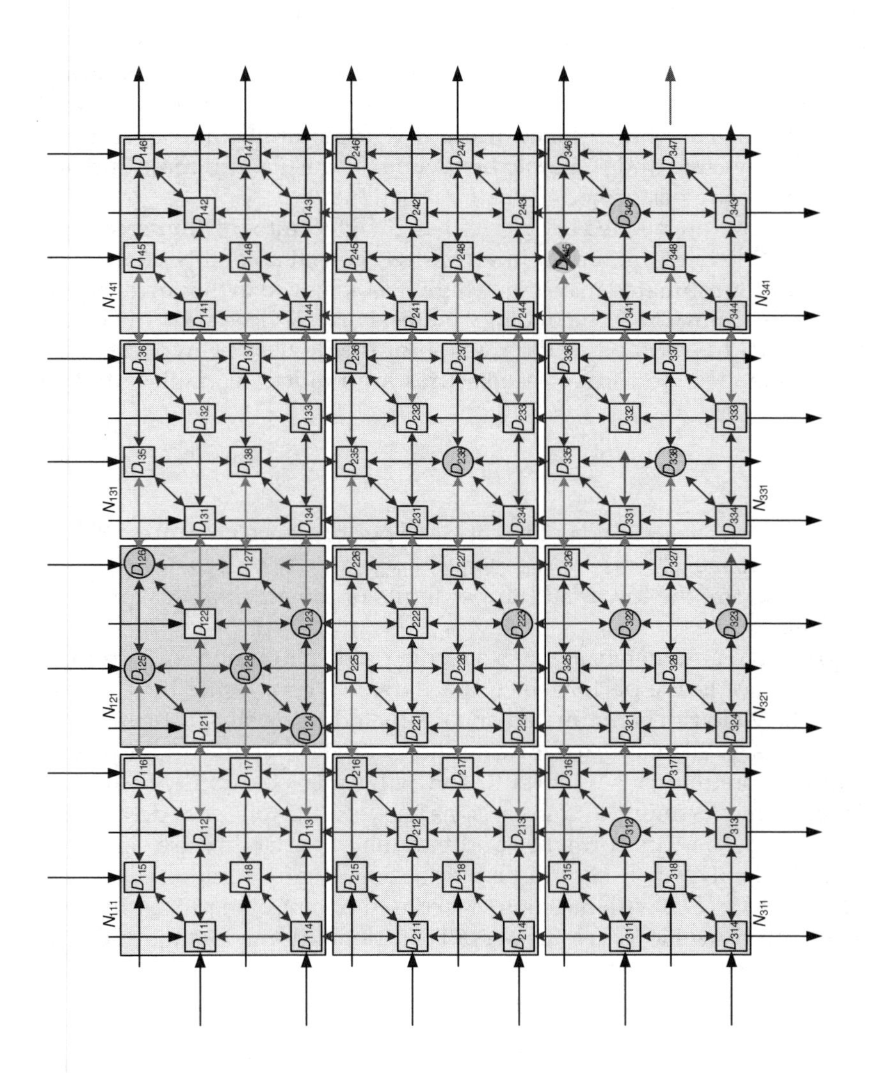

FIGURE 4.22
Aggregated ℵhypercells D_{ijk} within node *lattice*.

molecular electronics. However, owing to different device physics and technological limits, built-in self-test strategies (linear feedback shift registers are used to generate pseudorandom test patterns for synchronous and asynchronous circuits), response analyzers, observers, and field-programmable gate arrays (FPGAs) defect extraction may not be fully applicable to ^{M}ICs. For communication, configuration, combinational, and memory circuits, algorithms are different. Using the molecular hardware test logics, one can perform the hardware–software tests with evaluation and diagnostics features assuming the testability, controllability, and observability. In particular, the input–output mappings are utilized because any defects, faults, and misses lead to faulty steady-state and dynamic behaviors. The isolation, rerouting, and reconfiguration are performed to ensure functionality.

Novel concepts are needed to cope with high defect rates of $^{\aleph}$hypercells due to defective ^{M}devices, imperfect interconnect, partial assembly control, etc. The defect rate is estimated to be much higher as achieved in the current 65 nm and expected 45 nm CMOS technology nodes. For ^{M}devices and $^{\aleph}$hypercells, parametric yield is defined as the fraction of the devices or hypercells that are acceptable. The performance measures are random variables, and the yield is

$$Y = \Pr(\mathbf{r} \in R),$$

where $\mathbf{r} = [r_1, r_2, \ldots, r_{p-1}, r_p]$ is the vector of performance measures; R is the performance space, $R = \{\mathbf{r} \mid a_{i\min} \le r_i \le a_{i\max},\ i = 1, 2, \ldots, p-1, p\} \in \mid \mathbb{R}^p$; $a_{i\min}$ and $a_{i\max}$ are the lower and upper limits of the acceptability of the ith performance.

The primitive parameters are also specified. The mapping $\mathbf{x} \to \mathbf{r}$ from the parameter space to the performance space defines $\mathbf{r}(\mathbf{x})$. Using the indicator function, the yield can be estimated and evaluated using stochastic methods. As the synthesis will mature to fabricate and evaluate $^{\aleph}$hypercells, the probability distribution functions will result. This will lead to the ability to evaluate metrics enabling meaningful comparisons between the reference designs and synthesis yields. The performance and parameter variability are significant factors to be analyzed. It should be emphasized that the reported concept is applicable to ^{BM}PPs with neurons that consist of neuronal processing-and-memory primitives. However, the overall functionality, processing, and other features should be understood.

One needs to ensure defect isolation of defective or faulty primitives within a node *lattice* that may consist of a large number of random or partially-ordered self-assembled $^{\aleph}$hypercells. From the synthesis standpoints, a network of self-assembled primitives is formed using the *modular* $^{\aleph}$hypercell primitives as aggregated applying the *design rules*. The functionality can be ensured even if more than 25% of the $^{\aleph}$hypercells are faulty. This requires to execute failure-discontinue tasks ensuring defect isolation by means of hardware–software self-diagnostics, tests, evaluation, and reconfiguration. We estimate that for the interconnected three-terminal ^{ME}device the defect

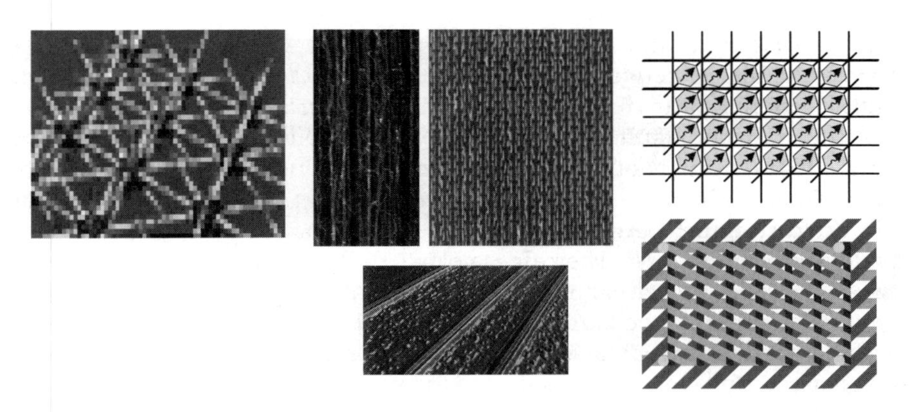

FIGURE 4.23
Schematics of $^{\aleph}$hypercells within a node *lattice* and its placement using carbon nanotubes and nanowires (3 nm wide parallel six-atom-wide erbium disilicide $ErSi_2$ nanowire, Hewlett-Packard, www.hp.com) within a crossbar topology.

rate will be larger that 1×10^{-5}, while for $^{\aleph}$hypercells the defect rate will be higher than 1×10^{-4}. The synthesis of multiterminal ^{ME}devices and $^{\aleph}$hypercells utilizing cyclic molecules is reported in Section 5.1. Using the *design rules*, the illustrative representation of $^{\aleph}$hypercells within a node lattice is shown in Figure 4.23. As possible technological solutions, aligned carbon nanotubes, nanowires, carbon cages, and other motifs can be examined. A crossbar fabrics, which utilizes the chemical-bond fabrics interconnect as, illustrated in Figure 4.23, is considered as one among other various promising concepts. However, the energy-based interconnect possesses greater advantages.

Faults and fault models should describe defects in the circuit. A defect, as a failure source, is the unintended difference between the ^{M}IC hardware and its intended design. The defects result due to the molecular fabrication inconsistency (missing connect, missing or faulty component, etc.), environmental (radiation-, temperature- and vibration-induced defects), as well as other hardware and physical imperfections. A fault is defined as a representation of the defect at the function level. A physical defect in a ^{M}IC can produce multiple faults, and a single test cannot detect all possible or actual defects. The problem of defect detection, localization, and determination should be solved. For molecular electronics, one may differentiate between *soft* (parametric, i.e., high delay, low speed, coupling, immunity, etc.) defects and *hard* (catastrophic) defects, which cause faults. There are a significant number of faults that may occur, and there are controllability, observability, detectability, equivalence, dominance, and other issues to be resolved.

Let $y(x)$ be the logic function of a combinational circuit C, where x is the input vector. Hence, $y(x)$ denotes the mapping realized by C. The presence of a fault ϕ changes C into a faulty circuit C^{ϕ} with $y^{\phi}(x)$. Taking note of the testing inputs t, the input test vector $T = \{t_1, t_2, \ldots, t_{n-1}, t_n\}$ provides a test sequence. One performs testing applying T that should detect the *detectable* faults and distinguish between them. A complete fault location test distinguishes between every pair of *distinguishable* faults. A complete fault location test can diagnose a fault within a *functionally equivalent class.*

A test t detects a fault ϕ iff $y(t) \neq y^{\phi}(t)$. A fault ϕ is *detectable* if there exists a test t that detects ϕ; otherwise, ϕ is *undetectable.* Two faults ϕ and g can be equivalent. Faults ϕ and g are called *functionally equivalent* iff $y^{\phi}(x) \equiv y^{g}(x)$. If T can distinguish between two faults ϕ and g, that is, $y^{\phi}(x) \neq y^{g}(x)$, these faults are *distinguishable*. There does not exist a t that can distinguish between two *functionally equivalent* faults. For test generation, it is sufficient to consider only one representative fault from every *functional equivalent class.*

For ^{M}ICs, we introduce a *functional fault model* to describe faults from a given arbitrary level of abstraction to the next higher level by means of the test generation design. Consider a Boolean function $y = f(x_1, x_2, \ldots, x_{n-1}, x_n)$ implemented by a ^{M}primitive in a circuit. Introduce a defect variable d that represents a given physical defect which affects y by changing the Boolean function to be $y = f^{d}(x_1, x_2, \ldots, x_{n-1}, x_n)$. We utilize a parametric function y_p that is a function of a defect variable d. In particular,

$$y_p = f_p(x_1, x_2, \ldots, x_{n-1}, x_n, d) = \bar{d}f \vee df^{d}$$

describes the behavior of the ^{M}primitive for both fault-free and faulty cases. For the faulty case, the value of the defect variable d is $d = 1$. While, for the fault-free case, $d = 0$. Hence

$$y_p = f^{d} \text{ if } d = 1, \text{ and } y_p = f \text{ if } d = 0.$$

The Boolean differential equation

$$B_d = (\partial y_p / \partial d) = 1$$

establishes the conditions that define the defect d as well as results in t. The parametric modeling of a given defect d allows one to perform: (1) defect-oriented fault simulation by verifying the condition $B_d = 1$, and, (2) defect-oriented test generation by solving the equation $B_d = (\partial y_p / \partial d) = 1$ when the defect d is activated and tested using the logic condition given by B_d. To find B_d for a given defect d one derives the corresponding logic expression for the faulty function f^{d} either by: (1) logical reasoning, (2) performing defect simulation, or (3) carrying out experiments to derive the physical behavior and f implemented.

Example 4.1

Consider a circuit shown in Figure 4.24, which implements a switching function $y = x_1x_2x_3 \vee x_4x_5$. A *short* defect d, shown in Figure 4.24, changes the circuit output to $y^{d} = (x_1 \vee x_4)(x_2x_3 \vee x_5)$. Thus, we have

$$\begin{aligned} B_d = \frac{\partial y_p}{\partial d} &= \frac{\partial[(x_1x_2x_3 \vee x_4x_5)\ \overleftarrow{d} \vee (x_1 \vee x_4)(x_2x_3 \vee x_5)d]}{\partial d} \\ &= x_1\ \overleftarrow{x}_2\ \overleftarrow{x}_4x_5 \vee x_1\ \overleftarrow{x}_3\ \overleftarrow{x}_4x_5 \vee\ \overleftarrow{x}_1x_2x_3x_4\ \overleftarrow{x}_5 = 1. \end{aligned}$$

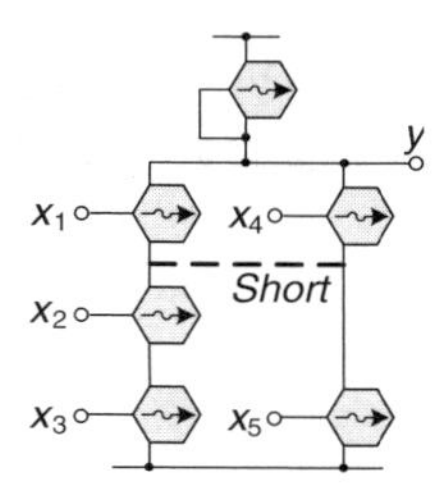

FIGURE 4.24
Circuit schematics.

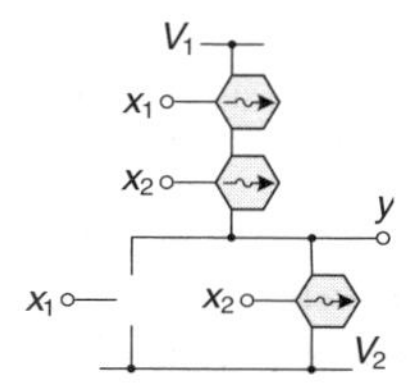

FIGURE 4.25
Molecular gate with an *open* fault.

The derived expression provides three values of t, and the test vector is found to be $T = \{10 \times 01, 1 \times 001, 01110\}$. Each t can be used as a test pattern for the given d.

Example 4.2

Consider a ^{M}NOR gate with an *open* fault as illustrated in Figure 4.25. The output retains its previous logic value. The considered combinational logic gate behaves as a dynamic memory element. The faulty function of the gate is $y^d = \overline{x}_1\overline{x}_2 \vee x_1\overline{x}_2 y_s$, where y_s is the output value stored at the output of the faulty ^{M}gate. We have $y_p = \overline{d}(\overline{x_1 \vee x_2}) \vee d(\overline{x}_1\overline{x}_2 \vee x_1\overline{x}_2 y_s) = \overline{x}_2(\overline{x}_1 \vee d y_s)$ and $B_d = \partial y_p/\partial d = x_1\overline{x}_2 y_s = 1$. The condition to activate the defect is $x_1 = 1$, $x_2 = 0$, and $y_s = 1$. Thus, for testing the fault one needs a test sequence of two patterns, for example, 00 (to obtain 1 on the output y) and then 11.

One can map the interconnect defects. Consider a ^{M}primitive (P) with a Boolean function $y = f(x_1, x_2, \ldots, x_{n-1}, x_n)$ and interconnect $I = \{x_{n+1}, \ldots, x_p\}$. We apply the defect variable d to represent physical defects in the circuit $C = (P, I)$. Let the defect d change the Boolean function $f(x_1, x_2, \ldots, x_{n-1}, x_n, x_{n+1}, \ldots, x_p)$ to $f^d(x_1, x_2, \ldots, x_{n-1}, x_n, x_{n+1}, \ldots, x_p)$. For modeling physical defects in C, we use the parametric function

$$y_p = f_p(x_1, x_2, \ldots, x_{n-1}, x_n, x_{n+1}, \ldots, x_p, d) = (\overline{d} \wedge f) \vee (d \wedge f^d),$$

which describes the behaviour of the circuit for the fault-free and faulty cases. For the faulty case $d = 1$, while for the fault-free case $d = 0$. Thus, $y_p = f^d$ if $d = 1$, and $y_p = f$ if $d = 0$. The solutions of the Boolean differential equation $B_d = (\partial y_p/\partial d) = 1$ allows one to perform the analysis and obtain the conditions that activate the defect d.

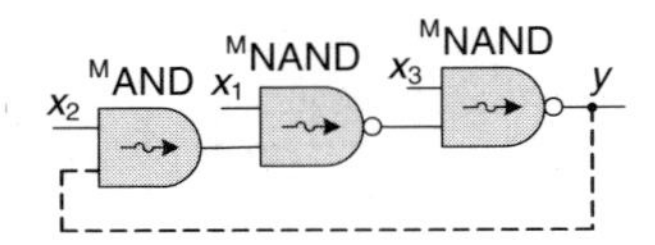

FIGURE 4.26
Molecular gates with a *short*.

Example 4.3

Consider a *short* in a circuit as shown in Figure 4.26. The parametric function is $y_p = \bar{d}(x_1x_2 \vee \bar{x}_3) \vee d(x_1x_2y \vee \bar{x}_3) = x_1x_2(\bar{d} \vee y_0)\bar{x}_3$, where y_0 denotes the previous value of y. The Boolean differential equation leads to $B_d = \partial y_p/\partial d = x_1x_2x_3\bar{y}_0 = 1$. Hence, one can test the *short* as follows: (1) Set the value $y = 0$ (for example, by assigning $x_3 = 0$); (2) Apply the test pattern 111 ($x_1 = 1$, $x_2 = 1$, and $x_3 = 1$).

The described method represents a general approach to map an arbitrary physical defect into a higher level. It was shown that the method of defining faults by logic condition $B_d = 1$ can be used both in fault simulation and in test generation. Consider a node k in a circuit. The output of a module M_k is y_k. For M_k, consider a set of faults $R_k = R_k^F \cup R_k^S$, where R_k^F is the subset of faults in M_k; R_k^S is the subset of structural faults (defects) in the network neighbourhood of M_k. In general, B_d allows one to examine conditions when the faults $d \in R_k$ change y_k. We denote by B_k^F the set of conditions B_d activating the defects $d \in R_k^F$, while B_k^S gives the set of conditions B_d activating the structural defects $d \in R_k^S$. Using B_k^F and B_k^S, one obtains a map of faults for test generation from a higher to a lower level, as well as for fault simulation and fault diagnostics tasks. In test generation, to map a lower level fault $d \in R_k$ to a higher level, we use $B_d = 1$. If $B_d = 1$ is guaranteed, the defect $d \in R_k$ changes y_k. For fault simulation and fault diagnostics, an erroneous y_k is

$$dy_k \rightarrow d_1B_{d1} \vee d_2B_{d2} \vee \cdots \vee d_nB_{dn}, d_i \in R_k.$$

For hierarchical testing, for each module M_k of the circuit, one studies R_k with the logical conditions B_d for each $d \in R_k$. The set of conditions B_k^F for the functional faults $d \in R_k^F$ of the module is found by low-level test generation for defects in the module. The set of conditions B_k^S for the structural faults $d \in R_k^S$ is to be derived from the Boolean differential analysis of the fault-free/faulty functions. For the concept under consideration, one considers $d = (d_1, d_2, \ldots, d_{i-1}, d_i)$.

For c17 circuits, represented in Figure 4.8a, by making use the concept reported, we obtain $T = \{10000\ 00101\ 01110\ 00110\ 00001\}$. The derive T one ensures more than 99% of possible fault detection.

Example 4.4

Consider a register and a gate level. The condition to detect the defect d on the observable Y is $B_D = \partial Y/\partial y_M \wedge \partial y_M/\partial y_G \wedge B_d = 1$, where y_M is the output variable of a logic-level module; y_G is the output of a logic gate with a physical defect d. In this equation, $\partial Y/\partial y_M$ gives the high-level fault propagation condition, $\partial y_M/\partial y_G$ is the fault propagation condition (Boolean derivative) at the gate level.

4.6.3 Design of Reconfigurable Molecular Processing Platforms

The overall objective can be achieved by guaranteeing the behavior (evolution, functionality, etc.) matching between the ideal ($\boldsymbol{C}_I$) and fabricated ($\boldsymbol{C}_F$) molecular platforms, subsystems, modules, or components. The molecular compensator ($\boldsymbol{C}_{F1}$) can be designed and implemented for a fabricated $\boldsymbol{C}_{F2}$ such that the response of the $\boldsymbol{C}_F$ will match the evolution of the $\boldsymbol{C}_I$; see Figure 4.27. Both $\boldsymbol{C}_{F1}$ and $\boldsymbol{C}_{F2}$ represent ^{M}ICs hardware. The $\boldsymbol{C}_I$ gives the reference ideal *model*, which provides the ideal input–output behavior. The compensator $\boldsymbol{C}_{F1}$ should modify the evolution of $\boldsymbol{C}_{F2}$ such that $\boldsymbol{C}_F$, as given by $\boldsymbol{C}_F = \boldsymbol{C}_{F1} \circ \boldsymbol{C}_{F2}$ for a series organization, matches the $\boldsymbol{C}_I$ behavior and functionality. Figure 4.27 illustrates the concept. The necessary and sufficient conditions for strong and weak evolution matching based on $\boldsymbol{C}_I$ and $\boldsymbol{C}_{F2}$ must be derived. We assume that the observability and controllability conditions are met. Our goal is to reconfigure $\boldsymbol{C}_{F2}$ to match $\boldsymbol{C}_I$. In general, $\boldsymbol{C}_{F2}$ can be given as $\boldsymbol{C}_{F2} = \boldsymbol{C}_{F2(1)} \circ \boldsymbol{C}_{F2(2)}$. For this case, the results described in the following should be slightly modified.

To address analysis, control, diagnostics, optimization, and design problems, the explicit models of ^{M}ICs, ^{M}PPs, and their units must be derived. There are different levels of abstraction in modeling, simulation, and analysis. High-level models can accept streams of instruction descriptions and memory references, while the low-level (device- or gate-level) modeling can be performed by making use of input–output mappings or examining device physics that results in nonlinear transient and steady-state analysis. The subsystem- or unit-level modeling (medium-level) also can be formulated

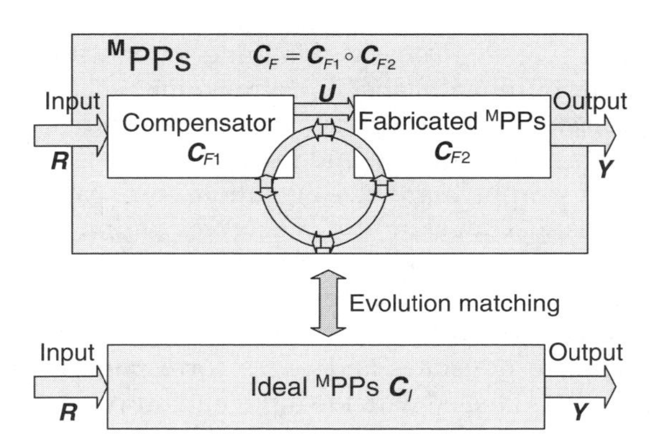

FIGURE 4.27
Molecular platform and evolution matching.

and performed. A subsystem can contain billions of ^{M}devices, and may not be modeled as queuing networks, difference equations, Boolean models, polynomials, information-theoretic models, and so forth. Different mathematical modeling concepts exist and have been developed for each level. We concentrate on the high-, medium-, and low-level systems modeling using the finite-state machine concept that is applicable to design adaptive (reconfigurable) defect-tolerant ^{M}ICs and ^{M}PPs.

Molecular processors and memories accept input information, process it according to the stored instructions, and produce the output. Any mathematical model is the mathematical idealization based on the abstractions, simplifications, and hypotheses made. It is virtually impossible to develop and apply the complete mathematical model because of complexities and uncertainties. Molecular platforms can be concurrently modeled using the six-tuple:

$$C = \{X, E, R, Y, F, X_0\},$$

where X is the finite set of states with initial and final states $x_0 \in X$ and $x_f \subseteq X$; E is the finite set of events (concatenation of events forms a string of events); R and Y are the finite sets of the input and output symbols (alphabets) or vectors; F are the transition functions mapping from $X \times E \times R \times Y$ to X (denoted as F_X), to E (denoted as F_E) or to Y (denoted as F_Y), $F \subseteq X \times E \times R \times Y$.

We assume that $F = F_X$, that is, the transition function defines a new state to each quadruple of states, events, references, and outputs, and F can be represented by a table listing the transitions or by a state diagram.

The evolution of a molecular platform is due to behavior of inputs, events, state evolutions, parameter variations, and so forth. A vocabulary (or an alphabet) A is a finite nonempty set of symbols (elements). A world (or sentence) over A is a string of finite length of elements of A. The empty (null) string does not contain symbols. The set of all words over A is denoted as A_w. A language over A is a subset of A_w. A finite-state machine with output $C_{FS} = \{X, A_R, A_Y, F_R, F_Y, X_0\}$ consists of a finite set of states S, a finite input alphabet A_R, a finite output alphabet A_Y, a transition function F_Y that assigns a new state to each state and input pair, an output function F_Y that assigns an output to each state and input pair, and initial state X_0.

Using the input–output map, the evolution of C can be expressed as $E_C \subseteq R \times Y$. That is, if C in state $x \in X$ receives an input $r \in R$, it moves to the next state $f(x,r)$ and produces the output $y(x,r)$. One can represent the molecular platform using the state tables that describe the state and output functions. In addition, the state transition diagram (direct graph whose vertices correspond to the states and edges correspond to the state transitions, and each edge is labeled with the input and output associated with the transition) can be used.

It should be emphasized that the quantum molecular platform is described by the seven-tuple $C_{\text{quantum}} = \{X, E, R, Y, H, U, X_0\}$, where H is the Hilbert space, and U is the unitary operator in the Hilbert space that satisfies the specific conditions.

The parameters set $\boldsymbol{P}$ should be used. Designing reconfigurable fault-tolerant platforms, sets $\boldsymbol{P}$ and $\boldsymbol{P}_0$ are integrated, and

$$C = \{\boldsymbol{X}, \boldsymbol{E}, \boldsymbol{R}, \boldsymbol{Y}, \boldsymbol{P}, \boldsymbol{F}, \boldsymbol{X}_0, \boldsymbol{P}_0\}.$$

Hence, the evolution of $\boldsymbol{C}$ depends on $\boldsymbol{P}$ and $\boldsymbol{P}_0$. Optimal performance can be achieved through testing, diagnostics, adaptation, and reconfiguration. For example, one can vary $\boldsymbol{F}$ and variable parameters $\boldsymbol{P}_v$ to attain the best possible performance. The evolution of states, events, outputs, and parameters is expressed as

$$(x_0, e_0, y_0, p_0) \overset{\text{evolution } 1}{\Longrightarrow} (x_1, e_1, y_1, p_1) \overset{\text{evolution } 2}{\Longrightarrow} \cdots$$
$$\overset{\text{evolution } j-1}{\Longrightarrow} (x_{j-1}, e_{j-1}, y_{j-1}, p_{j-1}) \overset{\text{evolution } j}{\Longrightarrow} (x_j, e_j, y_j, p_j).$$

The input, states, outputs, events, and parameter sequences are aggregated within the model as given by $\boldsymbol{C} = \{\boldsymbol{X}, \boldsymbol{E}, \boldsymbol{R}, \boldsymbol{Y}, \boldsymbol{P}, \boldsymbol{F}, \boldsymbol{X}_0, \boldsymbol{P}_0\}$. By taking note of defect (fault) set $\boldsymbol{D}$, we have

$$C = \{\boldsymbol{X}, \boldsymbol{E}, \boldsymbol{R}, \boldsymbol{Y}, \boldsymbol{P}, \boldsymbol{D}, \boldsymbol{F}, \boldsymbol{X}_0, \boldsymbol{P}_0, \boldsymbol{D}_0\}.$$

The concept reported allows us to find and apply the minimal but complete functional description of molecular processing and memory platforms. The minimal subset of state, event, output, and parameter evolutions (transitions) can be used. That is, the partial description $\mathbf{C}_{\text{partial}} \subset \mathbf{C}$ results, and every essential sixruple $(x_i, e_i, r_i, y_i, p_i, d_i)$ can be mapped by $(x_i, e_i, r_i, y_i, p_i, d_i)_{\text{partial}}$. This significantly reduces the complexity of diagnostics, evaluation, design, and reconfiguration problems.

Let the function $\boldsymbol{F}$ maps from $\boldsymbol{X} \times \boldsymbol{E} \times \boldsymbol{R} \times \boldsymbol{Y} \times \boldsymbol{P} \times \boldsymbol{D}$ to $\boldsymbol{X}$, that is, $\boldsymbol{F} : \boldsymbol{X} \times \boldsymbol{E} \times \boldsymbol{R} \times \boldsymbol{Y} \times \boldsymbol{P} \times \boldsymbol{D} \to \boldsymbol{X}$, $\boldsymbol{F} \subseteq \boldsymbol{X} \times \boldsymbol{E} \times \boldsymbol{R} \times \boldsymbol{Y} \times \boldsymbol{P} \times \boldsymbol{D}$. Thus, the transfer function $\boldsymbol{F}$ defines a next state $\boldsymbol{x}(t+1) \in \boldsymbol{X}$ based on the current state $\boldsymbol{x}(t) \in \boldsymbol{X}$, event $\boldsymbol{e}(t) \in \boldsymbol{E}$, reference $\boldsymbol{r}(t) \in \boldsymbol{R}$, output $\boldsymbol{y}(t) \in \boldsymbol{Y}$, parameter $\boldsymbol{p}(t) \in \boldsymbol{P}$, and defect $\boldsymbol{d}(t) \in \boldsymbol{D}$. Hence,

$\boldsymbol{x}(t+1) = \boldsymbol{F}[\boldsymbol{x}(t), \boldsymbol{e}(t), \boldsymbol{r}(t), \boldsymbol{y}(t), \boldsymbol{p}(t), \boldsymbol{d}(t)]$ for $\boldsymbol{x}_0(t) \in \boldsymbol{X}_0$, $\boldsymbol{e}_0(t) \in \boldsymbol{E}_0$, $\boldsymbol{r}_0(t) \in \boldsymbol{R}_0$, $\boldsymbol{y}_0(t) \in \boldsymbol{Y}_0$, $\boldsymbol{p}_0(t) \in \boldsymbol{P}_0$, and $\boldsymbol{d}_0(t) \in \boldsymbol{D}_0$.

Robust adaptation algorithms must be developed to ensure defect tolerance and reconfiguration. The control vector $\boldsymbol{u}(t) \in \boldsymbol{U}$ is integrated into the model. We have

$$C = \{\boldsymbol{X}, \boldsymbol{E}, \boldsymbol{R}, \boldsymbol{Y}, \boldsymbol{P}, \boldsymbol{D}, \boldsymbol{U}, \boldsymbol{F}, \boldsymbol{X}_0, \boldsymbol{P}_0, \boldsymbol{D}_0\},$$

and the problem is to design the compensator.

The strong evolutionary matching $\boldsymbol{C}_F = \boldsymbol{C}_{F1} \circ \boldsymbol{C}_{F2} = {}_B\boldsymbol{C}_I$ for given $\boldsymbol{C}_I$ and $\boldsymbol{C}_F$ is guaranteed if $E_{C_F} = E_{C_I}$. Here, $\boldsymbol{C}_F = {}_B\boldsymbol{C}_I$ means that the behaviors (evolution) of $\mathbf{C}_I$ and $\mathbf{C}_F$ are equivalent. The weak evolutionary matching $\mathbf{C}_F = \mathbf{C}_{F1} \circ \mathbf{C}_{F2} \subseteq {}_B\mathbf{C}_I$ for given $\mathbf{C}_I$ and $\mathbf{C}_F$ is guaranteed if $E_{C_F} \subseteq E_{C_I}$. Here, $\mathbf{C}_F \subseteq {}_B\mathbf{C}_I$ means that the evolution of $\mathbf{C}_F$ is contained in the behavior $\mathbf{C}_I$.

The problem is to derive a compensator $\mathbf{C}_{F1} = \{\mathbf{X}_{F1}, \mathbf{E}_{F1}, \mathbf{R}_{F1}, \mathbf{Y}_{F1}, \mathbf{F}_{F1}\}$ such that, for given $\mathbf{C}_I = \{\mathbf{X}_I, \mathbf{E}_I, \mathbf{R}_I, \mathbf{Y}_I, \mathbf{F}_I\}$ and $\mathbf{C}_{F2} = \{\mathbf{X}_{F2}, \mathbf{E}_{F2}, \mathbf{R}_{F2}, \mathbf{Y}_{F2}, \mathbf{P}_{F2}, \mathbf{D}_{F2}, \mathbf{F}_{F2}\}$, the following conditions:

$$\mathbf{C}_F = \mathbf{C}_{F1} \circ \mathbf{C}_{F2} = {}_B\mathbf{C}_I \quad \text{(strong behavior matching)}$$

or

$$\mathbf{C}_F = \mathbf{C}_{F1} \circ \mathbf{C}_{F2} \subseteq {}_B\mathbf{C}_I \quad \text{(weak behavior matching)}$$

are satisfied.

We assume that: (1) output sequences generated by $\mathbf{C}_I$ can be generated by $\mathbf{C}_{F2}$, and, (2) the $\mathbf{C}_I$ inputs match the $\mathbf{C}_{F1}$ inputs. The output sequences mean the state, event, output and/or parameters vectors.

If there exists the state-modeling representation $\gamma \subseteq \mathbf{X}_I \times \mathbf{X}_F$ such that $C_I^{-1} \subseteq {}^{\gamma}_{B}C_{F2}^{-1}$ (if $C_I^{-1} \subseteq {}^{\gamma}_{B}C_{F2}^{-1}$, then $C_I \subseteq {}^{\gamma}_{B}C_{F2}$), then the evolution matching problem is solvable. The compensator $\mathbf{C}_{F1}$ solves the strong-matching problem $\mathbf{C}_F = \mathbf{C}_{F1} \circ \mathbf{C}_{F2} = {}_B\mathbf{C}_I$ if there exist the state-modeling representations $\beta \subseteq \mathbf{X}_I \times \mathbf{X}_{F2}$, $(\mathbf{X}_{I\,0}, \mathbf{X}_{F2\,0}) \in \beta$ and $\alpha \subseteq \mathbf{X}_{F1} \times \beta$, $(\mathbf{X}_{F1\,0}, (\mathbf{X}_{I\,0}, \mathbf{X}_{F2\,0})) \in \alpha$ such that $C_{F1} = {}^{\alpha}_{B}C_I^{\beta}$ for $\beta \in \Gamma = \{\gamma | C_I^{-1} \subseteq {}^{\gamma}_{B}C_{F2}^{-1}\}$. The strong-matching problem is tractable if there exist C_I^{-1} and C_{F2}^{-1}.

The $\mathbf{C}$ can be decomposed using algebraic decomposition theory, which is based on the closed partition lattice. For example, consider the fabricated $\mathbf{C}_{F2}$ represented as $\mathbf{C}_{F2} = \{\mathbf{X}_{F2}, \mathbf{E}_{F2}, \mathbf{R}_{F2}, \mathbf{Y}_{F2}, \mathbf{P}_{F2}, \mathbf{D}_{F2}, \mathbf{F}_{F2}\}$. A partition on the state set for $\mathbf{C}_{F2}$ is a set $\{\mathbf{C}_{F2\,1}, \mathbf{C}_{F2\,2}, \ldots, \mathbf{C}_{F2\,i}, \ldots, \mathbf{C}_{F2\,k-1}, \mathbf{C}_{F2\,k}\}$ of disjoint subsets of the state set $\mathbf{X}_{F2}$ whose union is $\bigcup_{i=1}^{k} C_{F2\,i} = X_{F\,2}$ and $C_{F2\,i} \bigcap C_{F2\,j} = \varnothing$ for $i \neq j$. Hence, one designs and implements the compensators $\mathbf{C}_{F1\,i}$ for given $\mathbf{C}_{F2\,i}$.

4.7 Hardware–Software Design

Significant research activities have been focused on the software developments for efficient, robust, and homogeneous ^{M}PPs. The aforementioned activities must be supported by a broad spectrum of hardware–software codesign including technology-centric CAD developments toward SLSI. Hardware–software codesign, integration, and verification are important problems to be addressed. The synthesis of concurrent architectures and their organization (collection of functional hardware components, modules, subsystems, and systems that can be software programmable and adaptively reconfigurable) are among the most important tasks. The software depends

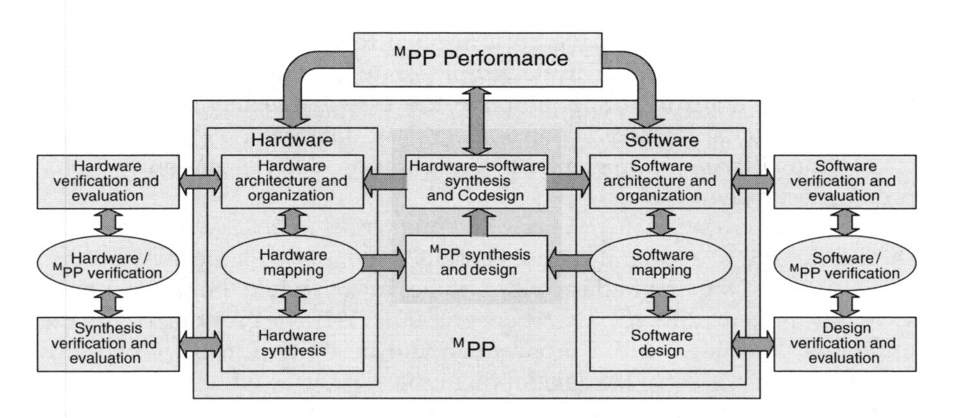

FIGURE 4.28
Hardware–software codesign for ^{M}PPs.

on hardware, and vice versa. The concurrency indicates hardware and software compliance and matching. It is impractical to fabricate high-yield ideal (defect-free) ^{M}ICs and ^{M}PPs. Furthermore, it is unlikely that the software can be developed for not-strictly-defined configurations that must be adapted, reconfigured, and optimized by using the design rules described in Section 4.6.2. Imperfect devices and interconnect show the importance of diagnostics, testing, evaluation, reconfiguration, and other tasks to be implemented through robust software. The flow chart for the systematic synthesis, analysis, optimization, and verification of hardware and software is illustrated in Figure 4.28.

Performance analysis, verification, evaluation, characterization, and other tasks can be formulated and examined only as the ^{M}PPs hardware is devised, synthesized, designed, and evaluated. It is important to start the design process from a high-level explicitly defining the abstraction domain that should:

1. Coherently capture the functionality and performance at all levels
2. Examine and verify the correctness of functionality, behavior and operation of devices, hypercells, nodes, modules, subsystems, and systems
3. Depict the specification of different organizations and architectures examining their adaptability, reconfigurability, performance, capabilities, and so forth.

System-level models describe ^{M}ICs and ^{M}PPs as a hierarchical collection of modules, subsystems, and systems. For example, steady state and dynamics of gates and modules are studied examining how these components perform and interact. The evolution of states, events, outputs, and parameters are of the designer's interest. Different discrete-event, process networks, Petri nets, and other methods have been applied to model ICs. Models, based on synchronous and asynchronous finite-state machine paradigm, ensure the meaningful features and describe the essential behavior in different abstraction domains. Mixed control, data flow, data processing (computing, encryption, filtering, coding, etc.), reconfiguration, defect isolation, and other processes can be modeled.

A program is a set of instructions that one writes to define what the circuit should do. For example, if the IC consists of *on* and *off* logic switches, one can assign that the first and second switches are *off*, while the third to eighth switches are *on* in order to have 8-bit signal 00111111. The program commands millions of switches, and a program should be written in the circuit-level language. For ICs, software developments have progressed to high-level programming languages. A high-level programming language allows one to use a vocabulary of terms, for example, `read`, `write`, or `do`, instead of creating the sequences of *on–off* switching that implements these functions. All high-level languages have their syntax, provide a specific vocabulary, and assign explicitly defined set of rules for using their vocabulary. A compiler is used to translate (interpret) the high-level language statements into machine code. The compiler issues error messages if the programmer uses the programming language incorrectly. This allows one to correct the error and perform other translation by compiling the program. Programming logic is an important issue because it involves executing various statements and procedures in the correct order to produce the desired results. One must use the syntax correctly and execute a logically constructed, sound program. Two commonly used approaches to write computer programs are procedural and object-oriented programming. Through procedural programming, one defines and executes computer memory locations (variables) to hold values and writes sequential steps to manipulate these values. The object-oriented programming is the extension of the procedural programming because it involves creating objects (program components) and creating applications that use these objects. Objects are made up of states that describe the characteristics of an object.

Specific hardware and software solutions must be developed and implemented. For example, ICs are designed by making use of the following hardware description languages (HDLs): Very high speed integrated circuit hardware description language (VHDL), Verilog, and so forth. The designer starts by interpreting the application requirements into organizational/architectural specifications. As the application requirements are examined, the designer translates the organizational/architectural specifications into behavior and structure domains. Behavior representation means the functionality required as well as the ordering of operations and completion of tasks in specified times. A structural description consists of a set of devices and their interconnection. Behavior and structure can be specified and studied using HDLs. These HDLs efficiently manage complex hierarchies that can include millions of logic gates. Another important feature is that HDLs are translated into netlists of library components using synthesis software.

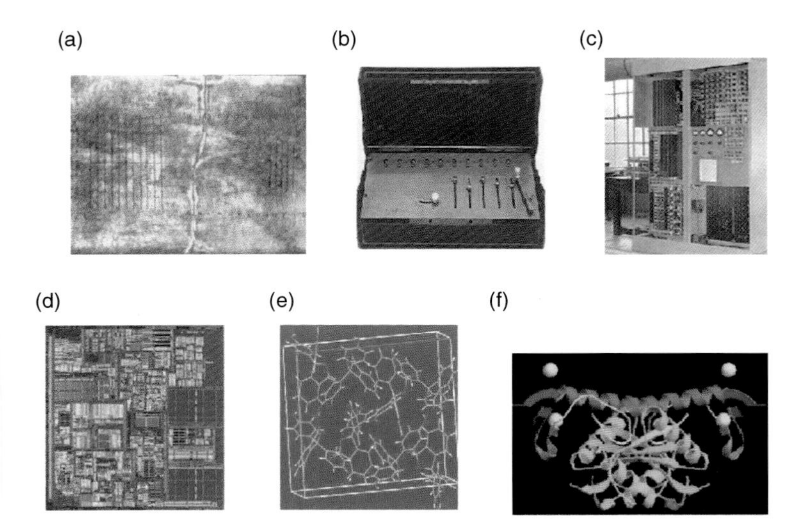

COLOR FIGURE 1.2
From the *abacus* (300 BC) to Thomas' "Arithmometer" (1820), from the electronic numerical integrator and computer (1946) to the 1.5 × 1.5 cm 478-pin Intel® Pentium® 4 processor with 42 million transistors (2002; http://www.intel.com/), and towards 3D *solid* and *fluidic* molecular electronics and processing.

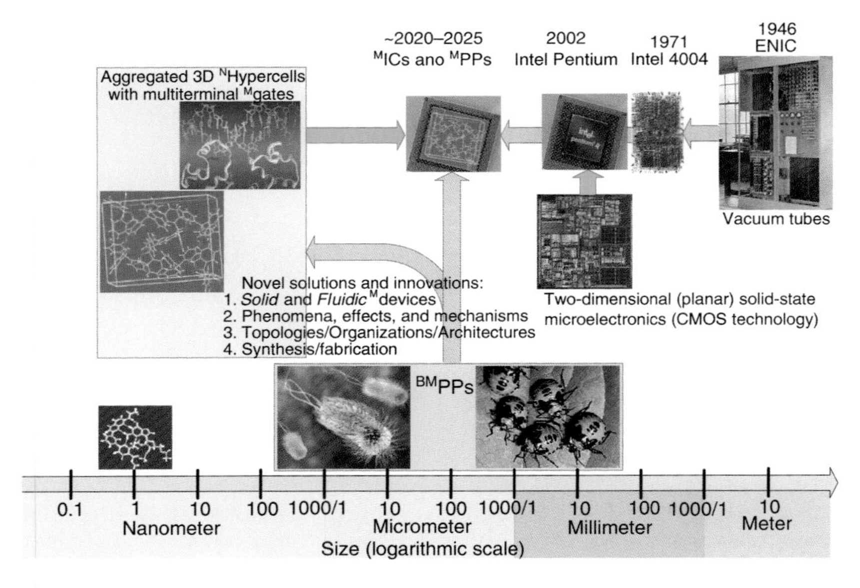

COLOR FIGURE 1.8
Envisioned roadmap: Toward super-high-performance ^{M}ICs and ^{M}PPs.

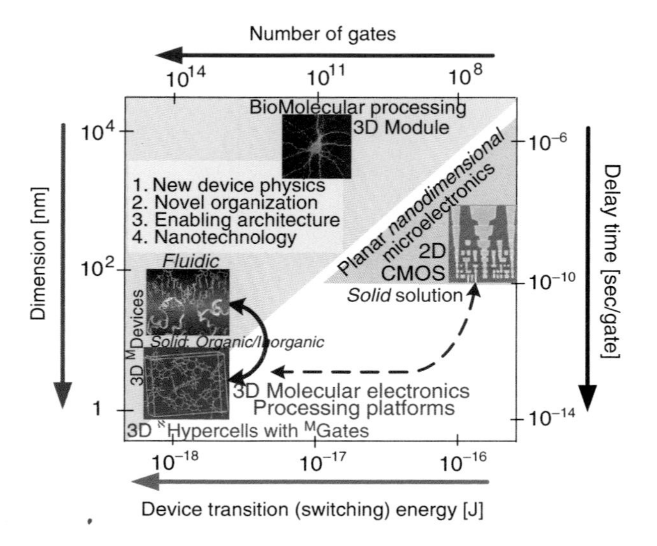

COLOR FIGURE 2.2
Toward molecular electronics and processing/memory platforms. Revolutionary advancements: From two-dimensional (2D) microelectronics to 3D molecular electronics. Evolutionary developments: From ^{BM}PPs to solid and fluidic molecular electronics and processing.

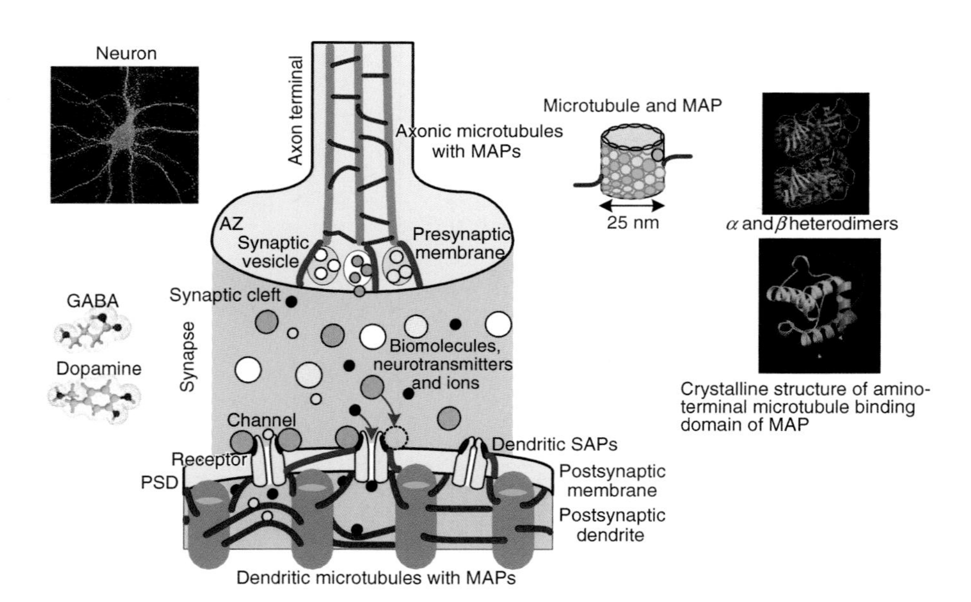

COLOR FIGURE 3.3
Schematic representation of the *axo-dendritic* organelles with presynaptic active zones (AZ) and postsynaptic density (PSD) protein assemblies. Electrochemomechanically induced transitions, interactions, and events in a *biomolecular processing hardware* may result to information processing and memory storage. For example, the *state* transitions results due to binding/unbinding of the *information carriers* (biomolecules and ions). 3D-topology lattice of synapse-associated proteins (SAPs) and microtubules with microtubule-associated proteins (MAPs) may ensure reconfigurable networking, and biomolecules can be utilized as the *routing carriers*.

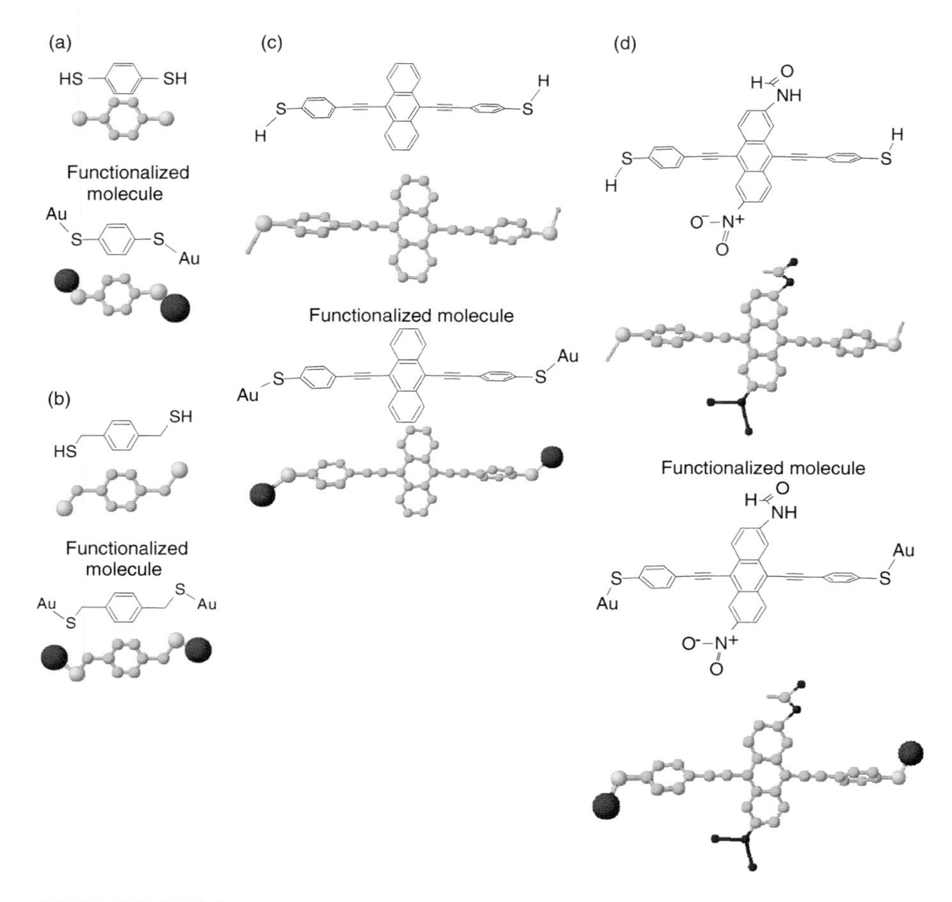

COLOR FIGURE 5.4
Molecules as potential two-terminal ^{ME}devices (atoms are colored as: H—green, C—cyan, N—blue, O—red, S—yellow, and Au—magenta): (a) 1,4-phenyledithiol molecule and functionalized 1,4-phenyledithiol molecule; (b) 1,4-phenylenedimethanethiol molecule; (c) 9,10-*bis*-((2′-para-mercaptophenyl)-ethinyl)-anthracene molecule; (d) 1,4-*bis*-((2′-para-mercaptophenyl)-ethinyl)-2-acetyl-amino-5-nitro-benzene molecule.

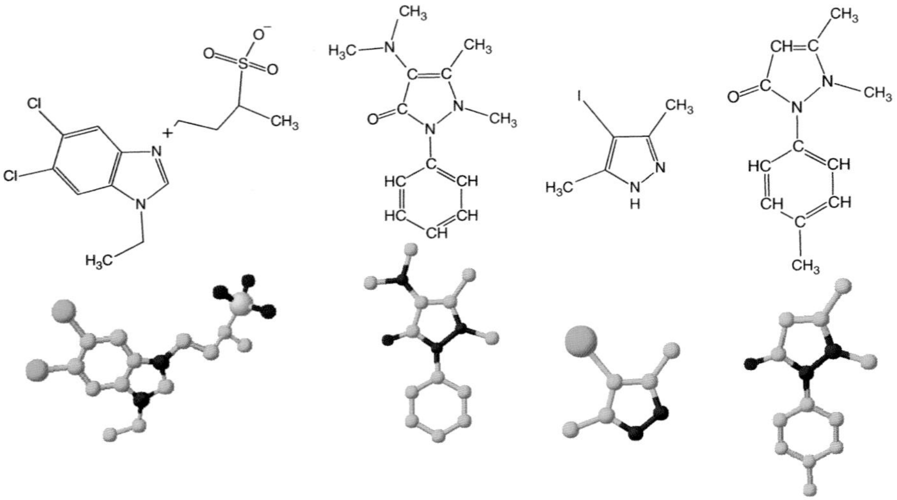

COLOR FIGURE 5.8

Multiterminal molecules, as illustrative ^{ME}devices, with *input*, *control*, and *output* terminals: (a) (2*S*)-4-(5,6-dichloro-1-ethyl-1*H*-3,1-benzimidazol-3-ium-3-yl)-butane-2-sulfonate molecule; (b) 4-(dimethylamino)-1,5-dimethyl-2-phenyl-1,2-dihydro-3*H*-pyrazol-3-one molecule; (c) 4-iodo-3,5-dimethyl-1*H*-pyrazole molecule; (d) 1,5-dimethyl-2-(4-methylphenyl)-1,2-dihydro-3*H*-pyrazol-3-one molecule.

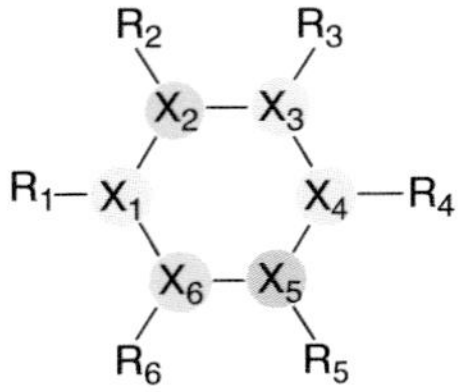

COLOR FIGURE 5.9

Monocyclic molecule as a multiterminal ^{ME}device.

The structural or behavioral representations are meaningful ways of describing a model. In general, HDLs can be used for design, verification, simulation, analysis, optimization, documentation, and so forth. For conventional ICs, VHDL and Verilog are among the standard design tools. In VHDL, a design is typically partitioned into blocks. These blocks are then integrated to form a complete design using the schematic capture approach. This is performed using a block diagram editor or hierarchical drawings to represent block diagrams. In VHDL, every portion of a VHDL design is considered as a block. Each block has an analogous to an off-the-shelf IC called an entity. The entity describes the interface to the block, schematics, and operation. The interface description is similar to a pin description and specifies the inputs and outputs to the block. A complete design is a collection of interconnected blocks. Consider a simple example of an entity declaration in VHDL. The first line indicates a definition of a new entity, while the last line marks the end of the definition. The lines in between, called the port clause, describe the interface to the design. The port clause provides a list of interface declarations. Each interface declaration defines one or more signals that are inputs or outputs to the design. Each interface declaration contains a list of names, mode, and type. As the interface declaration is accomplished, the architecture declaration is studied. As the basic building blocks, using entities and their associated organizations, one can combine them together to perform other designs. The structural description of a design is a textual description of a schematic. A list of components and their connections is called a netlist. In the data-flow domain, ICs are described by indicating how the inputs and outputs of built-in primitive components or pure combinational blocks are connected together. Thus, one describes how signals (data) flow through ICs. The architecture part describes the internal operation of the design. In the data-flow domain, one specifies how data flows from the inputs to the outputs. In VHDL this is accomplished with the signal assignment statement. The evaluation of the expression is performed substituting the values of the signals in the expression and computing the result of each operator in the expression. The scheme used to model a VHDL design is called discrete-event time simulation. When the value of a signal changes, this means that an event has occurred on that signal. The values of signals are only updated when discrete events occur. Because one event causes another, simulation proceeds in rounds. The simulator maintains a list of events that need to be processed. In each round, all events in a list are processed, and any new events produced are placed in a separate list (schedule) for processing in a later round. Each signal assignment is evaluated once, when simulation begins to determine the initial value of each signal to design ICs.

In general, one needs to develop new technology-centric HDLs by coherently integrating novel organizations, enabling architectures, device physics, *bottom-up* fabrication, and other distinctive features of molecular electronics. For ^{M}ICs and ^{M}PPs, there is a need to develop novel software environments that may be organizationally/architecturally neutral or specific. A single software tool unlikely can be used or will be functional to all classes of

^{M}ICs that utilize different hardware (solid, fluidic, or biomolecular) and distinct processing solutions—for example, analog versus digital, binary versus multiple-valued, quantum interactions versus quantum computing, and so forth. The software development and hardware–software codesign for molecular electronics, ^{M}ICs, and ^{M}PPs are formidable tasks to be addressed and solved.

References

1. S. Yanushkevich, V. Shmerko and S. E. Lyshevski, *Logic Design of NanoICs*, CRC Press, Boca Raton, FL, 2005.
2. S. E. Lyshevski, Three-dimensional molecular electronics and integrated circuits for signal and information processing platforms. In *Handbook on Nano and Molecular Electronics*, Ed. S. E. Lyshevski, CRC Press, Boca Raton, FL, pp. 6-1–6-104, 2007.
3. V. D. Malyugin, Realization of corteges of Boolean functions by linear arithmetical polynomials, *Auto. Telemekh.*, no. 2, pp. 114–121, 1984.
4. S. E. Lyshevski, Nanocomputers and nanoarchitectronics. In *Handbook of Nanoscience, Engineering and Technology*, Eds. W. Goddard, D. Brenner, S. Lyshevski and G. Iafrate, CRC Press, Boca Raton, FL, pp. 6.1–6.39, 2002.
5. W. Porod, Nanoelectronic circuit architectures. In *Handbook of Nanoscience, Engineering and Technology*, Eds. W. A. Goddard, D. W. Brenner, S. E. Lyshevski and G. J. Iafrate, CRC Press, Boca Raton, FL, pp. 5.1–5.12, 2003.
6. S. R. Williams and P. J. Kuekes, Molecular nanoelectronics, *Proc. Int. Symp. on Circuits and Systems*, Geneva, Switzerland, vol. 1, pp. 5–7, 2000.

5

Synthesis of Molecular Electronic Devices: Towards Molecular Integrated Circuits

5.1 Synthesis of Molecular Electronic Devices

The fundamentals of molecular electronics and circuits were covered in the previous chapters. For molecular electronics, device- and system-level research is very important. The functionality, performance, and characteristics of ^{M}devices and ^{M}ICs are defined by the phenomena exhibited, effects utilized, capabilities, and so forth. These ^{M}devices and ^{M}ICs must be tested, evaluated, and characterized. To ensure these tasks, ^{M}devices must be synthesized and high-yield *bottom-up* fabrication processes and technologies must be developed.

Molecular devices are comprised of functionalized aggregated molecules. In leaving organisms, biomolecules ultimately establish the *biomolecular processing hardware*. However, the formidable complexity and challenges result in unavailability to utilize *biomolecular hardware* as well as prototype and coherently mimic it. Therefore, we direct focused efforts on solid ^{ME}devices. The directly related problems, such as high-fidelity modeling and data-intensive analysis, are discussed in Chapter 6. It will be shown that, by applying quantum mechanics, one may examine electron transport in atomic complexes in order to evaluate the soundness of ^{ME}devices, assess device functionality, and analyze their performance characteristics. This chapter is aimed at covering the synthesis aspects.

All materials are composed of atoms and molecules. Lithographic microelectronic devices have been fabricated utilizing enhanced-functionality materials through photolithography, deposition, etching, doping, and other processes. In solid-state microelectronic devices, individual atoms and molecules have not been, and cannot be, utilized from the device physics perspective. At the device level, the key differences between molecular and microelectronic devices, as reported in Chapter 1, are: (1) device physics and phenomena exhibited; (2) effects, capabilities, and functionality utilized; (3) topologies and organizations attained; (4) fabrication processes and technologies used. For solid-state devices, using different composites, material science focuses largely on the top-down design in order to engineer enhanced-functionality materials (self-assembled thin films, templates, assemblies, etc.) with the overall goal to ensure the desired characteristics of microelectronic devices [1]. The scaling down of microelectronic devices results in performance degradation due to quantum effects (interference, inelastic scattering, vortices, resonance, etc.), discrete impurities, and other features [2]. In contrast, ^{ME}devices exhibit these phenomena but they are uniquely utilized, ensuring device functionality and guarantying superior capabilities. This ultimately results in novel device physics. One concludes that:

- In microelectronic devices, individual molecules and atoms do not depict the overall device physics and do not define the device performance, functionality, and capabilities.
- In molecular devices, individual molecules and atoms explicitly define the overall device physics depicting the device performance, functionality, capabilities, and topologies.

Reference [3] describes a D–σ–A molecular rectifier, with an electron donor moiety (D) bonded to an electron acceptor moiety (A) through an insulating saturated σ bridge. The small reverse current (for negative voltage) and large forward current (for the positive voltage) result. The nonlinear *I–V* characteristics result due to relative potentials arrangement and HOMOs-LUMOs-Fermi levels of two electrodes and two-terminal molecule leading to electron transport (flow). The first highly polar electronic excited state of D^+–σ–A^- becomes filled, and decays to the less-polar ground state D^0–σ–A^0 by inelastic tunneling through the molecule [3]. The electron flow can be enhanced by intramolecular charge transfer or intervalence transfer mixing of the donor and acceptor states. We assume the existence of an extra intramolecular charge transfer or intervalence transfer absorption band. If the D and A moieties are far apart (the σ bridge is too long), the number of electrons and the current decrease. If the moieties are close, a single mixed ground-state can form, qualitatively and quantitatively changing the *I–V* characteristic. The length of σ is ~2–10 carbon atoms or their equivalent. Different unimolecular rectifiers are documented in [4–9]. For example, [6] reports the experimental results for: (1) γ-hexadecylquinolinium tricyanoquinodimethanide **M**, and, (2) two thioacetyl derivatives of **M**—(*Z*)-α-cyano-β-[*N*-tetradecylthioacetylquinolin-4-ylium)-4-styryl-dicyanomethanide and (*Z*)-α-cyano-β-[*N*-hexadecylthioacetylquinolin-4-ylium)-4-styryl-dicyanomethanide. Other rectifiers were studied, for example: (1) 2,6-di[dibutylamino-phenylvinyl]-1-butylpyridinium iodide; (2) dimethylanilino-aza[C_{60}]-fullerene; (3) fullerene-*bis*-[4-diphenylamino-4″-(*N*-ethyl-*N*-2‴-ethyl)amino-1,4-diphenyl-1,3-butadiene] malonate. The experimental results for monolayers of these molecules, assembled between Au, Ti, Pt, or Al electrodes, exhibit asymmetric

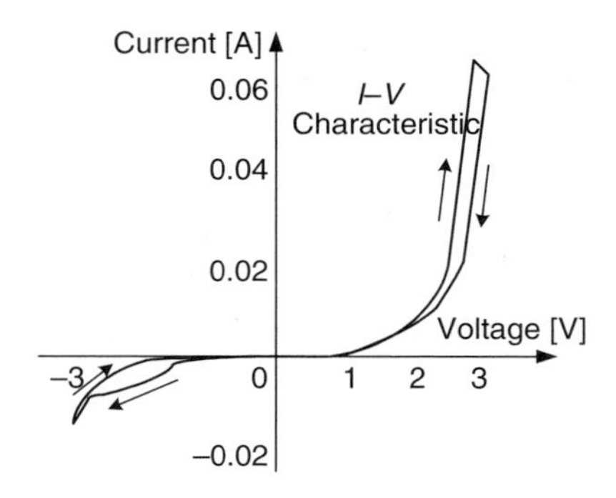

FIGURE 5.1
I–V characteristic for a Au–$C_{16}H_{33}$Q-3CNQ monolayer–Au.

I–V characteristics. For example, the *I–V* characteristics with a hysteresis for a two-terminal Au–monolayer of γ-hexadecylquinolinium tricyanoquinodimethanide molecules ($C_{16}H_{33}$Q-3CNQ)–Au assembly is illustrated in Figure 5.1 [6]. The reader can observe a high current due to the large area of electrodes, where the current flows through the millions of molecules.

For multiterminal ^{ME}devices, novel device solutions, and high-yield affordable fabrication technologies must be developed. One needs to synthesize not monolayer, stand-alone solid or fluidic ^{M}devices, but complex ^{M}ICs. Those ^{M}ICs can be synthesized utilizing molecular self-assembling and robust aggregations as reported in Sections 4.6.2 and 5.3.

The cyclic molecules ensure device physics soundness and provide the desired synthesis capabilities. An aromatic hydrocarbon is a cyclic compound with the sp^2-hybridized atoms in the ring. This molecule with a delocolized π-electron system has free *p*-orbitals ensuring conduction of π-electrons. There are some cyclic hydrocarbon molecules that have $(4n + 2)$ π-electrons, but these molecules are not aromatic because at least one of the carbon atoms within the ring is not sp^2-hybridized. For example, cycloheptatriene has six π-electrons; however, one of the seven carbon atoms is the sp^3-hybridized and the ring is not planar. The ring must be planar in order for the π-electrons to be delocalized in the ring. The planar structure ensures stability and rigidity. Benzene is the most commonly known aromatic hydrocarbon having six π-electrons with all six carbon atoms sp^2-hybridized, and therefore the ring is planar. In particular, the π-system of benzene is formed from six overlapping *p*-orbitals composing π-molecular orbitals with six π-electrons. In cyclic molecules, carbon atoms can be substituted. Figure 5.2 shows a structural and three-dimensional (3D) view of pyridine, pyrrole, furan, and thiophene. The well-known heterocyclic biomolecules, such as purine and pyrimidine, contain nitrogen and oxygen, see Figure 5.2. The derivatives of purine and purimidine can be utilized to synthesize modified nucleotides.

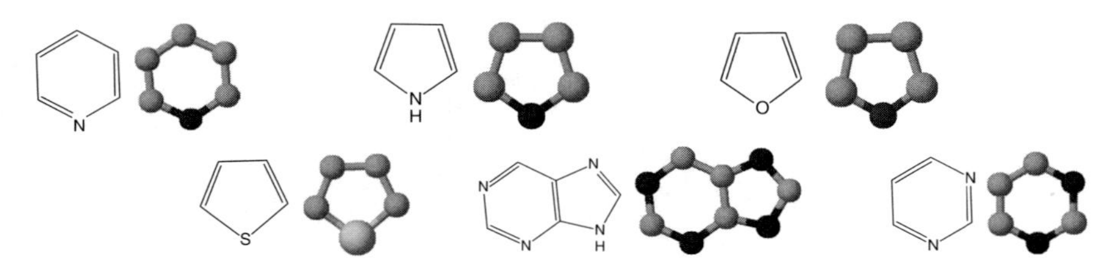

FIGURE 5.2
Pyridine, pyrrole, furan, thiophene, purine, and pyrimidine molecules.

Organic synthesis is the collection of procedures for the preparation of specific molecules and molecular aggregates. In planning the syntheses of desired molecules, the precursors must be selected. A great number of commercial and natural precursors are available. One carries out the retrosynthetic analysis as

```
Target Molecule ⇒ Precursors,
```

where the open arrow denotes "is made from."

Usually, more than one synthetic step is required. For example, one has

```
Target Molecule ⇒ Precursor 1 ⇒ ··· ⇒ Precursor Z
   ⇒ Starting Molecule.
```

A linear synthesis, which is adequate for simple molecules, is a series of sequential steps to be performed, resulting in synthetic intermediates. For complex molecules under our consideration, convergent or divergent synthesis is required. There are different procedures for synthesis of synthetic intermediates. For new synthetic intermediates, the discovery, development, optimization, and implementation steps are needed.

As an example, we report the Hantzsch pyridine (1,4-dihydropyridine dicarboxylate) synthesis as a multicomponent organic reaction between a formaldehyde (CH_2O), two molecules of an ethyl acetoacetate (Et denotes an ethyl C_2H_4), and an ammonium acetate (NH_4OAc) as a nitrogen donor. The initial reaction product is a dihydropyridine that can be oxidized in a subsequent step to a pyridine. Water is used as a reaction solvent, and ferric chloride ($FeCl_3$) leads to aromatization in the second reaction step as shown in Figure 5.3. A cyclic 1,4-dihydropyridine dicarboxylate molecule with side groups results.

Two-terminal molecular diodes and switches are reported in [4–11]. Devising multiterminal ^{ME}devices within a novel device physics is of a great importance. In multiterminal solid ^{ME}device, quantum effects (quantum interaction, quantum interference, quantum transition, vibration, Coulomb effect, etc.) could be used to ensure controlled *I–V* characteristics. The device physics, based on these and other phenomena and effects (electron spin, photon-electron-assisted transitions, etc.), must be coherently

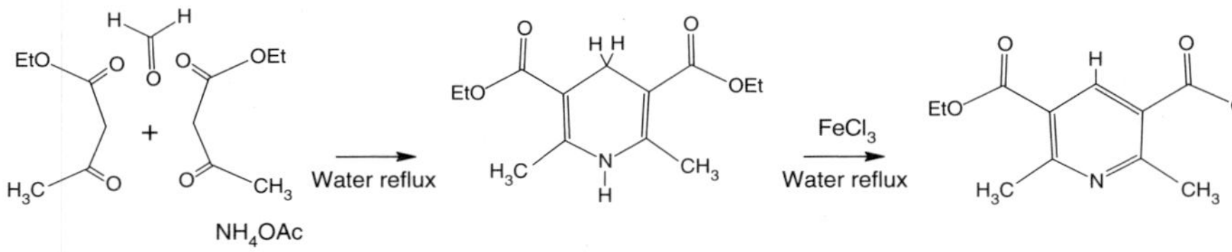

FIGURE 5.3
Synthesis of a cyclic 1,4-dihydropyridine dicarboxylate molecule.

complemented by the *bottom-up* synthesis of the molecular aggregates that exhibit those phenomena. We consider a 3D topology of two- and multiterminal solid ^{ME}devices for which one may utilize controlled electron transport, tunneling, interaction, and so forth.

Distinct solid ^{ME}devices have been proposed, ranging from resistors to multiterminal devices [4–11]. These ^{ME}devices are made of organic, inorganic, and biomolecules. Testing and characterization of some two-terminal ^{ME}devices are reported in [4–11]. Figure 5.4 illustrates different molecules that were thiol-functionalized so as to perform the characterizations measuring the *I–V* and *G–V* characteristics. Sulfur binds to the gold cluster consisting of usually four Au atoms at each binding cite, but Figure 5.4 schematically illustrates only one Au atom.

The density functional theory is used in [12] to examine the geometry, bonding, and energetics of thiol-functionalized molecules to the Au(111) surface. The gold electrodes comprise four-gold-atom clusters covalently bonded to S, and the schematics of the major structural motifs derived from energy minimization [12] are shown in Figure 5.5.

The aggregated molecules, examined in [12], are $^{2s+1}[Au_4{-}\mathbf{X}{-}Au_4]^q$, where $\mathbf{X}{=}S{-}C_8H_8{-}S$, $\mathbf{X}{=}S{-}CH_2{-}C_6H_4{-}CH_2{-}S$ (1,4-dithio-*p*-xylene), $\mathbf{X}{=}S{-}C_6H_4{-}S$ (1,4-phenyledithiol), $\mathbf{X}{=}S{-}C_2H_4{-}S$ (1,2-dithioethane), $\mathbf{X}{=}S{-}C_2H_2{-}S$ (1,2-dithioethylene), and $\mathbf{X}{=}S{-}C_2{-}S$ (dithoacetylene); Au_4 is the cluster of four Au atoms representing the electrode interconnect with the thiol-ended molecule **X**; s is the spin quantum number; q is the net charge on the complex. Here, $q = 0$ (neutral complex) with $s = 1$ and $s = 3$; $q = +1$ (cation) and $q = -1$ (anion) with $s = 2$.

Different types of geometric structures (geometrical motifs) were found in [12]. The derived gold cluster geometries are planar and developed from their initial tetrahedral Au_4 arrangement with the S atom on the three-fold axis equidistant from three Au atoms. The single Au—S and double Au=S bonds result in the Au_4—S complexes. The bond distances are usually in the following range: Au—Au from 2.6 to 3.1 Å; Au—S from 2.35 to 2.6 Å; S—C from 1.6 to 1.9 Å.

One faces significant challenges in the fabrication of experimental test beds (contact–$\sim$1 nm gap–contact) for multiterminal ^{ME}devices, as well as in their functionalization, testing, characterization, evaluation, and so forth. Only a limited number of molecules have been tested and characterized as

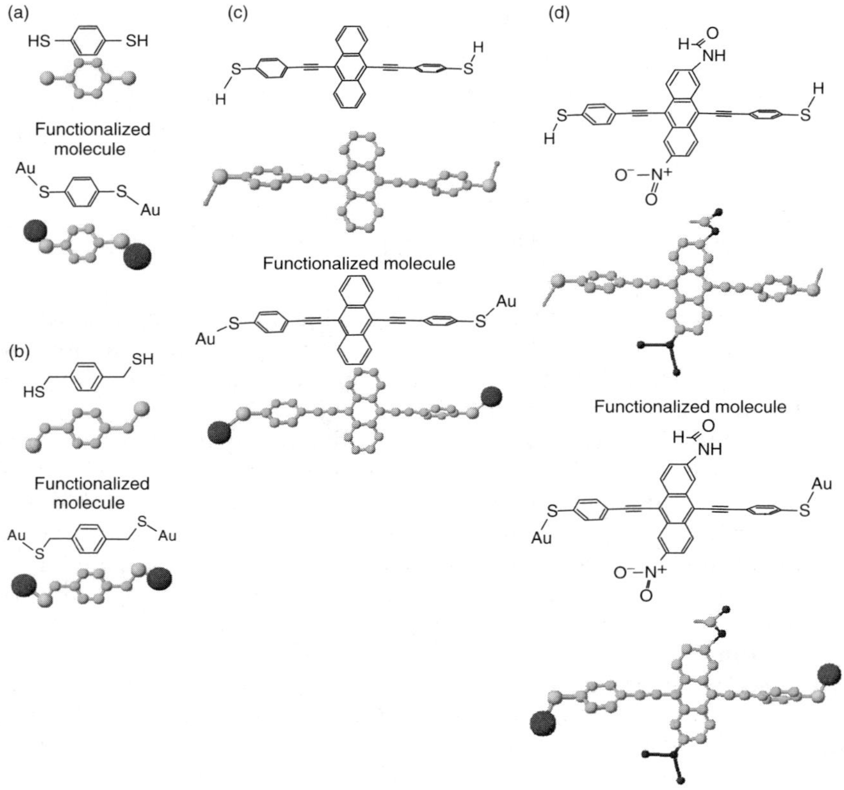

FIGURE 5.4
(See color insert following page 146.) Molecules as potential two-terminal ^{ME}devices (atoms are colored as: H—green, C—cyan, N—blue, O—red, S—yellow, and Au—magenta): (a) 1,4-phenyledithiol molecule and functionalized 1,4-phenyledithiol molecule; (b) 1,4-phenylenedimethanethiol molecule; (c) 9,10-*bis*-((2′-para-mercaptophenyl)-ethinyl)-anthracene molecule; (d) 1,4-*bis*-((2′-para-mercaptophenyl)-ethinyl)-2-acetyl-amino-5-nitro-benzene molecule.

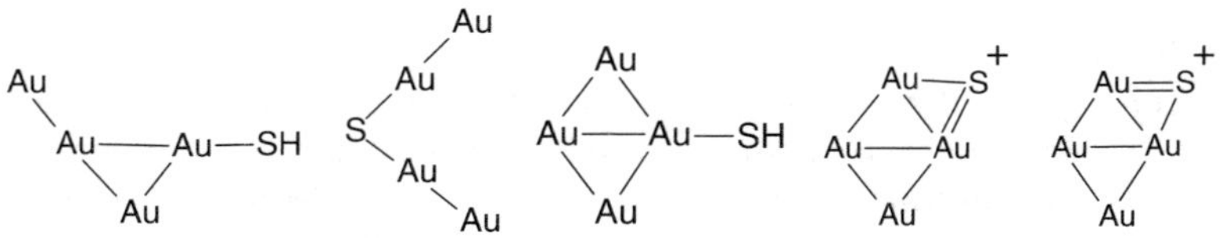

FIGURE 5.5
Gold cluster bonded to S motifs.

two-terminal devices. The major challenges are

1. Significant problems in functionalization of molecules and robust contact–molecule–contact interconnect. In fact, from the device

characterization viewpoint, not all molecules of interest can be thiol-functionalized by a thiol end-group that interacts with the Au(111) surface forming S–Au covalent bonds.

2. Difficulties in fabricating two-terminal (predominantly) test beds with ~1 nm gaps using microelectronic fabrication technologies.
3. *I–V* and other baseline steady-state and dynamic (switching) characteristics are significantly affected by the undesired functionalization and CMOS-fabrication-caused effects—for example, variations of the contact–molecule–contact interconnect (bond length, interbond angle, orbital overlap, etc.) and number of molecules functionalized.

Fabrication challenges do not allow one to test and characterize multi-terminal ^{ME}devices. As an illustration, Figure 5.6 shows a three-terminal 1,3,5-triazine-2,4,6-trithiol molecule.

To characterize a 1,3,5-triazinane-2,4,6-trione ($C_3N_3S_3^{3-}$) molecule (TMT), a functionalizable H_3TMT molecule should be used, as shown in Figure 5.7. A H_3TMT molecule can be prepared by treating the Na_3TMT · $9H_2O$

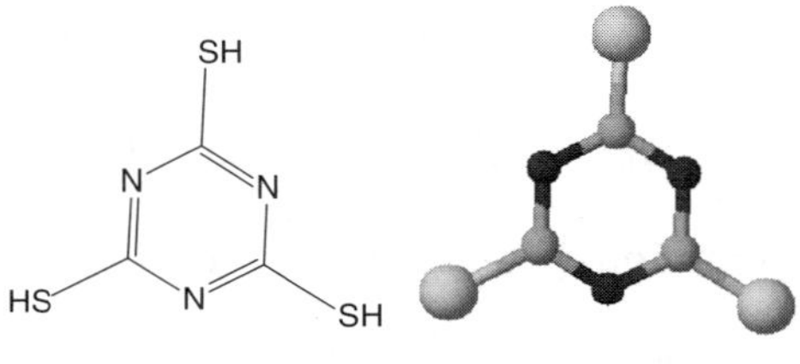

FIGURE 5.6
1,3,5-triazine-2,4,6-trithiol molecule.

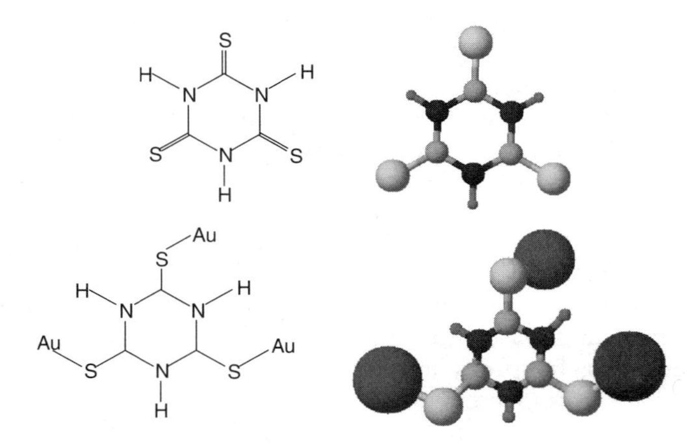

FIGURE 5.7
H_3TMT molecule and functionalized H_3TMT molecule with three Au—S bonds to form three terminals ensuring interconnect.

compound with concentrated hydrochloric acid in a 1:3 molar ratio. A 100 g of $Na_3TMT \cdot 9H_2O$ is dissolved in 350 mL of DI water with subsequent filtering. Then 60 mL of concentrated hydrochloric acid (12.1 *N*) is added to the filtrate. A yellow precipitate is formed immediately, and the mixture is stirred briefly. The precipitate is isolated by filtration, washed by copious amount of DI water, and dried first at room temperature and then at 110°C. The typical yield is ~40 g (91%), mp 230°C. The H_3TMT molecule can be examined by IR spectroscopy, elemental analysis, and x-ray diffraction (XRD) pattern. All reagents should have 95% purity. As the molecule is synthesized for a potential use as a ^{ME}device, testing and characterization of electronic characteristics are of a great importance. The molecule should be functionalized. The TMT molecule ensures stable complexes with transition metals. One can prepare divalent molecule–metal aggregates containing ligands—TMT^{3-}, $HTMT^{2-}$, and H_2TMT^{-}. Using Au and the thiol end-group, one obtains the three-terminal molecular complex, as shown in Figure 5.7. Hence, we depart form the two-terminal ^{ME}devices.

The electronic characteristics of many organic and inorganic molecules do not fully meet desired features because of insufficient controllability, symmetric *I–V* characteristics without the desired current saturation region, thermodynamic sensitivity, and so forth. Electron transport, tunneling, interactions, charge distributions, and other important features are modified by applying the potentials to three terminals. However, the *I–V* characteristics of the functionalized monocyclic H_3TMT and other symmetric molecule without side groups or asymmetry may not exhibit the desired characteristics, such as linear and saturation current regions, robustness, and so forth. Devising, engineering, and analyzing new functional ^{ME}devices are exceptionally important. We depart from the symmetric organic ^{ME}devices, proposing asymmetric multiterminal carbon-centered ^{ME}devices that comprise B, N, O, P, S, I, and other atoms. To ensure synthesis feasibility and practicality, these ^{ME}devices are engineered from cyclic molecules and their derivatives. This section emphasizes that the ^{ME}devices, formed from cyclic molecules with side groups, ensure the synthesis and aggregability features, while Chapter 6 reports that the overall device physics soundness is achieved. The concept is illustrated in Figure 5.8. The multiterminal molecules ensure the desired asymmetry of the *I–V* characteristics, while the saturation region or peaks-and-deeps should be examined.

Consider a multiterminal solid ^{ME}device with controlled electronic characteristics. Due to distinct device physics, one may find it unreasonable to employ the definitions and terminology of solid-state semiconductor transistors, where source–base–drain and emitter–base–collector terms are used for FETs and BJTs. To specify inputs, controls, and outputs, we propose to define the *input*, *output*, and *control* terminals. By applying the voltage to the *control* terminal, one varies the potential, regulates the charge and electromagnetic field, varies the interactions, and changes the tunneling affecting the electron transport. Hence, the input–output characteristics (*I–V* and *G–V*) can be controlled.

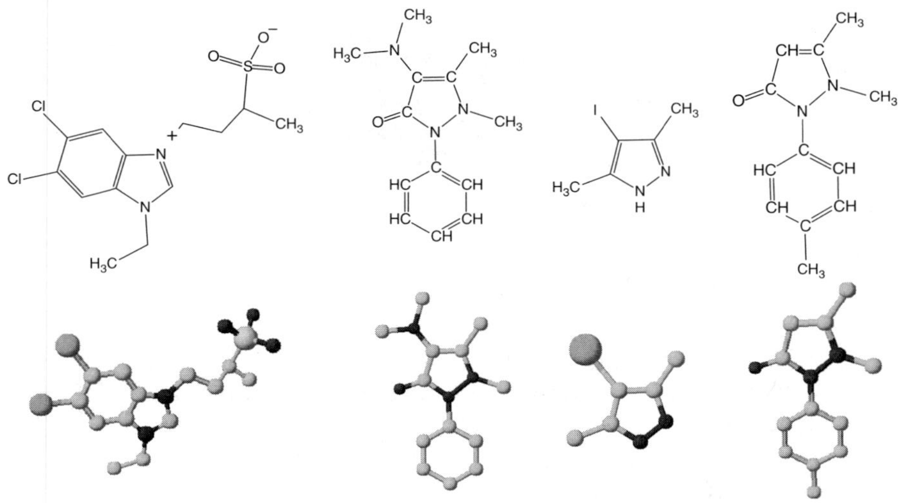

FIGURE 5.8
(See color insert following page 146.) Multiterminal molecules, as illustrative ^{ME}devices, with *input*, *control*, and *output* terminals: (a) (2*S*)-4-(5,6-dichloro-1-ethyl-1*H*-3,1-benzimidazol-3-ium-3-yl)-butane-2-sulfonate molecule; (b) 4-(dimethylamino)-1,5-dimethyl-2-phenyl-1,2-dihydro-3*H*-pyrazol-3-one molecule; (c) 4-iodo-3,5-dimethyl-1*H*-pyrazole molecule; (d) 1,5-dimethyl-2-(4-methylphenyl)-1,2-dihydro-3*H*-pyrazol-3-one molecule.

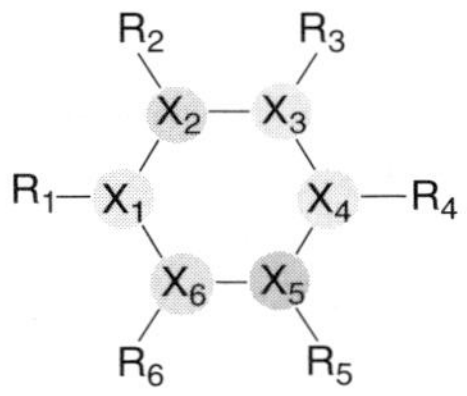

FIGURE 5.9
(See color insert following page 146.) Monocyclic molecule as a multiterminal ^{ME}device.

The monocyclic multiterminal molecule with a side groups is illustrated in Figure 5.9. Here, X_i denotes the specific atoms (B, C, N, O, Al, Si, P, S, Co, Br, etc.); R_i denotes the *input/control/output* terminals T_i and/or side groups: $R_i = (T_i, \text{Side Group}_i)$, $T_i = (T_{k\ \text{input}}, T_{l\ \text{control}}, T_{m\ \text{output}})$.

The use of specific atoms and side groups is defined by the device physics, synthesis, aggregability, and so forth. The aggregation and interconnect of *input/control/output* terminals can be accomplished within the carbon framework. The reported ^{ME}devices possess the quantum-effect device physics and exhibit and utilize quantum effects and transitions. For example:

1. The electron transport is predefined or significantly affected by X_i and side groups.

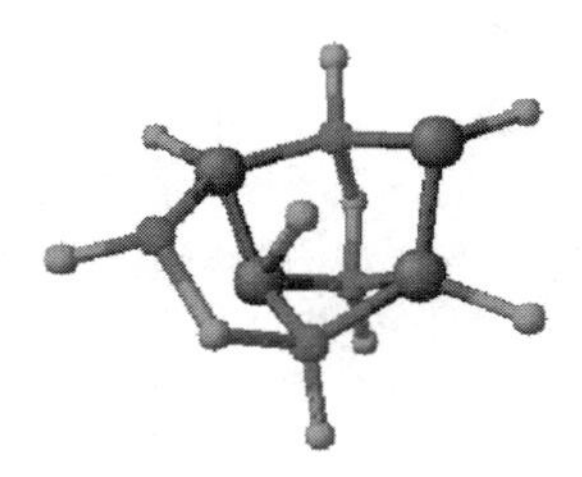

FIGURE 5.10
Molecular cage as a multiterminal ^{M}device.

2. The atomic structure of side groups can exhibit transitions or interactions under the external electromagnetic excitations and thermal gradient.
3. Side groups can be utilized as electron-*donating* and electron-*withdrawing* substituent groups, as well as interacting or interconnect groups.

The three-terminal quantum-effect ^{ME}device is documented in Section 6.9. The device aggregability and interconnect features are enhanced by utilizing side groups. The documented ^{ME}devices can be employed in solid, fluidic, and hybrid molecular electronics. A three-terminal monocyclic molecule was used to design ^{M}gates within the ^{ME}D–^{ME}D logic family, as shown in Figure 4.6.

Organometallic molecules, such as trimethylaluminum $(CH_3)_3Al$, triethylborane $(CH_3CH_2)_3B$, tetraethylstannane $(CH_3CH_2)_4Sn$, ethylmagnesium bromide CH_3CH_2MgBr, and so forth can be potentially utilized in molecular electronics. As an alternative solution, a 3D-topology molecular cage with carbon interconnects, as a multiterminal ^{M}device, is shown in Figure 5.10.

5.2 Testing and Characterization of Proof-of-Concept Molecular Electronic Devices

To date, some proof-of-concept two-terminal ^{ME}devices have been characterized and their *I–V* characteristics are measured [4–11]. To fabricate characterization test beds, conventional microelectronic fabrication techniques, processes, and materials are used. Horizontal and vertical gaps with separation between contacts in the range from ~one to tens of nanometers were fabricated using photolithography, deposition, etching, and other processes. High-resolution photolithography defines planar (two-dimensional) pattern and profile, thereby allowing one to achieve the specified patterns of insulator, metal, and other materials on the silicon wafer. Using photolithography, the mask pattern is transferred to a photoresist that is used to transfer the pattern to the substrate and distinct layers on it using sequential processes, including deposition and etching. Chemical and physical vapor deposition processes are used to deposit different insulators and conductors, while sputtering and evaporation are used to deposit Au, Pd, Ti, Cr, Al, and other metals. Wet chemical etching and dry etching are used to etch materials. Different etchants ensure desired vertical and lateral etching. Deep trenches and pits can be etched in a variety of materials, including silicon, silicon oxide, silicon nitride, and so forth. A combination of dry and wet etchings is integrated with materials ensuring etching selectivity, vertical (planar) and lateral (wall) profile control, etch rate ratio control, uniformity, and so forth. For the *anisotropic* etching one uses etchants (potassium hydroxide, sodium hydroxide, ethylene-diamine-pyrocatecol, etc.) which etch different crystallographic directions at different etch rates. In contrast, *isotropic* etching ensures the same (or close) etch rate in all directions. Different etch-stop materials are used, and these etch-stop layers can be sacrificial or structural. Shape, profile, thickness, and other features are controlled. The use of different *structural* and *sacrificial* materials, combined with etching and deposition processes, provides one with the opportunity to fabricate characterization test beds. Molecules to be examined must be functionalized with the metals forming robust contacts.

As a representative illustration, Figure 5.11 provides a cross-sectional view of a test bed to characterize two-terminal ^{ME}devices. Chromium and gold are sequentially evaporated on the insulators. The electron-beam gold evaporation with adhesion layer (Cr or Ti) is a well-established process for depositing a gold layer with the desired thickness and uniformity. Through the lateral etching of insulator 2, a "gap" is engineered. If needed, unwanted Cr (near gap) can be removed using Cr etchants. Figure 5.11 does not reflect the dimensionality as well as thickness of the insulators (silicon oxide, silicon dioxide, silicon nitride, aluminum oxide, zirconium oxide, or other high-*k* dielectrics), adhesive (Cr or Ti), and contact/pad (Au) layers. Distinct molecules, to be characterized as ^{ME}devices, can be functionalized to the evaporated gold or titanium layers using the thiol end group. A functionalized 1,4-phenylenedimethanethiol molecule is shown in Figure 5.11. The gap separation can be controlled by varying the processes (deposition and etching time, concentration, density, temperature, etc.). The separation between the

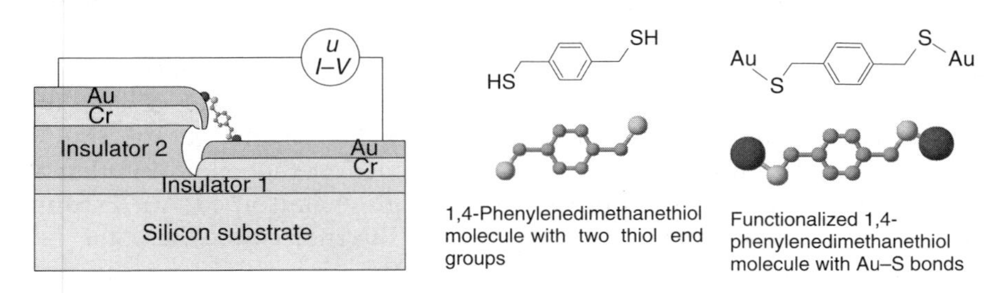

FIGURE 5.11
Characterization test bed with a 1,4-phenylenedimethanethiol molecule functionalized to Au contacts.

gold layers must match the functionalized molecule geometry and Au–**X**–Au length. For 1,4-benzenedimethane-thiols, the separation should be ≤1.2 nm to form a contact–molecule–contact assembly. After fabrication, a test bed is cleaned in the Ar/O_2 plasma, rinsed with ethanol, and stored in a glove box to avoid the oxidation of Au. The molecular deposition (functionalization) involves immersion of a test bed in a 1,4-phenylenedimethanethiol solution (1–10 mM in ethanol) and soaking for ~20 h. Following the ethanol rinse, the *I–V* and *G–V* characteristics are measured [10]. Many effects significantly affect the electronic characteristics, that is, attachment of multiple functionalized molecules to contacts, variation of the contact–S bonds, tunneling, leakage, electrostatic phenomena, and so forth. Molecules are attached by thiol (—SH) group that adsorbs to the gold lattice. The thiol group ensured conduction between metal and molecule. Although thiol is the most common end-group for the attachment of molecules to metals, it may not form the desired coherency for testing and characterization of ^{ME}devices. For example, the geometry of the S-orbitals may not ensure the conjugated π-orbitals from the molecule to interact strongly with the conduction orbitals of metal. The orbitals' mismatch creates an energy barrier at each bonding terminal, significantly impacting electron transport. One also should avoid or minimize the surface oxidation and side reactions effects.

Other processes to fabricate the so-called step junction were reported in [10] with the positive slope formed using the AZ-1518 photoresist. Chromium was used as a sacrificial layer. The electrodes are formed using Ti and Au.

Electromigration-induced, mechanical and electrical break "nanogaps" have been used. Alternative solutions are "nanopore," "nanoimprint," crossed wire, and so forth. The step and electromigration-induced gaps are relatively easy to fabricate and characterize using microscopy. An electromigration-induced break-junction technique at room temperature is reported in [11]. Photolithographically defined Au electrodes are evaporated on the oxide-coated silicon substrate silanized with (3-mercaptopropyl) trimethoxysilane. The subsequent electromigration procedure is carried out at room temperature to create a ~1–2 nm gap between two Au electrodes. The electromigration process is affected by the local Joule heating, melting, surface tension, migrating ions, electron forces, and so forth. The dissipated power per volume is estimated as $J^2\rho$, where J is the current density, which is a function of the cross-sectional area; ρ is the resistivity. The threshold current density is found to be from 1 to 2.5 A/μm^2 [11]. The test bed with the electromigration-induced break junction is cleaned in Ar/O_2 plasma and rinsed with ethanol to remove the oxides from Au. Finally, the substrate is immersed in a 1 mM solution of 1,4-phenylenedimethanethiol in ethanol. The Au–molecule contacts are formed through chemisorbed Au—S coupling, which forms contacts at both ends of the molecule. Then, the *I–V* characteristics are measured, as reported in [11].

The application of different biomolecules and modified biomolecules for molecular electronics was considered in [2,13]. The *I–V* characteristics of DNA were reported in [13]. Three different short (~5.4 nm) double-stranded DNA (dsDNA), functionalized using short oligonucleotide linkers and thiol end-groups, were examined. In particular, [13] documents the experimental results when 15 base-pair single-stranded oligonucleotides, **X**′-(CCGCGCGCCCGCCCG)-5′ with a complementary **X**′,**Y**3′-(CCGCGTTTTGCCCG)-5′ with **Y**′, and **Z** 3′-(GCCTCTCAACTCGTA)-5′ with **Z**′ were hybridized to form dsDNA. They were immobilized and functionalized to the gold electrodes using the $-(CH_2)_3SH$ and $-(CH_2)_6SH$ oligonucleotide linkers to their 3′ and 5′ ends. The electromigration-induced-break-gap test bed with ~10 nm gap was used to test the functionalized dsDNA. The uncertainties in the testing and characterization were emphasized. The quantitative results, which are of interest, indicate that for the applied voltage ±1.2 V, the current in the **X**–**X**′ dsDNA is ±0.35 nA, while the current in **Y**–**Y**′ dsDNA is ±0.065 nA. There was no current measured in the random paired **Z**–**Z**′ dsDNA.

It is difficult to make a conclusive assertion on the feasibility and soundness of DNA-centered ^{ME}devices; however, the use of double- or single-stranded DNA may not be a very promising direction owing to inadequate electronic characteristics and other limits emphasized in Section 3.9.4. The more promising direction is the use of cyclic or other multiterminal molecules with side groups, as emphasized in Section 5.1. The application of modified nitrogenous bases is an example of using the proposed concept. Our solution allows one to engineer functional ^{ME}devices with sound device physics ensuring desired characteristics, functionality, and aggregability features.

5.3 Molecular Integrated Circuits and Design Rules

This section extends the results of Section 4.6.2 by making use of the fabrication aspects covered. The synthesis of multiterminal carbon-centered ^{ME}devices was reported in Section 5.1. The *bottom-up* fabrication should provide techniques to engineer not only stand-alone ^{ME}devices but also ^{M}gates, *modular* $^{\aleph}$hypercells, $^{\aleph}$hypercells aggregates, node *lattices*, and ^{M}IC. In particular, combinational logics can be implemented using molecular multiplexers or molecular EXOR (^{M}EXOR) gates, while the memories can be realized applying ^{M}NAND or ^{M}NOR gates. There are procedures to synthesize complex molecular aggregates progressing from ^{M}gates to modular $^{\aleph}$hypercells. The ^{M}PPs are envisioned to be fabricated by aggregating modular $^{\aleph}$hypercells through the robust controlled synthesis and assembly, as reported in Section 4.6.2. For example, ^{M}ICs can be synthesized and implemented as cross-bar *fabrics*, as shown in Figure 4.23. The promising molecular interconnecting and interfacing solutions are under developments utilizing the *chemical-bond fabrics* and *energy-based* solutions.

The integration of molecular electronics and microelectronics is very important to ensure ICs–^{M}ICs–ICs interconnect and interface, as well as to test and characterize devices, modules, and systems. As was emphasized,

^{M}ICs will not employ microelectronic-centered interconnect solutions. There is no need to utilize the thiol and other end-groups and/or linkers to accomplish the contact–molecule–contact interconnect, as reported for a proof-of-concept device testing, characterization, and evaluation in Section 5.2. It is impractical to guarantee ICs–^{M}ICs–ICs interconnect and interface applying conventional technologies. For the envisioned ^{M}ICs, at the module and system levels, chemical, optical, electromagnetic, quantum, and other high-end I/O interconnect and interfacing paradigms and technologies are under the development, for example, the *chemical-bond fabrics* and *energy-based* ones.

Synthetic chemistry allows one to synthesize a wide range of complex molecules from atoms linked by covalent bonds. Utilizing noncovalent and covalent intermolecular interactions, as well as precisely controlling spatial (structural) and temporal (dynamic) features, supramolecular chemistry provides methods to synthesize even more complex atomic aggregates resulting in the ability to implement $^{\aleph}$hypercells. The molecular recognition is based on well-defined interaction patterns (hydrogen bonding arrays, sequences of donor and acceptor groups, ion coordination sites, etc.). One can design preorganized molecular receptors capable of binding specific substrates with high efficiency, robustness, and selectivity. The major features of the supramolecular noncovalent synthesis are

1. Molecular recognition based on molecular reactivity, catalysis, and transport.
2. Pairing and templating.
3. Controllable robust self-assembly that results in random, near-random, and directed ordering of molecular primitives.
4. Adaptive hierarchical self-organization (generation of well-defined. organized, and functional supramolecules by self-assembly).
5. Accurate entities positioning with the post-assembly modification through covalent bond formation.
6. Synthesis of interlocked molecular aggregates.
7. Programmable preorganization.
8. Recognition based on specific interaction patterns.
9. Self- and complementary-selection and compliance with self-recognition.

A self-organization process involves three main steps:

- Molecular recognition for the selective binding of the basic components.
- Growth through sequential and hierarchical binding of multiple components in the correct relative disposition.
- Termination of the process using a built-in feature.

The self-organization should be stable towards interfering interactions (metal coordination, van der Waals stacking, etc.) and robust towards modification of parameters (concentration and stoichiometry of the components, presence of other species, etc.). Multimode coordinated self-organization provides additional features. The ^{M}ICs and ^{M}PPs are envisioned to be synthesized utilizing the *bottom-up* fabrication and design rules.

Feasibility Analysis and Concluding Remarks: Biomolecular processing platforms and molecular electronics provide undisputable evidence of their superiority, surpassing any envisioned microelectronics solutions. In general, fluidic ^{M}devices offer a broader class of physics and phenomena to be utilized as compared with solid ^{ME}devices. However, taking into account existing and prospective technologies, from the fabrication viewpoint, it seems that in the near future solid molecular electronics may ensure a greater degree of feasibility. In biosystems, *biomolecular processing hardware*, which implements ^{BM}PPs, is synthesized through robustly controlled molecular assembling, which is far beyond even envisioned comprehension and synthesis capabilities. Although Chapter 5 was primarily focused on solid ^{ME}devices, the fluidic and biomolecular solutions are also of an importance taking into consideration the steady progress in biomolecular technologies and fundamental advances, as reported in Chapter 3.

References

1. S. M. Sze and K. K. Ng, *Physics of Semiconductor Devices*, John Wiley & Sons, NJ, 2007.
2. S. E. Lyshevski, Three-dimensional molecular electronics and integrated circuits for signal and information processing platforms. In *Handbook on Nano and Molecular Electronics*, Ed. S. E. Lyshevski, CRC Press, Boca Raton, FL, pp. 6-1–6-104, 2007.
3. A. Aviram and M. A. Ratner, Molecular rectifiers, *Chem. Phys. Letters*, vol. 29, pp. 277–283, 1974.
4. J. Chen, T. Lee, J. Su, et al., Molecular electronic devices. In *Handbook Molecular Nanoelectronics*, Eds. M. A. Reed and L. Lee, American Science Publishers, Stevenson Ranch, CA, 2003.
5. J. C. Ellenbogen and J. C. Love, Architectures for molecular electronic computers: Logic structures and an adder designed from molecular electronic diodes, *Proc. IEEE*, vol. 88, no. 3, pp. 386–426, 2000.
6. R. M. Metzger, Unimolecular electronics: results and prospects. In *Handbook on Molecular and NanoElectronics*, Ed. S. E. Lyshevski, CRC Press, Boca Raton, FL, pp. 3-1–3-25, 2007.
7. J. Reichert, R. Ochs, D. Beckmann, H. B. Weber, M. Mayor and H. V. Lohneysen, Driving current through single organic molecules, *Phys. Rev. Lett.*, vol. 88, no. 17, 2002.

8. J. M. Tour and D. K. James, Molecular electronic computing architectures, In *Handbook of Nanoscience, Engineering and Technology*, Eds. W. A. Goddard, D. W. Brenner, S. E. Lyshevski and G. J. Iafrate, CRC Press, Boca Raton, FL, pp. 4.1–4.28, 2003.
9. W. Wang, T. Lee, I. Kretzschmar and M. A. Reed, Inelastic electron tunneling spectroscopy of an alkanedithiol self-assembled monolayer, *Nano Lett.*, vol. 4, no. 4, pp. 643–646, 2004.
10. K. Lee, J. Choi and D. B. Janes, Measurement of *I–V* characteristic of organic molecules using step junction, *Proc. IEEE Conf. on Nanotechnology*, Munich, Germany, pp. 125–127, 2004.
11. A. K. Mahapatro, S. Ghosh and D. B. Janes, Nanometer scale electrode separation (nanogap) using electromigration at room temperature, *Proc. IEEE Trans. Nanotechnol.*, vol. 5, no. 3, pp. 232–236, 2006.
12. H. Basch and M. A. Ratner, Binding at molecule/gold transport interfaces. V. Comparison of different metals and molecular bridges, *J. Chem. Phys.*, vol. 119, no. 22, pp. 11926–11942, 2003.
13. A. K. Mahapatro, D. B. Janes, K. J. Jeong and G. U. Lee, 'Electrical behavior of nano-scale junctions with well engineered double stranded DNA molecules, *Proc. IEEE Conf. on Nanotechnology*, Cincinnati, OH, 2006.

6

Modeling and Analysis of Molecular Electronic Devices

6.1 Atomic Structures and Quantum Mechanics

Atomic and subatomic systems have been studied for centuries. Matter is made of atoms, and their properties and phenomena exhibited depend on the atomic structure. These are examined by applying the corresponding physical laws. Recalling Rutherford's structure of the atoms, we will view atoms omitting some details because only three subatomic *microscopic* particles (protons, neutrons, and electrons) have significance within the subject of this book. The nucleus of the atom bears the major mass, and it contains positively charged protons and neutral neutrons. It occupies a small atomic volume compared with the cloud of negatively charged electrons, which are attracted to the positively charged nucleus by the force that exists between the *microscopic* particles of opposite electric charge. In an atom of a particular element, the number of protons is the always same, but the number of neutrons may vary. An atom has no net charge due to the equal number of positively charged protons in the nucleus and negatively charged electrons around it. For example, carbon has six protons and six electrons. If electrons are lost or gained by the neutral atom in a chemical reaction, a charged particle, called an ion, is formed. The illustrative two-dimensional representation of the carbon atom with the orbiting electrons is depicted in Figure 6.1a. The orbitals shown do not define the electrons' pathway in three-dimensional (3D) space. However, the electron orbitals are three dimensional with the varying probability of being found in the given volume (position), as will be documented in this chapter.

There is an endless diversity of organic molecules, and the versatility of carbon in molecular assemblies is fascinating. Valence is the number of bonds an atom can form. The number of bonds is equal to the number of electrons required to complete the valence (outermost) electron shell. For $_6C$, the electron configuration is $1s^22s^22p^2$. Atoms with incomplete valence shells interact with certain other atoms to complete the valence shell by sharing electrons. A covalent bond is the sharing of a pair of valence electrons by

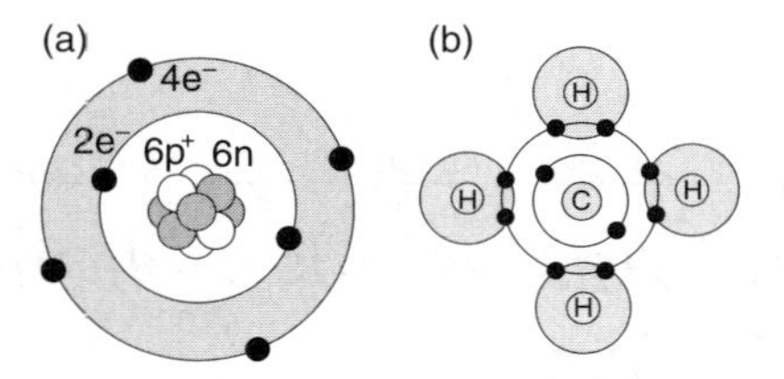

FIGURE 6.1
(a) Illustrative two-dimensional representation of electron configuration for a carbon atom ($_6$C): Six electrons (e^-, black dots), orbiting the nucleus, occupying two shells. Electrons in each energy level (filling electron shell) are represented as dots on concentric rings. Electrons in the first shell, closest to the nucleus, have the lowest energy. The four electrons in the second shell have higher energy. An electron can change its shell by absorbing or losing an amount of energy equal to the difference in potential energy between the original and final shell. Six protons (p^+, grey) and six neutrons (n, white) are in centrally located in a nucleus. (b) Methane is formed from one carbon and four hydrogen atoms forming covalent bonds.

two atoms. Hydrogen has one valence electron in the first shell, but the shell capacity is two electrons. Therefore, for example, four hydrogen atoms satisfy the valence (atoms share valence electrons) of one carbon atom forming methane (the molecular formula is H–C(H)(H)–H); see Figure 6.1b.

With six electrons in its second electron shell, oxygen ($_8$O), with the electron configuration $1s^2 2s^2 2p^4$, needs two more electrons to complete this valence shell. Two oxygen atoms form a molecule by sharing two pairs of valence electrons, and double covalent bonds are formed, that is, O=O.

Each atom sharing electrons has a bond capacity (number of covalent bonds that must be formed for the atom to have a full complement of valence electrons). The bond capacity is called the atom's valence. Carbon $_6$C has four valence electrons. Therefore, its bond capacity (or valence) is four. The valences for hydrogen, oxygen, nitrogen, and carbon are 1, 2, 3, and 4, respectively.

To study atoms and molecules, subatomic *microscopic* particles and systems are examined using quantum theory. This guarantees quantitative and qualitative analyses. Erwin Schrödinger in 1926 derived an equation that models and describes the wave nature of microscopic particles. The solution of the Schrödinger equation gives the wave functions and corresponding energies related to the orbitals. A collection of orbitals with the same principal quantum number, which describes the orbit, is called the electron shell. Each shell is divided into the number of subshells with the same principal quantum number. Each subshell consists of a number of orbitals. The Pauli exclusion principle states that each shell may contain only two electrons of the opposite spin. When the electron is in the lowest energy orbital, the atom is in its ground state. When the electron enters a higher orbital, the atom is in an excited state. To move the electron to the excited-state orbital, a photon of the appropriate energy should be absorbed.

When the size of the orbital increases, the electron spends more time farther from the nucleus, possesses more energy, and is less tightly bound to the nucleus. The outermost shell is the valence shell. The electrons that occupy it are referred to as valence electrons. Inner-shell electrons are called the core electrons. The valence electrons contribute to the bond formation between atoms when molecules are formed. When electrons are removed from the electrically neutral atom, a positively charged cation is formed. These electrons possess higher ionization energies (measures the ease of removing the electron from the atom) and occupy the energetically weakest orbital, which is the most remote orbital from the nucleus. The valence electrons removed from the valence shell become free electrons, transferring the physical quantities from one atom to another.

The electric conductivity of bulk media is predetermined by the density of free electrons. Good conductors have the free electron density $\sim 1 \times 10^{23}$ free electrons/cm^3. In contrast, in good insulators there is ~ 10 free electrons/cm^3. The free electron density of semiconductors is from 1×10^7 to 1×10^{15} free electron/cm^3 (the free electron concentration in silicon at 25°C and 100°C are 2×10^{10} and 2×10^{12}, respectively). The free electron density is determined by the energy gap between valence and conduction (free) electrons. That is, the properties of the media (conductors, semiconductors, and insulators) are determined by the atomic structure. However, the phenomena and effects exhibited by the bulk material and the molecules (or atoms) are profoundly different. For microscopic systems and particles (electron, atoms, molecules, molecular aggregates, etc.), classical physics and mechanics should be applied with a great deal of caution, as will be discussed. The dynamics of particles can be modeled utilizing distinct paradigms and applying various modeling concepts that result in different equations of motion. Newtonian and quantum mechanics employ different quantities, variables, and parameters. Microscopic particles exhibit wave dynamics. Waves are described using wavelength λ, angular frequency ω, amplitude A, velocity v, and so forth. These parameters are related to each other. For example, $\lambda = c/v$ and the energy of a wave is a function of v and A.

In 1900 Max Planck discovered the effect of quantization of energy, and the radiated (emitted) energy is given by the Planck's quantization law

$$E = nhv, \quad n = 0, 1, 2, \ldots,$$

where n is the non-negative integer, $n = 0, 1, 2, \ldots$; h is the Planck constant, $h = 6.62606876 \times 10^{-34}$ J sec or $h = 4.13566727 \times 10^{-15}$ eV sec; v is the frequency of radiation, $v = c/\lambda$; c is the speed of light, $c = 299{,}792{,}458$ m/sec; λ is the wavelength, $\lambda = c/v$.

The modified Planck constant, $\hbar = (h/2\pi) = 1.054571596 \times 10^{-34}$ J sec or $\hbar = 6.58211889 \times 10^{-16}$ eV sec, is frequently used.

The observation of discrete energy spectra suggests that each *microscopic* particle has the energy hv (radiation results due to N such particles), and such a particle is called a *photon*. Figure 6.2a demonstrates the photon emission

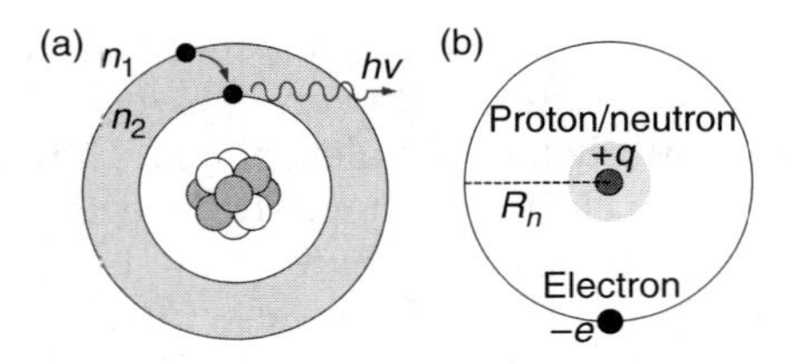

FIGURE 6.2
(a) An electron jumps emitting a photon; (b) hydrogen atom: uniform circular motion of a single electron.

when an electron jumps from the state (orbital) n_1 to the state n_2. The radiated energy is hv. The *photon* has the momentum

$$p = hv/c = h/\lambda.$$

Niels Bohr developed the model of the hydrogen atom. Figure 6.2b illustrates that the electron possess the planetary-type orbits. Electrons can be excited to an outer orbit and can "fall" to the inner orbits. Therefore, to carry out the analysis, Bohr postulated that the electron has certain stable circular orbits (the orbiting electron does not produce radiation because otherwise the electron would lose its energy and change its path); the electron changes orbitals with higher or lower energies by receiving or radiating a discrete amount of energy.

To attain the uniform circular motion, the electrostatic (Coulomb) force must be equal to the radial force. The radius R_{n} is found examining the energies.

The attractive Coulomb force is

$$F_{\mathrm{C}} = \frac{1}{4\pi\varepsilon_0}\frac{q_{\mathrm{electron}}q_{\mathrm{nucleus}}}{R_{\mathrm{n}}^2}.$$

For a single-electron hydrogen atom, taking into the account the centripetal acceleration v^2/R_{n}, one has

$$F_{\mathrm{C}} = \frac{1}{4\pi\varepsilon_0}\frac{e^2}{R_{\mathrm{n}}^2} = \frac{mv^2}{R_{\mathrm{n}}}.$$

The kinetic energy is

$$\Gamma = \frac{1}{2}mv^2 = \frac{e^2}{8\pi\varepsilon_0 R_{\mathrm{n}}}.$$

By making use the potential energy of the electron–nucleus system (Coulomb potential energy)

$$\Pi = -\frac{e^2}{4\pi\varepsilon_0 R_{\mathrm{n}}},$$

The total energy is

$$E = \Gamma + \Pi = -\frac{e^2}{8\pi\varepsilon_0 R_\text{n}}.$$

Analyzing the angular momentum of electron with respect to nucleus ($\mathbf{L} = \mathbf{r} \times \mathbf{p}$), one finds that $\mathbf{r}$ is perpendicular to $\mathbf{p}$. Hence, $L = rp = mvr$. Bohr postulated that $mvr = n\hbar$. Therefore, the kinetic energy is expressed as

$$\Gamma = \frac{1}{2}mv^2 = \frac{1}{2}m\left(\frac{n\hbar}{mR_\text{n}}\right)^2.$$

One obtains the following expression for the radius of the nth orbital:

$$R_\text{n} = \frac{4\pi\varepsilon_0\hbar^2}{me^2}n^2 = r_0 n^2.$$

Here, the Bohr radius is

$$r_0 = \frac{4\pi\varepsilon_0\hbar^2}{me^2} = 5.291772082 \times 10^{-11}\ \text{m}.$$

That is, the radius of the hydrogen atom is approximately 0.0529 nm.

One can substitute the R_n derived in the expression for E. Using n, the total energy of the electron in the nth orbit is

$$E_n = \Gamma + \Pi = -\frac{me^4}{32\pi^2\varepsilon_0^2\hbar^2 n^2}.$$

The application of quantum mechanics leads to the same expression for the quantized energy:

$$E_n = -\frac{me^4}{32\pi^2\varepsilon_0^2\hbar^2 n^2}.$$

It is obvious that E_n depends on the quantum number n. Figure 6.3 illustrates the quantized energy levels for different n. The MATLAB statement to perform the calculations and plot the results is given here (electronic mass is $9.10938188 \times 10^{-31}$ kg and charge is $1.602176462 \times 10^{-19}$ C). Using the conversion 1 eV $= 1.602176462 \times 10^{-19}$ J, the second plot provides the E_n in the eV unit, which is commonly used.

```
n=1:1:5; m=9.10938188e-31; e=1.602176462e-19;
eps=8.854187817e-12; h=1.054571596e-34;
En=-(m*e^4)./(32*pi*pi*eps*eps*h*h*n.*n);
stem(n,En); axis([0.9 5.1 -2.5e-18 0.5e-18]);
```

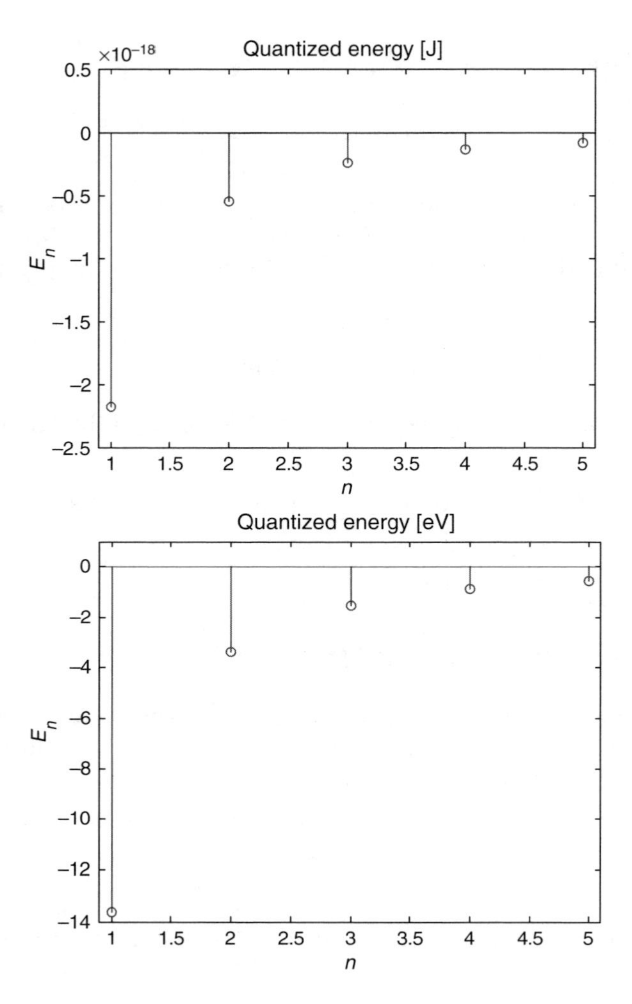

FIGURE 6.3
Quantized energy E_n for different n.

```
title('Quantized Energy, [J]','FontSize',14);
xlabel('\itn','FontSize',14);
ylabel('\itE_n','FontSize',14);
```

For the nucleus charge *Ze*, the derived equations can be modified as

$$E_n = -\frac{m(Ze^2)^2}{32\pi^2\varepsilon_0^2\hbar^2 n^2}.$$

One finds the energy difference between the orbitals as

$$\Delta E = E_{n1} - E_{n2} = \frac{me^4}{32\pi^2\varepsilon_0^2\hbar^2}\left(\frac{1}{n_2^2} - \frac{1}{n_1^2}\right).$$

The evaluation of the term $me^4/(32\pi^2\varepsilon_0^2\hbar^2)$ gives the energy levels differences $E_n = -13.6(Z^2/n^2)$ eV.

Using the difference between the energy levels, from $\Delta E = E_{n1} - E_{n2} = hv$, one finds

$$v = \frac{me^4}{64\pi^3\varepsilon_0^2\hbar^3}\left(\frac{1}{n_2^2} - \frac{1}{n_1^2}\right).$$

The wavelength of the emitted radiation is

$$\lambda = \frac{c}{v} = \frac{64\pi^3\varepsilon_0^2\hbar^3 c}{me^4}\left(\frac{n_1^2 n_2^2}{n_1^2 - n_2^2}\right) = \frac{1}{R_\infty}\left(\frac{n_1^2 n_2^2}{n_1^2 - n_2^2}\right),$$

where R_∞ is the Rydberg constant, $R_\infty = 1.097373 \times 10^7$ m^{-1}.

Example 6.1

Let the electron jumps from $n_1 = 3$ to $n_2 = 2$ and from $n_6 = 6$ to $n_5 = 5$. One can calculate the corresponding wavelengths. In particular,

$$\lambda = \frac{1}{1.097373 \times 10^7}\left(\frac{3^2 2^2}{3^2 - 2^2}\right) = 656 \text{ nm}$$

and

$$\lambda = \frac{1}{1.097373 \times 10^7}\left(\frac{6^2 5^2}{6^2 - 5^2}\right) = 7455.8 \text{ nm}.$$

Example 6.2

For a single photon of energy E, the momentum is $p = E/c$. The de Broglie formula relates the momentum and the wavelength λ as $p = h/\lambda$. The rest energy of electron is $m_e c^2 = 5.1 \times 10^5$ eV. For the electron with the kinetic energy Γ, if $\Gamma \ll m_e c^2$, one may use nonrelativistic formalism to find momentum as $p = \sqrt{2m_e\Gamma}$. Letting $\Gamma = 1$ eV, we have $p = 5.4 \times 10^{-25}$ kg m/sec, which gives $\lambda = 1.2$ nm. The frequency of radiation is $v = c/\lambda$.

6.2 Introduction to Modeling and Analysis

A great variety of molecules have been synthesized and examined for various applications other than electronics- or processing-centered ones. This section is devoted to introducing one to the topic of analysis of electron transport in ^{ME}devices. These ^{ME}devices, composed from atomic aggregates, should

- Guarantee functionality
- Ensure overall chemical synthesis soundness
- Exhibit quantum effects that can be utilized
- Guarantee desired performance characteristics

Molecular electronics devices should be examined by applying quantum mechanics. Coherent high-fidelity mathematical models are needed to examine behavior of molecular systems using various variables such as energy, velocity, momentum, charge variations. This ultimately results in the ability to examine the electron transport. Mathematical models should accurately describe the basic phenomena, be computationally tractable, and suit heterogeneous simulations as applied to carry out data-intensive analysis. Modeling and analysis of electronic devices are based on the Schrödinger equation, Green function, and other methods [1–5]. The kinetic energy, potentials, Fermi energy E_{F}, energy level broadening E_{B}, charge density, and other quantities, variables, and parameters are used.

Our goal is to study the equations of motion to analyze *microscopic* particles and *microscopic* system behavior. The dynamic and steady-state behavior of all dynamic systems, including ^{M}devices, is analyzed using the state variables and quantities, for example, position, velocity, field, charge, radiation, polarization. Using classical mechanics, the evolution of particles is given using physical variables, that is, the *canonically coupled variables*. Quantum mechanics offers a different framework. In general, when a system is "in motion" due to different $\left\{\begin{matrix}\text{"force"}\\ \text{"potential"}\end{matrix}\right\}$, the system behavior is found solving $\left\{\begin{matrix}\text{Newton/Lagrange/Hamilton}\\ \text{Schrödinger}\end{matrix}\right\}$ equations. The $\left\{\begin{matrix}\text{variable}\\ \text{wave function}\end{matrix}\right\}$ is continuous across the boundary if $\left\{\begin{matrix}\text{"force"}\\ \text{"potential"}\end{matrix}\right\}$ is finite. Quantum mechanics describes *microscopic* particles (for example, electrons) and systems (^{M}devices) using the wave function $\Psi(t, \mathbf{r})$. The physical quantities of these particles and systems can be derived by making use of $\Psi(t, \mathbf{r})$. There is a direct correspondence between the physical quantities of particles, for example, energy E, momentum $\mathbf{p}$, position $\mathbf{r}$, frequency v, wave vector $\mathbf{k}$. We recall the de Broglie relation $\mathbf{p} = \hbar\mathbf{k}$, Planck's quantization law $E = nhv$, $k = 2\pi/\lambda$, and so forth.

Example 6.3

One may derive the Schrödinger equation using the familiar conservation energy concept and de Broglie's equation. From the conservation of energy concept, one has $E = \Gamma + \Pi$.

The kinetic energy is $\Gamma = \frac{1}{2}mv^2 = \frac{1}{2}(p^2/m)$.

By taking note of the de Broglie equation $p = \hbar k = \hbar(2\pi/\lambda)$, we have

$$\Gamma = \frac{1}{2}\frac{\hbar^2 k^2}{m},$$

where p is the momentum; k is the wave number.

Using the results from the classical mechanics and electromagnetics (string, membrane, Maxwell's equation, etc.), one can define the free-particle deBroglie wave as $\Psi(t,x) = A\sin(kx - \omega t)$. For $t = 0$, the differentiation of $\Psi(t,x) = A\sin kx$ gives

$$\frac{d^2\Psi}{dx^2} = -k^2\Psi = -\frac{2m}{\hbar^2}\Gamma\Psi = -\frac{2m}{\hbar^2}(E - \Pi)\Psi.$$

One obtains the Schrödinger equation

$$-\frac{\hbar^2}{2m}\frac{d^2\Psi}{dx^2} + \Pi\Psi = E\Psi.$$

Figure 6.4a schematically illustrates 3D-topology multiterminal and two-terminal ^{ME}devices.

It has been reported that by using quantum mechanics one can derive the dimensionless transmission probability of electron tunneling $T(E)$, which is a function of energy E, and $0 \le T(E) \le 1$. The conductance of molecular wires and some two-terminal ^{ME}devices were examined in [1–5]. A linear conductance that neglects thermal relaxation and other effects can be estimated by applying the so-called Landauer [6] or Landauer–Buttiker [7] expression:

$$g(E) = \frac{e^2}{\pi\hbar}T(E),$$

where the transmission coefficient $T(E)$ is evaluated at the energy E equal to the Fermi energy E_F at zero voltage bias.

The so-called quantum conductance is defined to be

$$g_0 = (e^2/\pi\hbar) = 7.75 \times 10^{-5}\ \Omega^{-1}.$$

The constant $(e^2/3\pi^2\hbar^2)$ in defining the expression for conductance was originally reported in [6], where electron transport in an electric field was studied. By making the use of the acceleration of electrons $(d\mathbf{k}/dt) = -(e\mathbf{E}/\hbar)$,

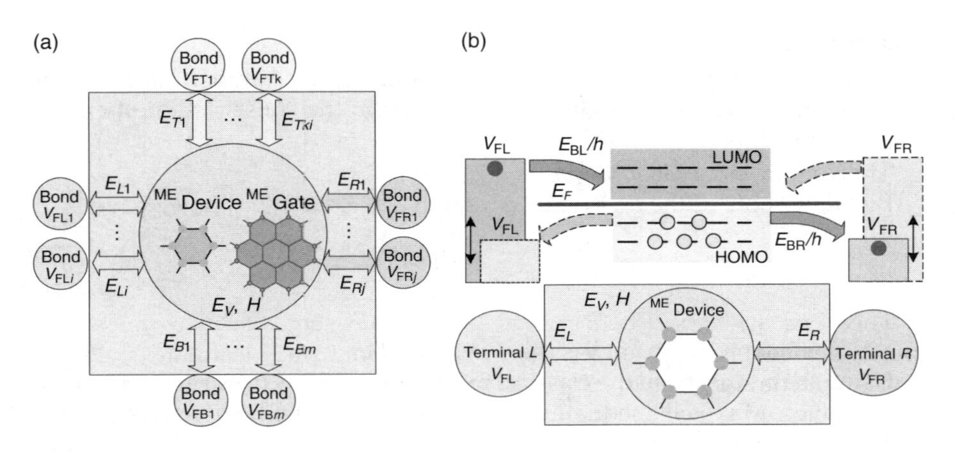

FIGURE 6.4
Molecular electronic devices: Multiterminal ^{ME}device with the left (L), right (R), top (T), and bottom (B) bonds forming *input*, *control*, and *output* terminals; (b) Two-terminal ^{ME}device with Hamiltonian H, single energy potential E_V, and varying left/right potentials V_{FL} and V_{FR}.

the expression for conductivity was obtained. In particular, assuming the equilibrium condition, it is stated that [6]: "For our isotropic band structure and isotropic background scattering the conductivity ... is given by $\sigma_{\mathrm{B}} = (\tau_{\mathrm{B}}/3\pi^2)(e^2/\hbar^2)k^2(\mathrm{d}U/\mathrm{d}k)$." Here, τ_{B} is the relaxation time.

Assuming the applicability of the Fermi–Dirac distribution, the current–voltage (I–V) characteristics for two-terminal electronic devices (see Figure 6.4b) are commonly found by applying the following equation [1–5]:

$$I(E) = \frac{2e}{h}\int_{-\infty}^{+\infty} T(E)[f(E_V, V_{\mathrm{FL}}) - f(E_V, V_{\mathrm{FR}})]\mathrm{d}E,$$

where $f(E_V, V_{\mathrm{FL}})$ and $f(E_V, V_{\mathrm{FR}})$ are the Fermi–Dirac distribution functions, $f(E_V, V_{\mathrm{FL}}) = (1+\mathrm{e}^{(E_V - V_{\mathrm{FL}}/kT)})^{-1}$ and $f(E_V, V_{\mathrm{FR}}) = (1+\mathrm{e}^{(E_V - V_{\mathrm{FR}}/kT)})^{-1}$; E_V is the single energy potential that depends on the charge density $\rho(E)$ or the number of electrons N, $E_V = E_{V0} + V_{\mathrm{SC}}$; V_{SC} is the self-consistent potential to be determined by solving the Poisson equation using the charge density, $V_{\mathrm{SC}} = f_\rho(\rho)$ or $V_{\mathrm{SC}} = V(N - N_0)$; N is the electron concentration; N_0 is the number of electrons at the equilibrium, $N_0 = 2f(E_{V0}, E_{\mathrm{F}}) = 2(1 + \mathrm{e}^{(E_{V0} - E_{\mathrm{F}}/kT)})^{-1}$; V_{FL} and V_{FR} are the left and right electrochemical potentials related to the Fermi levels.

The electrochemical potentials V_{FL} and V_{FR} vary, and there is no electron transport if $V_{\mathrm{FL}} = V_{\mathrm{FR}}$. The highest occupied molecular orbital (HOMO) and lowest unoccupied molecular orbital (LUMO) orbitals, as well as the Fermi level, are documented in Figure 6.4b. Depending on the HOMO and LUMO levels, as well as E_{F}, electron transport takes place trough particular orbitals. The electron transport rates E_{BL}/h and E_{BR}/h are functions of the broadening energies E_{BL} and E_{BR}. One may estimate the number of electrons and current as [3–5]:

$$N = 2\frac{E_{\text{BL}}f(E_V, V_{\text{FL}}) + E_{\text{BR}}f(E_V, V_{\text{FR}})}{E_{\text{BL}} + E_{\text{BR}}}$$

and

$$I = \frac{eNE_{\text{BR}}}{h}.$$

The approach reported here is well-suited for semiconductor microelectronic devices. For ^{ME}devices, many assumptions and postulates valid for solid-state devices (Fermi–Dirac distribution, carrier velocity, continuous energy, etc.) may not be ensured, or may not be sound or applicable from the device physics standpoints.

For example, Section 2.1.2 reports the distribution statistics with the applied assumptions. Correspondingly, other methods should be applied. Quantum mechanics will be utilized to examine the functionality, performance, and baseline characteristics of ^{ME}devices. The wave function $\Psi(t, \mathbf{r})$, allowed energies, potentials, and other quantities must be studied to qualitatively and quantitatively examine temporal and spatial evolution of quantum system (^{M}device) states. This ensures coherent analysis of behavior and phenomena such as electron transport. For example, the transmission coefficient and expectation values of system variables are derived using the wave function obtained by solving the Schrödinger equation.

6.3 Heisenberg Uncertainty Principle

We apply quantum theory and perform some analysis from the experimental perspective by employing the Heisenberg uncertainty principle. The Heisenberg uncertainty principle specifies that no experiment can be performed to furnish uncertainties below the limits defined by the uncertainty relationship. For a perturbed particle, using complementary observable variables A and B, the generalized uncertainty principle is given as

$$\sigma_A^2 \sigma_B^2 \geq \left(\frac{1}{2i} \langle [\hat{A}, \hat{B}] \rangle \right)^2,$$

where σ_A and σ_B are the standard deviations; $[\hat{A}, \hat{B}]$ is the commutator of two Hermitian operators $\hat{A}$ and $\hat{B}$, where $[\hat{A}, \hat{B}] = \hat{A}\hat{B} - \hat{B}\hat{A}$.

We conclude that it is impossible to measure simultaneously two complementary observable variables with arbitrary accuracy. One may use the observable position x, for which $\hat{A} = x$, as well as the momentum p with the corresponding operator $\hat{B} = -i\hbar(\partial/\partial x)$. By taking note of the canonical commutation relation $[\hat{x}, \hat{p}] = i\hbar$, we obtain the position–momentum uncertainty principle as

$$\sigma_x^2 \sigma_p^2 \geq \left(\frac{1}{2i} i\hbar\right)^2 = \left(\tfrac{1}{2}\hbar\right)^2 \quad \text{or} \quad \sigma_x \sigma_p \geq \tfrac{1}{2}\hbar.$$

The energy–time uncertainty principle is

$$\sigma_E \sigma_t \geq \tfrac{1}{2}\hbar.$$

Notations Δx, Δp, ΔE, and Δt are frequently used to define the standard deviations as uncertainties. In books on quantum mechanics, ΔE gives the energy difference between the quantum states. Hence, in covering the Heisenberg uncertainty principle, we use the notation $\Delta \hat{E}$, which is not ΔE.

One defines the uncertainties ΔA and ΔB in the measurement of A and B by their dispersions:

$$(\Delta A)^2 = \langle(\hat{A} - \langle \hat{A} \rangle)^2\rangle = \langle \hat{A}^2 \rangle - \langle \hat{A} \rangle^2$$

and

$$(\Delta B)^2 = \langle(\hat{B} - \langle \hat{B} \rangle)^2\rangle = \langle \hat{B}^2 \rangle - \langle \hat{B} \rangle^2,$$

or

$$\Delta A = \sqrt{\langle \hat{A}^2 \rangle - \langle \hat{A} \rangle^2}$$

and

$$\Delta B = \sqrt{\langle \hat{B}^2 \rangle - \langle \hat{B} \rangle^2}.$$

The uncertainty relation is

$$\Delta A \Delta B \geq \tfrac{1}{2} |\langle [\hat{A}, \hat{B}] \rangle|.$$

The position–momentum and energy–time uncertainty principles are rewritten as

$$\Delta x \Delta p_x \geq \tfrac{1}{2}\hbar, \quad \Delta y \Delta p_y \geq \tfrac{1}{2}\hbar, \quad \Delta z \Delta p_z \geq \tfrac{1}{2}\hbar$$

and

$$\Delta \hat{E} \Delta t \geq \tfrac{1}{2}\hbar.$$

Example 6.4

Consider the position–momentum uncertainty relation $\Delta x \Delta p_x \geq (1/2)\hbar$. The subscript x is used for the momentum p_x to indicate that $\Delta x \Delta p_x \geq (1/2)\hbar$ applies to motion of particle in a given direction and relates the uncertainties in position x and momentum p_x in that direction only. The relationship $\Delta x \Delta p_x \geq (1/2)\hbar$ gives an estimate (one cannot do better) of the minimum uncertainty that can result from any experiment, and measurement of the position and momentum of a particle will give uncertainties Δx and Δp_x.

Hence, the Heisenberg uncertainty principles indicates that, if the x-component of the momentum of a particle is measured with uncertainty Δp_x, then its x-position cannot be measured more accurately than

$$\Delta x \geq \frac{\hbar}{2\Delta p_x}.$$

Thus, it is impossible simultaneously measure two observable variables with an arbitrary accuracy.

As documented, there is a limit on the accuracy. One cannot perform experiments better than imposed by

$$\Delta x \Delta p_x \geq \tfrac{1}{2}\hbar, \quad \Delta y \Delta p_y \geq \tfrac{1}{2}\hbar, \quad \Delta z \Delta p_z \geq \tfrac{1}{2}\hbar$$

and

$$\Delta \hat{E} \Delta t \geq \tfrac{1}{2}\hbar,$$

no matter which measuring hardware is used.

Example 6.5

Discussions on What Δt is in the Heisenberg Uncertainty Principle

The particle position, momentum, and energy are dynamic variables (measurable characteristics of the system or device) at any given time. In contrast, time is the independent variable of which the dynamic quantities are functions. That is, in

$$\Delta \hat{E} \Delta t \geq \tfrac{1}{2}\hbar,$$

Δt is the time it takes the system to change substantially.

For example, Δt represents the amount of time it takes the expectation value of E to change by one standard deviation in order to ensure the observability of E. It should be emphasized again that $\Delta \hat{E}$ defines the standard deviations as uncertainties in measurement of E.

It is obvious that Δt does not define or allows one to estimate transition time, switching time, bandwidth, or other system (device) characteristics.

Example 6.6

Derive the position uncertainties Δx for a 9.1×10^{-31} kg electron (*microscopic* particle) and a 9.1×10^{-3} kg bullet (*macroscopic* particle) if the speed 1000 m/sec is measured with uncertainty 0.001%.

From $p = mv$, one finds $\Delta p = m\Delta v$.

Using $\Delta x \geq (\hbar/2\Delta p_x)$, for an electron one obtains $\Delta x \geq 0.00577$ m, while for a bullet we have $\Delta x \geq 5.77 \times 10^{-31}$ m.

For the electron, taking note of the atomic radius of the silicon atom, which is 117 pm, one concludes that the position uncertainty Δx is 2.47×10^7 larger than the diameter of Si atom.

The dimension of a 1 cm bullet is 1.73×10^7 times larger than Δx. This guarantees no restrictions on measurements for a bullet.

Conclusions on the Role and Use of Heisenberg Uncertainty Principle: The reported results related to the Heisenberg uncertainty principle impose constraints and limits on testing, evaluation, and characterization of quantum systems, including ^{M}devices. The ability to conduct measurements for particular devices depends on the device physics, functionality, phenomena, carriers (photons, electrons, or ions), and so forth. The Heisenberg uncertainty principle imposes limits as utilized in the experimental studies, testing, and data analysis. The uncertainty principle does not define or imply the dimensionality, switching time, power dissipation, switching energy, carrier velocity, and other *achievable* device characteristics. Those quantities must be found coherently applying other concepts reported in this chapter.

6.4 Particle Velocity

For ^{ME}devices, it is important to examine how wave packets evolve in time and space to provide an answer on motion of *microscopic* particles in space. The velocity of the group of matter waves is equal to the particle velocity whose motion they are governing. For the wave packets propagating in the x-direction, in order to examine the time evolution, we apply the following equation:

$$\Psi(t, x) = \frac{1}{\sqrt{2\pi}} \int_{-\infty}^{\infty} \phi(k) e^{i(kx - \omega t)} \, dk,$$

where $\phi(k)$ is the magnitude of the wave packet; k is the wave number; ω is the angular frequency.

Examining the time evolution of the wave packet, the group and phase velocities are given as

$$v_g = \frac{d\omega(k)}{dk} = v_{ph} + k\frac{dv_{ph}}{dk} = v_{ph} + p\frac{dv_{ph}}{dp}$$

and

$$v_{ph} = \frac{\omega(k)}{k}.$$

The group velocity represents the velocity of motion of the group of propagating waves that compose the wave packet. The phase velocity is the velocity of propagation of the phase of a single mth harmonic wave $e^{ik_m(x-v_{ph}t)}$. The wave packet travels with the group velocity.

Taking note of $E = \hbar\omega$ and $p = \hbar k$, one obtains

$$v_g = \mathrm{d}E(p)/\mathrm{d}p$$

and

$$v_{ph} = E(p)/p.$$

From $E = (p^2/2m) + \Pi$, assuming that $\Pi = \text{const}$, we have

$$v_g = \mathrm{d}E(p)/\mathrm{d}p = p/m = v$$

and

$$v_{ph} = E(p)/p = p/2m + \Pi/p.$$

Thus, the group velocity of the wave packet is equal to the particle velocity v.

For a free electron, the energy is

$$E = \frac{p^2}{2m} = \frac{\hbar^2 k^2}{2m} = \hbar\omega.$$

One finds

$$v_g = \frac{\mathrm{d}\omega}{\mathrm{d}k} = \frac{\hbar k}{m} = \frac{p}{m} = v.$$

Consider a free electron in the electric field with the intensity E_E. We have

$$\mathrm{d}E = eE_E\,\mathrm{d}x = eE_E\frac{\mathrm{d}x}{\mathrm{d}t}\,\mathrm{d}t = eE_E v\,\mathrm{d}t$$

and

$$\mathrm{d}E = \hbar\,\mathrm{d}\omega = \hbar\frac{\mathrm{d}\omega}{\mathrm{d}k}\,\mathrm{d}k = \hbar v\,\mathrm{d}k.$$

Thus, one finds $qE_E = \hbar(\mathrm{d}k/\mathrm{d}t)$.

The time derivative of the electron velocity $v = (\mathrm{d}\omega/\mathrm{d}k) = (1/\hbar)(\mathrm{d}E/\mathrm{d}k)$ gives the acceleration of the electron, and

$$a = \frac{\mathrm{d}v}{\mathrm{d}t} = \frac{1}{\hbar}\frac{\mathrm{d}^2E}{\mathrm{d}k\,\mathrm{d}t} = \frac{1}{\hbar}\frac{\mathrm{d}^2E}{\mathrm{d}k^2}\frac{\mathrm{d}k}{\mathrm{d}t} = \frac{1}{\hbar^2}\frac{\mathrm{d}^2E}{\mathrm{d}k^2}eE_E.$$

The force acting on electron is $F = (\mathrm{d}p/\mathrm{d}t) = \hbar(\mathrm{d}k/\mathrm{d}t)$ or $F = eE_E$. Hence,

$$a = \frac{1}{\hbar^2}\frac{d^2E}{dk^2}F.$$

The expression $F = \hbar^2(\mathrm{d}^2E/\mathrm{d}k^2)^{-1}(\mathrm{d}v/\mathrm{d}t)$ is used in solid-state semiconductor devices to introduce the so-called *effective* mass of electron which is $m_{\text{eff}} = \hbar^2(\mathrm{d}^2E/\mathrm{d}k^2)^{-1}$.

In *solid* ^{ME}devices, the device physics and 3D-topology must be coherently integrated. The derived expressions for the particle velocity can be used to obtain the *I–V* and *G–V* characteristics, estimate propagation delays, analyze switching speed, and examine other characteristics.

Example 6.7

Consider a wave packet corresponding to a relativistic particle. The energy and momentum are

$$E = mc^2 = \frac{m_0c^2}{\sqrt{1 - v^2/c^2}}$$

and

$$p = mv = \frac{m_0v}{\sqrt{1 - v^2/c^2}},$$

where m_0 is the rest mass of the particle.

From $E = c\sqrt{p^2 + m_0^2c^2}$, one obtains

$$v_g = \frac{\mathrm{d}E}{\mathrm{d}p} = \frac{\mathrm{d}\left(c\sqrt{p^2 + m_0^2c^2}\right)}{\mathrm{d}p} = \frac{pc}{\sqrt{p^2 + m_0^2c^2}} = v \quad \text{and} \quad v_{ph} = \frac{E}{p} = \frac{c^2}{v}.$$

Example 6.8

Consider an electron as a not relativistic particle. From $E = mv^2/2$, one has $v = \sqrt{(2E/m)}$. Let $E = 0.1$ eV $= 0.1602176462 \times 10^{-19}$ J. For a nonrelativistic electron, we find $v = 1.88 \times 10^5$ m/sec. The time it takes to electron to travel 1 nm distance is $t = L/v = 5.33 \times 10^{-15}$ sec.

The particle (electron) traversal time is of interest to analyze the device performance [2]. For a one-dimensional case, for a particle with an energy E in $\Pi(x)$, one has [2,8]:

$$\tau(E) = \int_{x_0}^{x_f} \sqrt{\frac{m}{2[\Pi(x) - E]}}\,dx.$$

For a one-dimensional rectangular barrier with Π_0 and width L,

$$\tau(E) = \sqrt{\frac{m}{2(\Pi_0 - E)}}\,L.$$

By using the transmission probabilities of two particle states $T_1(E)$ and $T_2(E)$, we have [9]

$$\tau(E) = \lim_{\lambda \to 0}\left(\frac{\hbar}{|\lambda|}\sqrt{\frac{T_2(E)}{T_1(E)}}\right).$$

Example 6.9

Let $(\Pi_0 - E) = 0.1$ eV $= 0.16 \times 10^{-19}$ J and $L = 1$ nm. One finds $\tau = 5.33 \times 10^{-15}$ sec. The estimated τ agrees with the results reported for $\tau(E)$ in [9], where the transmission probabilities are used. As will be documented in Section 6.5, 6.6, and 6.9, using the Schrödinger equation, we can find the wave function, energy, momentum, and other variables. This enables one to obtain $\tau(E)$.

6.5 Schrödinger Equation

6.5.1 Introduction and Fundamentals

The time-invariant (time-independent) Schrödinger equation for a particle in the Cartesian coordinate system is given as

$$-\frac{\hbar^2}{2m}\nabla^2\Psi(x,y,z) + \Pi(x,y,z)\Psi(x,y,z) = E(x,y,z)\Psi(x,y,z),$$

where ∇^2 is the Laplacian, $\nabla^2 = (\partial^2/\partial x^2) + (\partial^2/\partial y^2) + (\partial^2/\partial z^2)$; $\Pi(x,y,z)$ is the potential energy function; $E(x,y,z)$ is the total energy.

The Hamiltonian is $H = -(\hbar^2/2m)\nabla^2 + \Pi$.

Hence, $H(x,y,z)\Psi(x,y,z) = E(x,y,z)\Psi(x,y,z)$ or $H(\mathbf{r})\Psi(\mathbf{r}) = E(\mathbf{r})\Psi(\mathbf{r})$.

The time-dependent Schrödinger equation is

$$-\frac{\hbar^2}{2m}\nabla^2\Psi(t,x,y,z) + \Pi(t,x,y,z)\Psi(t,x,y,z) = i\hbar\frac{\partial\Psi(t,x,y,z)}{\partial t},$$

or

$$-\frac{\hbar^2}{2m}\nabla^2\Psi(t,\mathbf{r}) + \Pi(t,\mathbf{r})\Psi(t,\mathbf{r}) = i\hbar\frac{\partial\Psi(t,\mathbf{r})}{\partial t}.$$

The Schrödinger equation has the following attributes:

- Consistent with the de Broglie postulates $p = h/\lambda$ and $v = E/h$
- Consistent with total, kinetic, and potential energies, that is, $E = p^2/2m + \Pi$
- Linear in $\Psi(t,\mathbf{r})$

The Schrödinger equation should be solved using normalizing, boundary, and continuity conditions in order to find the wave function $\Psi(t,\mathbf{r})$. The probability of finding a particle within a volume V is $\int_V \Psi^*(t,\mathbf{r})\Psi(t,\mathbf{r})\,\mathrm{d}V$, where $\Psi^*(t,\mathbf{r})$ is the complex conjugate of $\Psi(t,\mathbf{r})$.

Hence, the wave function is normalized as

$$\int_{-\infty}^{\infty} \Psi^*(t,\mathbf{r})\Psi(t,\mathbf{r})\,\mathrm{d}V = 1,$$

where in the Cartesian coordinate system $\mathrm{d}V = \mathrm{d}x\,\mathrm{d}y\,\mathrm{d}z$.

The system behavior and time evolution of the system states are defined by the wave function. The basic connection between the properties of $\Psi(t,\mathbf{r})$ and the particle behavior is expressed by the probability density $P(t,\mathbf{r})$. For example, the quantity $P(t,x)$ specifies the probability, per unit length, of finding the particle near x at time t. Thus,

$$P(t,x) = \Psi^*(t,x)\Psi(t,x).$$

For a physical observable C that has an associated operator $\hat{C}$, the average expectation value of the observable is

$$\langle C\rangle = \int \Psi^*(t,\mathbf{r})\hat{C}\Psi(t,\mathbf{r})\,\mathrm{d}V.$$

The following momentum and energy operators are applied

$$p \leftrightarrow -i\hbar\frac{\partial}{\partial x} \text{ and } E \leftrightarrow i\hbar\frac{\partial}{\partial t}.$$

In general, for a momentum one has $p \leftrightarrow -i\hbar\nabla$.

For a given probability density $P(t,x)$, the expected values of any function of x can be derived. In particular,

$$\langle f(x)\rangle = \int_{-\infty}^{\infty} f(x)P(t,x)\,\mathrm{d}x = \int_{-\infty}^{\infty} \Psi^*(t,x)f(x)\Psi(t,x)\,\mathrm{d}x.$$

For example, the expectation values of x and x^2 are

$$\langle x \rangle = \int_{-\infty}^{\infty} xP(t,x)\,\mathrm{d}x = \int_{-\infty}^{\infty} \Psi^*(t,x)x\Psi(t,x)\,\mathrm{d}x$$

and

$$\langle x^2 \rangle = \int_{-\infty}^{\infty} x^2P(t,x)\,\mathrm{d}x = \int_{-\infty}^{\infty} \Psi^*(t,x)x^2\Psi(t,x)\,\mathrm{d}x.$$

For a one-dimensional case the expectation values of the momentum and total energy are

$$\langle p \rangle = \int_{-\infty}^{\infty} \Psi^*(t,x)\left(-i\hbar\frac{\partial}{\partial x}\right)\Psi(t,x)\,\mathrm{d}x = -i\hbar\int_{-\infty}^{\infty} \Psi^*(t,x)\frac{\partial\Psi(t,x)}{\partial x}\,\mathrm{d}x$$

and

$$\langle E \rangle = \int_{-\infty}^{\infty} \Psi^*(t,x)\left(i\hbar\frac{\partial}{\partial t}\right)\Psi(t,x)\,\mathrm{d}x = i\hbar\int_{-\infty}^{\infty} \Psi^*(t,x)\frac{\partial\Psi(t,x)}{\partial t}\,\mathrm{d}x$$

$$= \int_{-\infty}^{\infty} \Psi^*(t,x)\left(-\frac{\hbar^2}{2m}\frac{\partial^2}{\partial x^2} + \Pi(t,x)\right)\Psi(t,x)\,\mathrm{d}x.$$

For $f(p)$, we have

$$\langle f(p) \rangle = \int_{-\infty}^{\infty} \Psi^*(t,x)f\left(-i\hbar\frac{\partial}{\partial x}\right)\Psi(t,x)\,\mathrm{d}x.$$

For example, one finds

$$\langle p^2 \rangle = \int_{-\infty}^{\infty} \Psi^*(t,x)\left(-i\hbar\frac{\partial}{\partial x}\right)^2\Psi(t,x)\,\mathrm{d}x = -\hbar^2\int_{-\infty}^{\infty} \Psi^*(t,x)\frac{\partial^2\Psi(t,x)}{\partial x^2}\,\mathrm{d}x.$$

For any dynamic quantity that is a function of x and p, for example, $f(t,x,p)$, the expectation value is

$$\langle f(t,x,p) \rangle = \int_{-\infty}^{\infty} \Psi^*(t,x)f\left(t,x,-i\hbar\frac{\partial}{\partial x}\right)\Psi(t,x)\,\mathrm{d}x.$$

As an illustration, for a potential $\Pi(t,x)$, we have

$$\langle \Pi(t,x) \rangle = \int_{-\infty}^{\infty} \Psi^*(t,x)\Pi(t,x)\Psi(t,x)\,\mathrm{d}x.$$

Example 6.10

Let the wave function for the lowest energy state for a free particle is

$$\Psi(t,x) = \begin{cases} A\cos\dfrac{\pi x}{L}\mathrm{e}^{-(iE/\hbar)t} & \text{for } -\frac{1}{2}L < x < \frac{1}{2}L, \\ 0 & \text{for } x \le -\frac{1}{2}L,\ x \ge \frac{1}{2}L. \end{cases}$$

As will be documented later, we are considering a particle in a one-dimensional potential well with $\Pi(x) = 0$ in $-L/2 < x < L/2$, and $\Pi(x) = \infty$ otherwise.

One finds the total energy E by using the Schrödinger equation, which is

$$-\frac{\hbar^2}{2m}\frac{\partial^2\Psi}{\partial x^2} = i\hbar\frac{\partial\Psi}{\partial t} \quad \text{for } -L/2 < x < L/2.$$

The expressions for the spatial and time derivatives are

$$\frac{\partial\Psi}{\partial x} = -\frac{\pi}{L}A\sin\frac{\pi x}{L}\mathrm{e}^{-(iE/\hbar)t},$$

$$\frac{\partial^2\Psi}{\partial x^2} = -\frac{\pi^2}{L^2}A\cos\frac{\pi x}{L}\mathrm{e}^{-(iE/\hbar)t} = -\frac{\pi^2}{L^2}\Psi$$

and

$$\frac{\partial\Psi}{\partial t} = -\frac{iE}{\hbar}A\cos\frac{\pi x}{L}\mathrm{e}^{-(iE/\hbar)t} = -\frac{iE}{\hbar}\Psi.$$

Thus, the Schrödinger equation gives

$$\frac{\hbar^2}{2m}\frac{\pi^2}{L^2}\Psi = -i\hbar\frac{iE}{\hbar}\Psi.$$

Therefore, one has $E = (\pi^2\hbar^2/2mL^2)$.

The expectation values of x and x^2 are found by making use of

$$\langle x\rangle = \int_{-\infty}^{\infty} xP(t,x)\,\mathrm{d}x = \int_{-\infty}^{\infty} \Psi^*(t,x)x\Psi(t,x)\,\mathrm{d}x$$

and

$$\langle x^2\rangle = \int_{-\infty}^{\infty} x^2P(t,x)\,\mathrm{d}x = \int_{-\infty}^{\infty} \Psi^*(t,x)x^2\Psi(t,x)\,\mathrm{d}x.$$

Taking note of $\Psi(t,x)$, we have

$$\langle x\rangle = \int_{-(1/2)L}^{(1/2)L} A\cos\frac{\pi x}{L}\mathrm{e}^{(iE/\hbar)t}xA\cos\frac{\pi x}{L}\mathrm{e}^{-(iE/\hbar)t}\,\mathrm{d}x = A^2\int_{-(1/2)L}^{(1/2)L} x\cos^2\frac{\pi x}{L}\,\mathrm{d}x = 0,$$

and

$$\langle x^2\rangle = \int_{-(1/2)L}^{(1/2)L} A\cos\frac{\pi x}{L}\mathrm{e}^{(iE/\hbar)t}x^2 A\cos\frac{\pi x}{L}\mathrm{e}^{-(iE/\hbar)t}\,\mathrm{d}x = A^2\int_{-(1/2)L}^{(1/2)L} x^2\cos^2\frac{\pi x}{L}\,\mathrm{d}x$$

$$= 2A^2\int_0^{(1/2)L} x^2\cos^2\frac{\pi x}{L}\,\mathrm{d}x = 2A^2\frac{L^3}{\pi^3}\int_{-(1/2)L}^{(1/2)\pi}\left(\frac{\pi x}{L}\right)^2\cos^2\frac{\pi x}{L}\,\mathrm{d}\frac{\pi x}{L}$$

$$= A^2\frac{L^3}{24\pi^2}(\pi^2-6).$$

The wave function should be normalized, and the amplitude A can be found. One has

$$\int_{-\infty}^{\infty}\Psi^*(t,x)\Psi(t,x)\,\mathrm{d}x = A^2\int_{-(1/2)L}^{(1/2)L}\cos^2\frac{\pi x}{L}\,\mathrm{d}x = 2A^2\frac{L}{\pi}\int_0^{(1/2)\pi}\cos^2\frac{\pi x}{L}\mathrm{d}\frac{\pi x}{L}$$

$$= 2A^2\frac{L}{\pi}\frac{\pi}{4} = \frac{1}{2}A^2 L.$$

By normalizing the wave function as $\int_{-\infty}^{\infty}\Psi^*(t,x)\Psi(t,x)\,\mathrm{d}x = 1$, we obtain $A = \sqrt{2/L}$.

Hence, $\langle x^2\rangle = (2/L)(L^3/24\pi^2)(\pi^2-6) = (L^2/12\pi^2)(\pi^2-6)$, which gives the fluctuations of particle about the average, and the root-mean-square value is $\sqrt{\langle x^2\rangle}$.

From

$$\langle p^2\rangle = -\hbar^2\int_{-\infty}^{\infty}\Psi^*(t,x)\frac{\partial^2\Psi(t,x)}{\partial x^2}\,\mathrm{d}x,$$

one has

$$\langle p^2\rangle = \hbar^2\frac{\pi^2}{L^2}\int_{-\infty}^{\infty}\Psi^*(t,x)\Psi(t,x)\,\mathrm{d}x = \frac{\hbar^2\pi^2}{L^2}.$$

Thus, the root-mean-square momentum is $\sqrt{\langle p^2\rangle} = (\pi\hbar/L)$, and $\sqrt{\langle p^2\rangle}$ represents the average momentum fluctuations about the average $\langle p\rangle = 0$.

By making use of $E = (\pi^2\hbar^2/2mL^2)$, from $p = \pm\sqrt{2mE}$, one concludes that the magnitude of momentum is $\pi\hbar/L$.

Example 6.11

Let

$$\Psi(x) = \begin{cases} 2a\sqrt{a}x\mathrm{e}^{-ax} & \text{for } x \ge 0, \\ 0 & \text{for } x < 0. \end{cases}$$

The peak of $P(x) = |\Psi(x)|^2$ occurs at

$$\frac{\mathrm{d}P(x)}{\mathrm{d}x} = 4a^3 \frac{\mathrm{d}(x^2\,\mathrm{e}^{-2ax})}{\mathrm{d}x} = 0.$$

By making use of $x(1 - ax)\mathrm{e}^{-2ax} = 0$, we have $x = 1/a$.
The expected values for x and x^2 are

$$\langle x \rangle = \int_0^\infty x(4a^3x^2\,\mathrm{e}^{-2ax})\,\mathrm{d}x = \frac{1}{4a}\int_0^\infty y^3\,\mathrm{e}^{-y}\,\mathrm{d}y = \frac{3!}{4a} = \frac{3}{2a}$$

and

$$\langle x \rangle^2 = \int_0^\infty x^2(4a^3x^2\,\mathrm{e}^{-2ax})\,\mathrm{d}x = \frac{4!}{8a^2} = \frac{3}{a^2}.$$

In addition to wave function, the probability current density $J(t, \mathbf{r})$ is frequently applied. For a 3D problem, we have

$$\frac{\partial P(t,\mathbf{r})}{\partial t} + \nabla \cdot \mathbf{J}(t,\mathbf{r}) = 0.$$

Here, the probability density and probability current density are

$$P(t,\mathbf{r}) = \Psi^*(t,\mathbf{r})\Psi(t,\mathbf{r})$$

and

$$\mathbf{J}(t,\mathbf{r}) = \frac{i\hbar}{2m}[\Psi(t,\mathbf{r})\nabla\Psi^*(t,\mathbf{r}) - \Psi^*(t,\mathbf{r})\nabla\Psi(t,\mathbf{r})].$$

Example 6.12

Discussion on the Probability Current Density and Current Density Nomenclature

The probability current density $\mathbf{J}(t, \mathbf{r})$ and the current density $\mathbf{j}$ are entirely different physical quantities. In semiconductor devices, one of the basic equation is

$$\mathbf{j} = Q\mathbf{v},$$

where Q is the charge density; $\mathbf{v}$ is the velocity of the charge carrier (electron or hole), which is found by making use of the applied potential, electric field and other quantities.

Taking note of the volume charge density ρ_V, one has $\mathbf{j} = \rho_V\mathbf{v}$.

Electric charges in motion constitute a current. As charged particles move from one region to another within a *conducting* path, electric potential energy is transformed. The current through the closed surface is

$$I = \oint_S \mathbf{j} \cdot \mathrm{d}\mathbf{s},$$

and

$$I = \mathrm{d}Q/\mathrm{d}t.$$

The current density in electronic devices is the number of electrons crossing a unit area per unit time $N_s \bar{v}_x$ (the unit for N_s is electrons/cm^2) multiplied by the electron charge. For a one-dimensional case

$$j_x = -eN\bar{v}_x \quad \text{or} \quad j_x = -e\sum_i \bar{v}_{xi}.$$

The average net velocity is found using the average momentum per electron: $\bar{v}_x = \bar{p}_x/m$.

In contrast, in quantum mechanics, $\mathbf{J}(t, \mathbf{r})$ represents the rate of probability changes, allowing one to estimate $\langle p \rangle$ which is found using $\Psi(t, \mathbf{r})$.

6.5.2 Application of the Schrödinger Equation: One-Dimensional Potentials

In one-dimensional case, the time-invariant (time-independent) and time-dependent Schrödinger equations are

$$-\frac{\hbar^2}{2m}\frac{\partial^2 \Psi(x)}{\partial x^2} + \Pi(x)\Psi(x) = E(x)\Psi(x)$$

and

$$-\frac{\hbar^2}{2m}\frac{\partial^2 \Psi(t,x)}{\partial x^2} + \Pi(t,x)\Psi(t,x) = i\hbar\frac{\partial \Psi(t,x)}{\partial t}.$$

These Schrödinger equations will be solved to obtain the expression for the wave function using the boundary, continuity and normalization conditions. For a one-dimensional problem, the probability current density $J(t,x)$ is given as

$$J(t,x) = \frac{i\hbar}{2m}\left(\Psi(t,x)\frac{\partial \Psi^*(t,x)}{\partial x} - \Psi^*(t,x)\frac{\partial \Psi(t,x)}{\partial x}\right).$$

The probability of finding a particle in the region $a < x < b$ at time t is $P_{ab}(t) = \int_a^b \Psi^*(t,x)\Psi(t,x)\,\mathrm{d}x$, and $(\mathrm{d}P_{ab}/\mathrm{d}t) = J(t,a) - J(t,b)$. For the probability density $P(t,x) = \Psi^*(t,x)\Psi(t,x)$, one has the following continuity equation

$$\frac{\partial P(t,x)}{\partial t} + \frac{\partial J(t,x)}{\partial x} = 0.$$

Example 6.13

Let the solution of the Schrödinger equation be $\Psi(t,x) = \mathrm{e}^{-i(E/\hbar)t}\Psi(x)$.

The probability density does not depend on time, $\mathrm{d}P_{ab}/\mathrm{d}t = 0$, and $J(t,x) = \text{const}$.

If $\Psi(x) = A\ \mathrm{e}^{ikx}$, we have $P_{ab} = |A|^2(b-a)$ and $P = |A|^2$. Hence, $J = (\hbar k/m)|A|^2 = (\hbar k/m)P$.

Example 6.14

Let $\Psi(x) = A\ \mathrm{e}^{ikx} + B\ \mathrm{e}^{-ikx}$.

We have $\Psi^*(x) = A^*\ \mathrm{e}^{-ikx} + B^*\ \mathrm{e}^{ikx}$ and $(\mathrm{d}\Psi(x)/\mathrm{d}x) = ik(A\ \mathrm{e}^{ikx} - B\ \mathrm{e}^{-ikx})$.

Thus

$$J(x) = \frac{\hbar k}{m}\left(|A|^2 - |B|^2\right) = \frac{p}{m}\left(|A|^2 - |B|^2\right).$$

For $\Psi(x) = A\ \mathrm{e}^{(iD(x)/\hbar)}$, one finds $J(x) = (1/m)A^2(\mathrm{d}D(x)/\mathrm{d}x)$.

Consider a particle in a finite one-dimensional potential $\Pi(x)$ for $x \to \pm\infty$ with $\Pi(-\infty) = \Pi_1$ and $\Pi(+\infty) = \Pi_2$ as shown in Figure 6.5. Let the potential has one minimum $\Pi_{\min} < \Pi_1 < \Pi_2$. Bound states (states whose wave functions are finite or zero at $x \to \pm\infty$) occur because the particle with energy $\Pi_{\min} < E < \Pi_1$ cannot move to infinity, that is, the particle is confined (bound) at all energies to move within a finite and limited region. The Schrödinger equation admits only a discrete solution, for example, infinite square well and harmonic oscillator problems. Unbound states (continuous spectrum) occur when the motion of particle is not confined, that is, a particle is free. In particular, if $\Pi_1 < E < \Pi_2$, the particle moves towards $-\infty$, for example, particle moves between x_1 to $-\infty$. The energy spectrum is continuous and none of

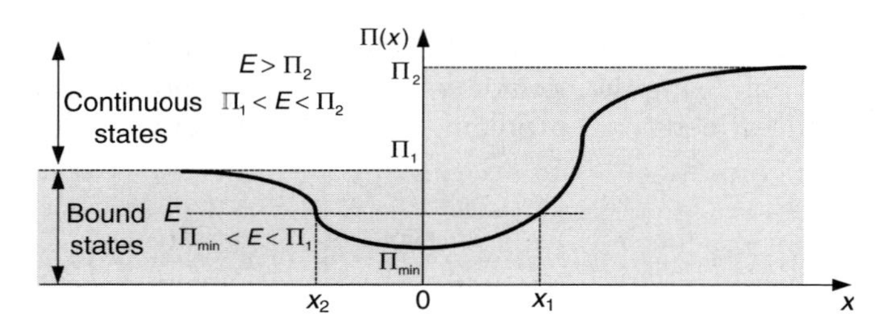

FIGURE 6.5
One-dimensional potential.

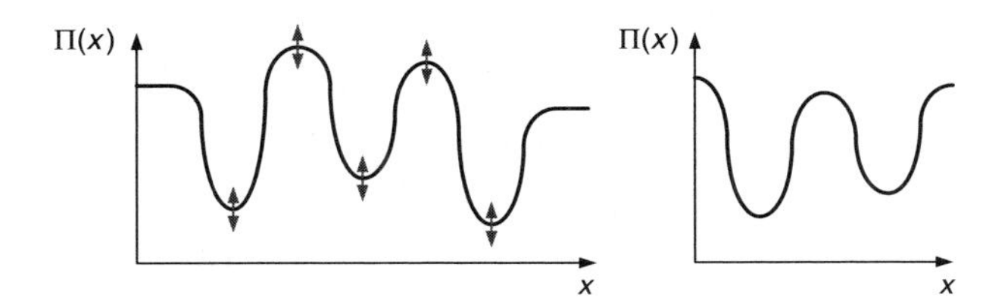

FIGURE 6.6
One-dimensional metastable potentials.

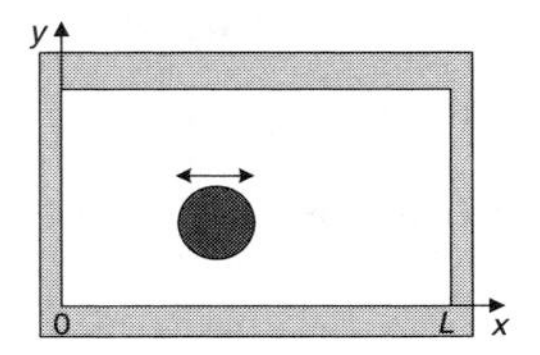

FIGURE 6.7
Particle in the box: Particle motion is confined in $0 \leq x \leq L$.

the energy eigenvalues is degenerate. If $E > \Pi_2$, the energy spectrum is continuous, and particle motion is infinite in $\pm\infty$. The energy levels are doubly degenerate. It should be emphasized that the mixed spectrum corresponds to potentials that confine the particle for some energies only, for example, Coulomb and molecular potentials.

Let a particle be trapped in a metastable potential well [10] as shown in Figure 6.6. Due to thermodynamic fluctuations and electromagnetic fields, the particle can gain the energy from the environment or control apparatus to escape, transmit, or tunnel. Theoretical results reported in the literature provide one with various details and contradictory results. Quantum theory can be applied to ^{ME}devices emphasizing the engineering solutions that are based on solid theoretical fundamentals. Taking note of $\Pi(x)$, the Wentzel–Kramers–Brillouin approximation for the transmission $T(E)$ is given as

$$T(E) \cong e^{-(2/\hbar)\int_{x_0}^{x_f} \sqrt{2m[\Pi(x)-E]}\mathrm{d}x}.$$

Example 6.15

We apply the Schrödinger equation studying the *microscopic* particle if a one-dimensional motion as confined by [0 *L*]; see Figure 6.7. For rigid "walls" and elastic collisions, we have an infinite square-well potential.

The Schrödinger equation to be solved is

$$-\frac{\hbar^2}{2m}\frac{\partial^2\Psi(x)}{\partial x^2} + \Pi(x)\Psi(x) = E(x)\Psi(x).$$

The Hamiltonian is

$$H(x,p) = \frac{p^2(x)}{2m} + \Pi(x) = -\frac{\hbar^2}{2m}\frac{d^2}{dx^2} + \Pi(x).$$

The particle moves in x ∈ [0 L], and the potential energy is

$$\Pi(x) = \begin{cases} 0 & \text{for } 0 < x < L, \\ \infty & \text{for } x \leq 0 \text{ and } x \geq L. \end{cases}$$

The motion of the particle is bounded within the infinite square-well potential, and

$$\Psi(x) = \begin{cases} \text{continuous} & \text{if } 0 < x < L, \\ 0 & \text{if } x \leq 0 \text{ and } x \geq L. \end{cases}$$

In $0 < x < L$, the potential energy is zero, and we have

$$-\frac{\hbar^2}{2m}\frac{d^2\Psi(x)}{dx^2} = E\Psi(x).$$

The solution of the resulting second-order differential equation

$$\frac{d^2\Psi(x)}{dx^2} + k^2\Psi(x) = 0, \quad k^2 = \frac{2mE}{\hbar^2}$$

is

$$\begin{aligned}\Psi(x) &= A\,e^{ikx} + B\,e^{-ikx} = A(\cos kx + i\sin kx) + B(\cos kx - i\sin kx) \\ &= C\sin kx + D\cos kx.\end{aligned}$$

The solution can be verified by substituting $\Psi(x)$ in the left-hand side of the differential equation $-(\hbar^2/2m)(d^2\Psi(x)/dx^2) = E\Psi(x)$ obtaining $E\Psi(x) = E\Psi(x)$.

The kinetic energy of the particle is $p^2/2m$, where $p = kh$, and the wave number k is $k = 2\pi/\lambda$.

Using the boundary conditions, the wave function is equal to zero at $x = 0$ and $x = L$. We have

$$\Psi(x)|_{x=0} = \Psi(0) = 0 \quad \text{and} \quad \Psi(x)|_{x=L} = \Psi(L) = 0.$$

From $\Psi(0) = C\sin kx + D\cos kx = 0$, we conclude that $D = 0$.

Using $C\sin kL = 0$, one finds the expression for kL. In particular, taking note of $C \neq 0$, we have

$$kL = n\pi, \quad n = 1, 2, 3, \ldots,$$

where n is the integer (if $n = 0$, the wave function vanishes everywhere, and thus, $n \neq 0$).

Hence, $k = n\pi/L$, and the wavelength is $\lambda = 2\pi/k = 2L/n$, $n = 1, 2, 3, \ldots$.

From $k = \sqrt{2mE/\hbar^2}$, by making use of $kL = n\pi$, we obtain the expression for the allowed energy levels. In particular, discrete values of the energy derived solving the Schrödinger equation are

$$E_n = \frac{\hbar^2\pi^2}{2mL^2}n^2, \quad n = 1, 2, 3, \ldots.$$

Thus, we derived the allowable energy levels for a *microscopic* particle. Each energy level has its quantum number n and the corresponding wave function. For example, if $n = 1$ and $n = 2$, we have

$$E_{n=1} = \frac{\hbar^2\pi^2}{2mL^2}$$

(the lowest possible energy, called the ground state, and usually denoted as E_0) and

$$E_{n=2} = \frac{2\hbar^2\pi^2}{mL^2}.$$

The energy at the ground state (state of lowest energy) is $E_{n=1} = (\hbar^2\pi^2)/(2mL^2)$, which is different from a classical particle at rest with $p = 0$ and $\Pi(x) = 0$, that is, the sum of the kinetic and potential energy is zero.

We demonstrated that the energy levels for the particle is quantized. Furthermore, the wave function is

$$\Psi_n(x) = C \sin kx + D \cos kx = C \sin\left(\frac{n\pi}{L}x\right).$$

The constant C is found using the normalization condition $\int_{-\infty}^{\infty} \Psi^*(t,x) \times \Psi(t,x)\mathrm{d}x = 1$, which gives $\int_{-\infty}^{+\infty} \Psi_n^2(x)\mathrm{d}x = 1$. This equation indicates that the probability of finding the particle along the x axis is 1. The wave function is equal to zero except $0 < x < L$. Using the probability density, we normalize the wave function as

$$\int_0^L \Psi_n^2(x)\,\mathrm{d}x = C^2 \int_0^L \sin^2\left(\frac{n\pi}{L}x\right)\mathrm{d}x = C^2\frac{L}{n\pi}\int_0^{n\pi} \sin^2 z\,\mathrm{d}z = 1, \quad z = \frac{n\pi}{L}x.$$

From

$$C^2\frac{L}{n\pi}\frac{n\pi}{2} = C^2\frac{L}{2} = 1,$$

we have $C = \sqrt{2/L}$.

One obtains

$$\Psi_n(x) = \sqrt{\frac{2}{L}}\sin\left(\frac{n\pi}{L}x\right)$$
$$= \frac{1}{i\sqrt{2L}}\left(e^{i(n\pi/L)x} - e^{-i(n\pi/L)x}\right), \quad 0 \le x \le L, \ n = 1, 2, 3, \ldots.$$

For $n = 1$ and $n = 2$, we have

$$\Psi_1(x) = \sqrt{\frac{2}{L}}\sin\left(\frac{\pi}{L}x\right) \quad \text{and} \quad \Psi_2(x) = \sqrt{\frac{2}{L}}\sin\left(\frac{2\pi}{L}x\right).$$

The probability density $P(t, x)$ specifies the probability, per unit length, of finding the particle near x at time t. Taking note of $P(t, x) = \Psi^*(t, x)\Psi(t, x)$, we have

$$P_n(x) = \frac{2}{L}\sin^2\left(\frac{n\pi}{L}x\right).$$

The probability of finding a particle between $x = 0$ and $x = l$ is given by

$$P = \int_0^l \Psi^*(x)\Psi(x)\mathrm{d}x = \frac{2}{L}\int_0^l \sin^2\left(\frac{n\pi}{L}x\right)\mathrm{d}x = \frac{1}{L}\left[x - \frac{L}{2n\pi}\sin\left(\frac{2n\pi}{L}x\right)\right]\Bigg|_0^l$$
$$= \frac{1}{L}\left[l - \frac{L}{2n\pi}\sin\left(\frac{2n\pi}{L}l\right)\right].$$

The derived probability can be analyzed. If $l = L$, we have $P = 1$, and $P = 0.5$ when $l = L/2$.

One of the special properties of eigenfunctions are that the solutions are orthogonal (although the solutions are related to each other, their overlap is zero).

We may derive the expectation (average value) of variables. The average momentum is

$$\langle p\rangle = \int_0^L \Psi^*(x)\frac{\hbar}{i}\frac{\partial(\Psi(x))}{\partial x}\mathrm{d}x = \frac{2}{L}\int_0^L \sin\left(\frac{n\pi}{L}x\right)\frac{\hbar}{i}\frac{\partial[\sin((n\pi/L)x)]}{\partial x}\mathrm{d}x$$
$$= \frac{2\hbar n\pi}{iL^2}\int_0^L \sin\left(\frac{n\pi}{L}x\right)\cos\left(\frac{n\pi}{L}x\right)\mathrm{d}x = \frac{i\hbar}{L}\left[\sin^2\left(\frac{n\pi}{L}x\right)\right]_0^L = 0.$$

The average momentum is zero because the particle will travel to the right and to the left with equal probability, resulting in no net momentum.

In contrast, the average square of the momentum is not zero:

$$\langle p^2\rangle = \int_0^L \Psi^*(x)\left(\frac{\hbar}{i}\frac{\partial}{\partial x}\right)^2\Psi(x)\,\mathrm{d}x = \frac{\hbar^2\pi^2n^2}{L^2} = \frac{h^2n^2}{4L^2} \neq 0.$$

We also recall that $p^2 = 2mE$, and we obtain $E_n = (\hbar^2\pi^2/2mL^2)n^2$.

The expectation value of position is

$$\langle x \rangle = \int_0^L \Psi^*(x)x\Psi(x)\,\mathrm{d}x = \frac{2}{L}\int_0^L \sin\left(\frac{n\pi}{L}x\right) x \sin\left(\frac{n\pi}{L}x\right)\,\mathrm{d}x$$
$$= \frac{2}{L}\int_0^L \sin^2\left(\frac{n\pi}{L}x\right) x\,\mathrm{d}x = \frac{1}{2}L.$$

We conclude that the particle has an average position at $L/2$, that is, the particle is most likely to be in the middle of the well.

Figure 6.8 shows the wave functions $\Psi_n(x)$ and probability densities $P_n(x)$. For the illustrative purpose, the wave function and probability density are given in the *arbitrary unit* (regularized to 1) by dividing by the maximum value. For the ground state ($n = 1$) and ($n = 5$) the results are reported

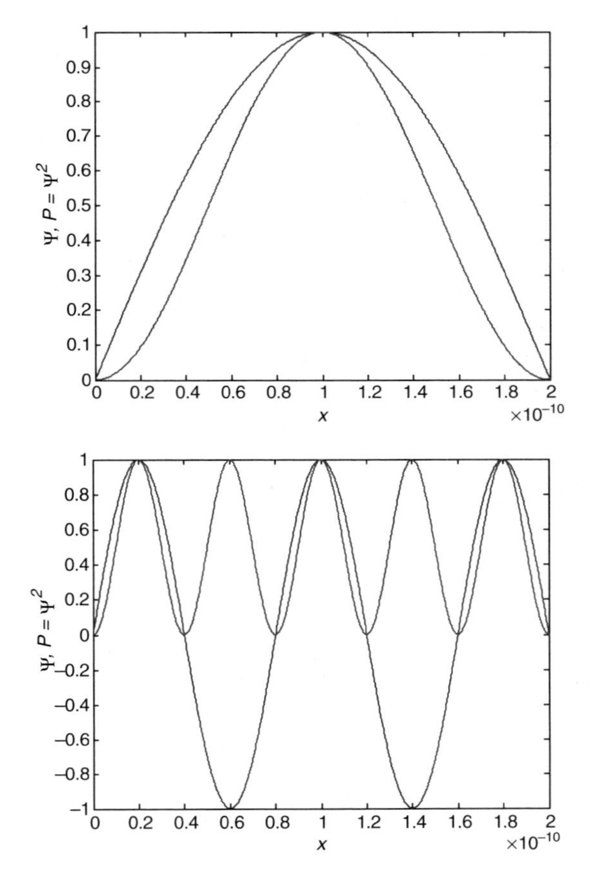

FIGURE 6.8
Wave functions and probability densities for $n = 1$ and $n = 5$.

letting $L = 2 \times 10^{-10}$ m. The equations $E_n = (\hbar^2\pi^2/2mL^2)n^2$ and $\Psi_n(x) = \sqrt{2/L}\sin(n\pi/Lx)$ are solved in MATLAB using the following statement:

```
L=2e-10;  m=9.10938188e-31;  x=0:L/1000:L;
h=1.054571596e-34;  n=5;  En=h*h*pi*pi*n*n/(2*m*L*L);
P=sqrt(0.5*L)*sin(n*pi*x/L);
plot(x,P/max(P),x,P.*P/max(P)^2);
title('Wave Function and Probability Density','FontSize',14);
xlabel('\itx','FontSize',14);
ylabel('\it\Psi, \itP=\Psi^2','FontSize',14);
```

For $n = 1$ and $n = 5$, the allowed energy levels are E_n are $E_1 = 1.5062 \times 10^{-18}$ J and $E_5 = 3.7654 \times 10^{-17}$ J, respectively. The states for which $n > 1$ are called the excited states. For $n = 1$, the probability density is maximum at $L/2$, and the probability density falls to zero at the edges. In contrast, the classical mechanics gives equal probability at any position. For $n = 5$, the probability density is maximum at $L/10$, $3L/10$, $5L/10$, $7L/10$, and $9L/10$ with zero probability density at 0, $L/5$, $2L/5$, $3L/5$, $4L/5$, and L. The *microscopic* particle moves and can be found (with maximum probabilities) at $L/10$, $3L/10$, $5L/10$, $7L/10$, and $9L/10$, and cannot be found at positions 0, $L/5$, $2L/5$, $3L/5$, $4L/5$, and L. How can the particle travel for $L/10$ to $3L/10$ and not being found at $L/5$? Quantum physics examines behavior, phenomena, and effects in terms of waves and not particle position.

The potential examined, E_n obtained, and wave function derived for standing waves are not directly related to the electron transport problem in the finite potentials. To study electron transport, finite potentials that correspond to realistic potentials in atomic complexes should be studied. However, the considered example is related to insulation and noise immunity problems important in ^{ME}devices. For an infinite potential, the difference between the energy levels $(E_n - E_{n-1})$ is proportional to $1/L^2$. That is, a small width L leads to high $(E_n - E_{n-1})$, distinguishing molecular (nano) electronics (L is in the range of Å) and microelectronics.

Example 6.16

An electron is trapped in a one-dimensional region (infinite square well) with dimensions $x = 0$ to $L = 1 \times 10^{-10}$ m. One can find the external energy needed to excite the electron from the ground state to the first and second exited states ($n = 2$ and $n = 3$). In Example 6.15 we derived the expression $E_n = (\hbar^2\pi^2/2mL^2)n^2, n = 1, 2, 3, \ldots$.

Hence, one obtains $E_0 = (\hbar^2\pi^2/2mL^2)$, and

$$E_0 = \frac{\hbar^2\pi^2}{2mL^2} = \frac{(1.055 \times 10^{-34})^2 3.14^2}{2(9.1 \times 10^{-31})(1 \times 10^{-10})^2} = 6 \times 10^{-18}\ \text{J} = 37\ \text{eV}.$$

In the first and second excited state, the values are 4 and 9 times E_n due to the term n^2. Correspondingly, the external energy needed to excite the electron from the ground state to the first ($n = 2$) and second ($n = 3$) excited states are

$$E_2 = 4E_0 - E_0 = 3E_0 = 111 \text{ eV}$$

and

$$E_3 = 9E_0 - E_0 = 8E_0 = 296 \text{ eV}.$$

The probability of finding a *microscopic* particle between $[x_1\ x_2]$, if $n = 1$, is

$$P = \int_{x_1}^{x_2} \Psi^2(x)\,\mathrm{d}x = \frac{2}{L}\int_{x_1}^{x_2} \sin^2\left(\frac{\pi}{L}x\right)\mathrm{d}x = \frac{1}{L}\left[x - \frac{L}{2\pi}\sin\left(\frac{2\pi}{L}x\right)\right]\Bigg|_{x_1}^{x_2}.$$

In the first exited state ($n = 2$), the probability to find the electron in $[0\,0.25 \times 10^{-10}]$ is

$$P = \int_{x_1}^{x_2} \Psi^2(x)\,\mathrm{d}x = \frac{2}{L}\int_{x_1}^{x_2} \sin^2\left(\frac{2\pi}{L}x\right)\mathrm{d}x = \frac{1}{L}\left[x - \frac{L}{4\pi}\sin\left(\frac{4\pi}{L}x\right)\right]\Bigg|_{x_1}^{x_2} = 0.25.$$

Example 6.17

Consider a *microscopic* particle in a quantum well as illustrated in Figure 6.9. This quantum well represents a finite potential well. Outside of the well the potential is Π_0. In the studied three regions (I, II, and III), the potential is given as

$$\Pi(x) = \begin{cases} 0 & \text{for } -L \le x \le L, \\ \Pi_0 & \text{for } |x| > L. \end{cases}$$

Hence, at $x = \pm L$, the particle encounters a barrier with the potential Π_0. Our goal is to find the wave function and derive the expression for the quantized energy as functions of Π_0 and width L.

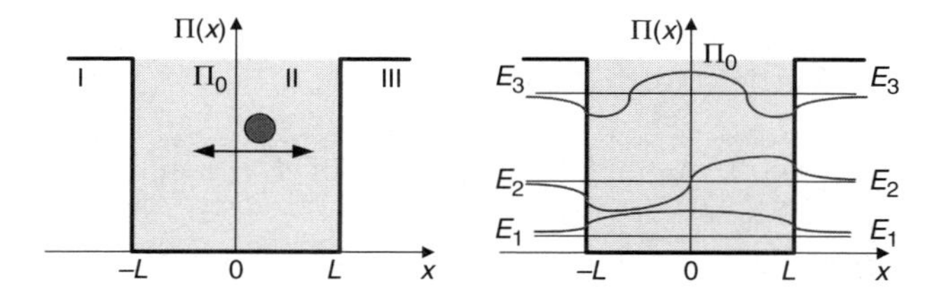

FIGURE 6.9
Particle in quantum well with a square-well finite potential, and energy level diagram with the wave functions for three bound states.

In Newtonian mechanics, a particle trapped (localized) in a potential well can vibrate with periodic motion if $E < \Pi_0$. We apply the quantum mechanics. The symmetry of the potential leads to the reflection symmetry (smooth joining conditions are satisfied for even and odd eigenfunctions) and symmetry of the Hamiltonian. Correspondingly, the wave function satisfies $\Psi(x) = \Psi(-x)$ for even, and $\Psi(x) = -\Psi(-x)$ for odd case. Hence, we may solve the problem for $0 \le x < \infty$, and then derive the solution in $-\infty < x < \infty$.

In the Schrödinger equation

$$\left\{\frac{\mathrm{d}^2}{\mathrm{d}x^2} + \frac{2m}{\hbar^2}[E - \Pi(x)]\right\}\Psi(x) = 0,$$

we denote

$$k^2 = \frac{2mE}{\hbar^2} \quad \text{and} \quad \kappa^2 = \frac{2m}{\hbar^2}(\Pi_0 - E) = \frac{2m\Pi_0}{\hbar^2} - k^2.$$

Therefore, the Schrödinger equation in two regions (II and III) are

$$\left(\frac{\mathrm{d}^2}{\mathrm{d}x^2} + k^2\right)\Psi_{\mathrm{II}} = 0, \quad 0 \le x \le L \text{ (inside the well)}$$

and

$$\left(\frac{\mathrm{d}^2}{\mathrm{d}x^2} - \kappa^2\right)\Psi_{\mathrm{III}} = 0, \quad x > L \text{ (outside the well)}.$$

A finite solution for $x \to \infty$ (in region III) is

$$\Psi_{\mathrm{III}} = A_{\mathrm{III}}\,\mathrm{e}^{-\kappa x}, \quad \kappa > 0.$$

In region II, we have two solutions for even and odd cases

$$\Psi_{\mathrm{II}} = B_{\mathrm{II}} \cos kx \quad \text{and} \quad \Psi_{\mathrm{II}} = C_{\mathrm{II}} \sin kx.$$

The constants A_{III}, B_{II}, and C_{II} are found using the boundary and normalization conditions.

For the even case, letting $x = L$, we have

$$B_{\mathrm{II}} \cos kL = A_{\mathrm{III}}\,\mathrm{e}^{-\kappa L} \quad \text{and} \quad kB_{\mathrm{II}} \sin kL = \kappa A_{\mathrm{III}}\,\mathrm{e}^{-\kappa L}.$$

Dividing $kB_{\mathrm{II}} \sin kL = \kappa A_{\mathrm{III}}\,\mathrm{e}^{-\kappa L}$ by $B_{\mathrm{II}} \cos kL = A_{\mathrm{III}}\,\mathrm{e}^{-\kappa L}$, we obtain

$$k \tan(kL) = \kappa = \sqrt{\frac{2m\Pi_0}{\hbar^2} - k^2} = \sqrt{\frac{2m}{\hbar^2}(\Pi_0 - E)}.$$

For the odd case, using $C_{II} \sin kL = A_{III}\, e^{-\kappa L}$ and $kC_{II} \cos kL = -\kappa A_{III}\, e^{-\kappa L}$, the division gives

$$k\text{ctn}(kL) = -\kappa = -\sqrt{\frac{2m\Pi_0}{\hbar^2} - k^2} = -\sqrt{\frac{2m}{\hbar^2}(\Pi_0 - E)}.$$

We obtain two transcendental equations that result in positive values of the wave number that corresponds to the energy levels of the particle. In fact, using k and κ, we have the equation for a circle $k^2 + \kappa^2 = (2m\Pi_0/\hbar^2)$. By multiplying the left and right sides by L^2, one obtains $(k^2 + \kappa^2)L^2 = (2m\Pi_0/\hbar^2)L^2$ with $r^2 = (2m\Pi_0 L^2/\hbar^2)$.

Equations

$$kL \tan(kL) = \kappa L = \sqrt{\frac{2m\Pi_0}{\hbar^2}L^2 - k^2L^2}, \quad -kL\text{ctn}(kL) = \kappa L = \sqrt{\frac{2m\Pi_0}{\hbar^2}L^2 - k^2L^2}$$

and

$$(k^2 + \kappa^2)L^2 = \frac{2m\Pi_0}{\hbar^2}L^2$$

provide algebraic equations with two variables kL and κL. The numeric, analytic, and graphic solutions can be obtained. Figure 6.10 shows plots for positive values kL and κL. The unknown constants A_{III}, B_{II}, and C_{II} are found as the intersections as shown in Figure 6.10. Here, the dimensionless parameter $A = (2m\Pi_0/\hbar^2)L^2$ is used. For distinct m, L, and Π_0, one can draw the circle to derive A_{III}, B_{II}, and C_{II}. The resulting transcendental equations can be solved using other methods.

By making use of $k^2 = 2mE/\hbar^2$, the quantized energy is found to be

$$E_n = \frac{\hbar^2 k_n^2}{2m}, \quad n = 1, 2, 3, \ldots.$$

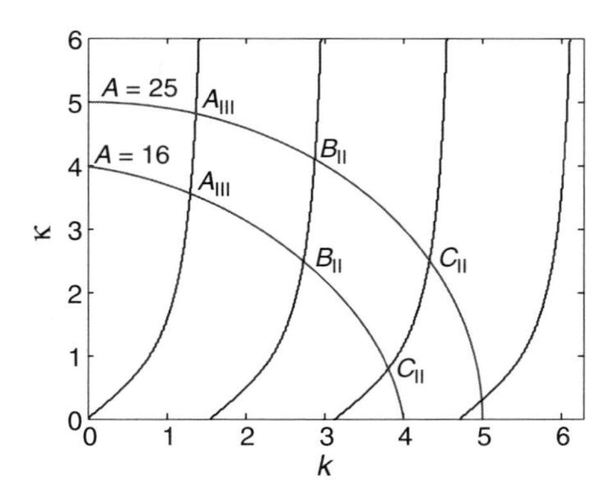

FIGURE 6.10
Graphical solutions of transcendental equations: The interceptions determine the allowed energy levels in a quantum well with potential Π_0 and width $2L$. The dimensionless parameter A is assumed to be 16 and 25.

In the case of a quantum well of infinite depth ($\Pi_0 \to \infty$), one has $k_n = n\pi/2L$. Hence,

$$E_n = \frac{\hbar^2\pi^2}{2m(2L)^2}n^2, \quad n = 1, 2, 3, \ldots.$$

When the width L increases, the energy decreases. When $L \to \infty$, the particle becomes free and has continuous values of energy $E = \Pi_0 + \hbar^2 k^2/2m$.

Example 6.18

Consider a finite square well of width L with three regions (I, II, and III) similar to the potential barrier as documented in Figure 6.9. Let

$$\Pi(x) = \begin{cases} 0 & \text{for } x < \frac{1}{2}L, \\ -\Pi_0 & \text{for } -\frac{1}{2}L \le x \le \frac{1}{2}L, \\ 0 & \text{for } x > \frac{1}{2}L. \end{cases}$$

The potential admits bound states ($E < 0$) and scattering states with $E > 0$. That is, for $E > 0$, the particle is unconfined and scattered by the potential.

Outside (if $|x| > \frac{1}{2}L$) and inside (for $-\frac{1}{2}L \le x \le \frac{1}{2}L$) the quantum well, the Schrödinger equations are

$$-\frac{\hbar^2}{2m}\frac{d^2\Psi}{dx^2} = E\Psi \quad \text{or} \quad \frac{d^2\Psi}{dx^2} + k^2\Psi = 0, \quad \text{where } k^2 = \frac{2m}{\hbar^2}E,$$

and

$$-\frac{\hbar^2}{2m}\frac{d^2\Psi}{dx^2} - \Pi_0\Psi = E\Psi \quad \text{or} \quad \frac{d^2\Psi}{dx^2} + \kappa^2\Psi = 0, \quad \text{where } \kappa^2 = \frac{2m}{\hbar^2}(E + \Pi_0).$$

The general solutions are

$$\Psi_{\mathrm{I}}(x) = A_{\mathrm{I}}\,e^{ikx} + B_{\mathrm{I}}\,e^{-ikx}, \quad x < -\tfrac{1}{2}L,$$
$$\Psi_{\mathrm{II}}(x) = A_{\mathrm{II}}\,e^{i\kappa x} + B_{\mathrm{II}}\,e^{-i\kappa x}, \quad -\tfrac{1}{2}L \le x \le \tfrac{1}{2}L,$$
$$\Psi_{\mathrm{III}}(x) = A_{\mathrm{III}}\,e^{ikx} + B_{\mathrm{III}}\,e^{-ikx}, \quad x > \tfrac{1}{2}L,$$

where A_i and B_i are the unknown constants that can be derived using the boundary and continuity conditions.

The Schrödinger differential equations are valid in all three regions. In order to simplify the solution, we are using the particular solution taking into account the convergence (decaying) of $\Psi(x)$, that is, real exponentials can be used instead of complex exponentials. Taking note of $\Psi_{II}(x) = A_{II}\sin(\kappa x) + B_{II}\cos(\kappa x)$, the continuity of $\Psi(x)$ and $d\Psi(x)/dx$ at $x = -L/2$ gives

$$A_I\,e^{-(1/2)ikL} + B_I\,e^{(1/2)ikL} = -A_{II}\sin\left(\tfrac{1}{2}\kappa L\right) + B_{II}\cos\left(\tfrac{1}{2}\kappa L\right)$$

and

$$ik(A_I\,e^{-(1/2)ikL} - B_I\,e^{(1/2)ikL}) = \kappa\left[A_{II}\cos\left(\tfrac{1}{2}\kappa L\right) + B_{II}\sin\left(\tfrac{1}{2}\kappa L\right)\right],$$

while at $x = L/2$, taking into account $B_{III} = 0$, one obtains

$$A_{II}\sin\left(\tfrac{1}{2}\kappa L\right) + B_{II}\cos\left(\tfrac{1}{2}\kappa L\right) = A_{III}\,e^{(1/2)ikL}$$

and

$$\kappa[A_{II}\cos\left(\tfrac{1}{2}\kappa L\right) - B_{II}\sin\left(\tfrac{1}{2}\kappa L\right)] = ikA_{III}\,e^{(1/2)ikL}.$$

Here, A_I, B_I and A_{III} are the incident, reflected and transmitted amplitudes. With the ultimate objective to study the transmission coefficient, we express

$$A_{III} = (e^{-ikL}/(\cos(\kappa L) - i(k^2 + \kappa^2/2k\kappa)\sin(\kappa L)))A_I.$$

Correspondingly, the transmission coefficient is found as

$$T(E) = \frac{|A_{III}|^2}{|A_I|^2} = \left[1 + \frac{\Pi_0^2}{4E(E+\Pi_0)}\sin^2\left(\frac{L}{\hbar}\sqrt{2m(E+\Pi_0)}\right)\right]^{-1}.$$

The transmission coefficient is a periodic function. The maximum achievable transmission, that is, the *total* transmission $T(E) = 1$, is guaranteed if $L\sqrt{2m(E+\Pi_0)}/\hbar = n\pi,\ n = 1, 2, 3, \ldots$. The energies for a *total* transmission are related as $E_n + \Pi_0 = (n^2\pi^2\hbar^2/2mL^2)$.

Denoting $K = \frac{1}{2}L\kappa$ and $K_0 = (L/\sqrt{2}\hbar)\sqrt{m\Pi_0}$, the transcendental equation, that defines K and E, is $\tan K = \sqrt{(K_0^2/K^2) - 1}$. This equation can be solved analytically and numerically.

For shallow and narrow quantum wells, there is a limited number of bound states. There is always one bound state, and for $K_0 < \frac{1}{2}\pi$ only one state remains. Having solved the transcendental equation, one uses E to find the transmission coefficient, which is a function of energy. The potential $\Pi(x)$, mass and well width result in variations of $T(E)$.

We examine two quantum wells. Let $\Pi_0 = 0.3$ eV, $L = 14$ nm and $L = 0.14$ nm, effective masses are $0.1m_e$, $0.5m_e$, and m_e (semiconducting heterogeneous structure) as well as $0.5 \times 10^4 m_e$ and $1 \times 10^4 m_e$. The well width $L = 0.14$ nm corresponds to the one-dimensional analysis of the electron transport in organic molecules for which the bond lengths C—C and

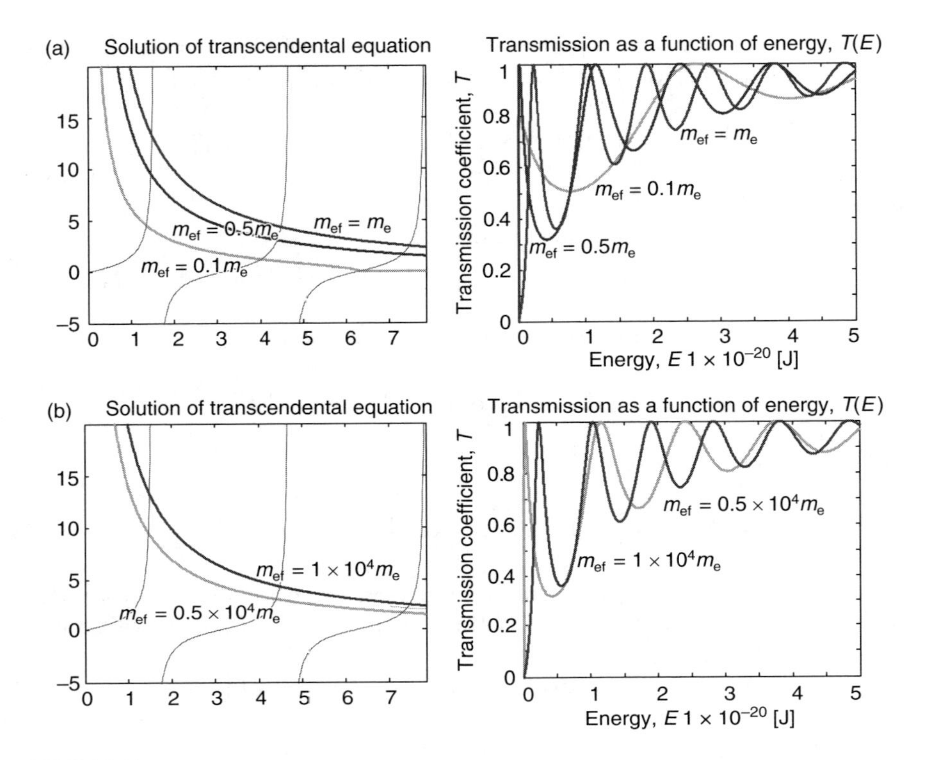

FIGURE 6.11
Solution of the transcendental equation and transmission coefficient $T(E)$ for a finite quantum wells: (a) $\Pi_0 = 0.3$ eV and $L = 14$ nm; (b) $\Pi_0 = 0.3$ eV and $L = 0.14$ nm.

C—N are approximately 0.14 nm. Figures 6.11a and b document the numerical solutions for the studied heterogeneous structure and atomic complex. As E increases, the resonance becomes broader.

Example 6.19

Consider a one-dimensional scattering problem for a particle of mass m that moves from the left to a rectangular potential barrier with

$$\Pi(x) = \begin{cases} 0 & \text{for } x < 0, \\ \Pi_0 & \text{for } 0 \le x \le L, \\ 0 & \text{for } x > L. \end{cases}$$

We consider two cases when $E > \Pi_0$ and $E < \Pi_0$.

The Schrödinger equation results in two differential equations for two distinct regions when $\Pi(x) = 0$ or $\Pi(x) = \Pi_0 \neq 0$. In particular,

$$-\frac{\hbar^2}{2m}\frac{d^2\Psi(x)}{dx^2} = E\Psi(x) \quad \text{or} \quad \frac{d^2\Psi(x)}{dx^2} + k^2\Psi(x) = 0, \quad \text{where } k^2 = \frac{2m}{\hbar^2}E,$$

and

$$-\frac{\hbar^2}{2m}\frac{d^2\Psi(x)}{dx^2} + \Pi_0\Psi(x) = E\Psi(x) \quad \text{or} \quad \frac{d^2\Psi(x)}{dx^2} + \kappa_i^2\Psi(x) = 0,$$

where κ_i (κ_1 or κ_2) depends on the amplitudes of the incident particle energy E and potential Π_0.

For $E > \Pi_0$, the general solutions in three regions are

$$\Psi_{\text{I}}(x) = A_{\text{I}}\,e^{ikx} + B_{\text{I}}\,e^{-ikx}, \quad x < 0,$$
$$\Psi_{\text{II}}(x) = A_{\text{II}}\,e^{i\kappa_1 x} + B_{\text{II}}\,e^{-i\kappa_1 x}, \quad 0 \le x \le L,$$
$$\Psi_{\text{III}}(x) = A_{\text{III}}\,e^{ikx} + B_{\text{III}}\,e^{-ikx}, \quad x > L,$$

where

$$\kappa_1^2 = \frac{2m}{\hbar^2}(E - \Pi_0).$$

While, for $E < \Pi_0$, one defines $\kappa_2^2 = \frac{2m}{\hbar^2}(\Pi_0 - E)$, and

$$\Psi_{\text{I}}(x) = A_{\text{I}}\,e^{ikx} + B_{\text{I}}\,e^{-ikx}, \quad x < 0,$$
$$\Psi_{\text{II}}(x) = A_{\text{II}}\,e^{-\kappa_2 x} + B_{\text{II}}\,e^{\kappa_2 x}, \quad 0 \le x \le L,$$
$$\Psi_{\text{III}}(x) = A_{\text{III}}\,e^{ikx} + B_{\text{III}}\,e^{-ikx}, \quad x > L.$$

For $E > \Pi_0$, applying classical considerations, the particle with momentum $p_1 = \sqrt{2mE}$ entering the potential slows to $p_2 = \sqrt{2m(E - \Pi_0)}$, and gains momentum as $x = L$ regaining p_1 and keeping p_1 for $x > L$. Hence, in regions $x < 0$ and $x > L$ we have a *total* transmission. The application of quantum mechanics leads to the solution of the Schrödinger equation with different results for three distinct regions. For $E > \Pi_0$, one obtains

$$T(E) = \frac{|A_{\text{III}}|^2}{|A_{\text{I}}|^2} = \left[1 + \frac{\Pi_0^2}{4E(E - \Pi_0)}\sin^2\left(\frac{L}{\hbar}\sqrt{2m(E - \Pi_0)}\right)\right]^{-1}$$

which is usually written as

$$T(E) = \left[1 + \frac{1}{4(E/\Pi_0)((E/\Pi_0) - 1)}\sin^2\left(\frac{L}{\hbar}\sqrt{2m\Pi_0\left(\frac{E}{\Pi_0} - 1\right)}\right)\right]^{-1}.$$

The *total* transmission occurs when the incident energy of a particle is

$$E_n = \Pi_0 + \frac{n^2\pi^2\hbar^2}{2mL^2}, \quad n = 1, 2, 3, \ldots.$$

The maxima of the $T(E)$ coincides with the energy eigenvalues of the infinite square-well potential known as resonances, which do not appear in classical physics consideration. This resonance phenomenon is due to interference between the incident and reflected waves observed in the atomic structures in low-energy ($E \sim 0.1$ eV) scattering phenomena such as Ramsauer–Townsend and other effects.

If $E \gg \Pi_0$, $T(E) \approx 1$ and $R(E) \approx 0$.

The tunneling problem is focused on the analysis of propagation of particles through regions (barrier) where the particle energy is smaller than the potential energy, that is, $E < \Pi(x)$. For tunneling, $E < \Pi_0$, and one has

$$T(E) = \left[1 + \frac{\Pi_0^2}{4E(\Pi_0 - E)} \sinh^2\left(\frac{L}{\hbar}\sqrt{2m(\Pi_0 - E)}\right)\right]^{-1}$$

or

$$T(E) = \left[1 + \frac{1}{4(E/\Pi_0)(1 - (E/\Pi_0))} \sinh^2\left(\frac{L}{\hbar}\sqrt{2m\Pi_0\left(1 - \frac{E}{\Pi_0}\right)}\right)\right]^{-1}$$

And

$$R(E) = \frac{\Pi_0^2}{4E(\Pi_0 - E)} \sinh^2\left(\frac{L}{\hbar}\sqrt{2m(\Pi_0 - E)}\right) T(E).$$

For $E \ll \Pi_0$, taking note of the following approximation $\sinh(z) \approx e^z/2$, one finds

$$T(E) \approx \frac{16E}{\Pi_0}\left(1 - \frac{E}{\Pi_0}\right) e^{-(2L/\hbar)\sqrt{2m(\Pi_0 - E)}}.$$

If $E \approx \Pi_0$, we obtain

$$T(E) = (1 + (mL^2\Pi_0/2\hbar^2))^{-1} \text{ and } R(E) = (1 + (2\hbar^2/mL^2\Pi))^{-1}.$$

Example 6.20

In the analysis, the scattering and transfer matrices can be used.

For a rectangular potential barrier we have

$$\Pi(x) = \begin{cases} 0 & \text{for } x < -L, \\ \Pi_0 & \text{for } -L \le x \le L, \\ 0 & \text{for } x > L. \end{cases}$$

The Schrödinger equation when $\Pi(x) = 0$ (in regions $x < -L$ and $x > L$) is

$$-\frac{\hbar^2}{2m}\frac{d^2\Psi(x)}{dx^2} = E\Psi(x) \quad \text{or} \quad \frac{d^2\Psi(x)}{dx^2} + k^2\Psi(x) = 0, \quad \text{where } k^2 = \frac{2m}{\hbar^2}E,$$

In the region where $\Pi(x) = \Pi_0 \neq 0$ $(-L \leq x \leq L)$, one has

$$-\frac{\hbar^2}{2m}\frac{\mathrm{d}^2\Psi(x)}{\mathrm{d}x^2} + \Pi_0\Psi(x) = E\Psi(x) \quad \text{or} \quad \frac{\mathrm{d}^2\Psi(x)}{\mathrm{d}x^2} + \kappa^2\Psi(x) = 0,$$

where $\kappa^2 = (2m/\hbar^2)(\Pi_0 - E)$ for $E < \Pi_0$.

The expressions for wave functions are

$$\Psi_\mathrm{I}(x) = A_\mathrm{I}\,\mathrm{e}^{ikx} + B_\mathrm{I}\,\mathrm{e}^{-ikx}, \quad x < -L,$$
$$\Psi_\mathrm{II}(x) = A_\mathrm{II}\,\mathrm{e}^{-\kappa x} + B_\mathrm{II}\,\mathrm{e}^{\kappa x}, \quad -L \leq x \leq L,$$
$$\Psi_\mathrm{III}(x) = A_\mathrm{III}\,\mathrm{e}^{ikx} + B_\mathrm{III}\,\mathrm{e}^{-ikx}, \quad x > L.$$

Using the continuity of $\Psi(x)$ at $x = -L$, one obtains $A_\mathrm{I}\,\mathrm{e}^{-ikL} + B_\mathrm{I}\,\mathrm{e}^{ikL} = A_\mathrm{II}\,\mathrm{e}^{-\kappa L} + B_\mathrm{II}\,\mathrm{e}^{\kappa L}$, while continuity of $\mathrm{d}\Psi/\mathrm{d}x$ at $x = -L$ gives $A_\mathrm{I}\,\mathrm{e}^{-ikL} - B_\mathrm{I}\,\mathrm{e}^{ikL} = (i\kappa/k)(A_\mathrm{II}\,\mathrm{e}^{\kappa L} - B_\mathrm{II}\,\mathrm{e}^{-\kappa L})$.

Hence, we have

$$\begin{bmatrix} A_\mathrm{I} \\ B_\mathrm{I} \end{bmatrix} = M_1 \begin{bmatrix} A_\mathrm{II} \\ B_\mathrm{II} \end{bmatrix},$$

where $M_1 \in \mathbb{R}^{2\times 2}$ is the matrix

$$M_1 = \begin{bmatrix} \dfrac{k + i\kappa}{2k}\mathrm{e}^{(ik+\kappa)L} & \dfrac{k - i\kappa}{2k}\mathrm{e}^{(ik-\kappa)L} \\ \dfrac{k - i\kappa}{2k}\mathrm{e}^{-(ik-\kappa)L} & \dfrac{k + i\kappa}{2k}\mathrm{e}^{-(ik+\kappa)L} \end{bmatrix}.$$

Continuity conditions for $\Psi(x)$ and $\mathrm{d}\Psi/\mathrm{d}x$ at boundary $x = L$ result in

$$\begin{bmatrix} A_\mathrm{II} \\ B_\mathrm{II} \end{bmatrix} = M_2 \begin{bmatrix} A_\mathrm{III} \\ B_\mathrm{III} \end{bmatrix},$$

where $M_2 \in \mathbb{R}^{2\times 2}$ is the matrix

$$M_2 = \begin{bmatrix} \dfrac{k - i\kappa}{2k}\mathrm{e}^{(ik+\kappa)L} & \dfrac{k + i\kappa}{2k}\mathrm{e}^{-(ik-\kappa)L} \\ \dfrac{k + i\kappa}{2k}\mathrm{e}^{(ik-\kappa)L} & \dfrac{k - i\kappa}{2k}\mathrm{e}^{-(ik+\kappa)L} \end{bmatrix}.$$

The transfer matrix, which relates the amplitudes of wave functions in the regions I, II, and III, is $M = M_1 M_2$. This transfer matrix, which provides the relationship between the incident, reflected, and transmitted wave functions, is straightforwardly applied to derive $T(E)$. In particular, we have

$$\begin{bmatrix} A_{\mathrm{I}} \\ B_{\mathrm{I}} \end{bmatrix} = \begin{bmatrix} \left(\cosh 2\kappa L + \frac{1}{2} i \frac{\kappa^2 - k^2}{k\kappa} \sinh 2\kappa L\right) e^{2ikL} & \frac{1}{2} i \frac{\kappa^2 + k^2}{k\kappa} \sinh 2\kappa L \\ -\frac{1}{2} i \frac{\kappa^2 + k^2}{k\kappa} \sinh 2\kappa L & \left(\cosh 2\kappa L - \frac{1}{2} i \frac{\kappa^2 - k^2}{k\kappa} \sinh 2\kappa L\right) e^{-2ikL} \end{bmatrix} \begin{bmatrix} A_{\mathrm{III}} \\ B_{\mathrm{III}} \end{bmatrix}.$$

Hence

$$\frac{A_{\mathrm{III}}}{A_{\mathrm{I}}} = \frac{e^{-2ikL}}{\cosh 2\kappa L + (1/2) i (\kappa^2 - k^2/k\kappa) \sinh 2\kappa L}.$$

Assuming $\kappa L \gg 1$, and denoting $z = 2\kappa L$, one may apply the following approximation $\cosh(z) \approx \sinh(z) \approx e^z/2$. Hence, we have

$$T(E) = \frac{|A_{\mathrm{III}}|^2}{|A_I|^2} \approx 16\, e^{-4\kappa L} \left(\frac{k\kappa}{k^2 + \kappa^2}\right)^2.$$

For the narrow and high barrier, we have $\Pi_0 \gg E$, $\kappa \gg k$ and $\kappa L \ll 1$. Terms $\Pi_0 L$ and $\kappa^2 L$ are finite. One obtains

$$T(E) = (|A_{\mathrm{III}}|^2/|A_I|^2) \approx (k^2/k^2 + \kappa^4 L^2) = (E/E + (2m/\hbar^2)\Pi_0^2 L^2).$$

The results in deriving $T(E)$ are enhanced by applying the Wentzel–Kramers–Brillouin approximation. For an integrable, continuous, slow-varying potential $\Pi(x)$, one has

$$T(E) \cong e^{-(2/\hbar)\int_{x_0}^{x_f} \sqrt{2m[\Pi(x) - E]} dx}.$$

This equation is obtained by making use of the Schrödinger equation

$$-\frac{\hbar^2}{2m}\frac{d^2\Psi}{dx^2} + \Pi(x)\Psi = E\Psi,$$

which is rewritten as

$$\frac{d^2\Psi(x)}{dx^2} + \frac{p^2(x)}{\hbar^2}\Psi(x) = 0,$$

where $p^2(x) = 2m[E - \Pi(x)]$.

The general approximate solution is $\Psi(x) \cong (A/\sqrt{p(x)})e^{\pm(1/\hbar)\int |p(x)|dx}$, and $|\Psi(x)|^2 \cong |A|^2/p(x)$.

In the *classical* region with $E > \Pi(x)$, $p(x)$ is real, while for the tunneling problem $p(x)$ is imaginary because $E < \Pi(x)$. The $\Psi(x)$ amplitudes are used to derive the Wentzel–Kramers–Brillouin expression for $T(E)$. As an alternative, the potential $\Pi(x)$ can be approximated by a number of steps $\Pi_j(x)$.

Example 6.21

Consider an electron with mass m and energy E in an external time-invariant electric field E_E. The potential barrier that corresponds to the scattering of electrons (cold emission of electron from metal with the work function Π_0) is

$$\Pi(x) = \begin{cases} 0 & \text{for } x \leq 0, \\ (\Pi_0 - eE_E x) = (\Pi_0 - fx) & \text{for } x > 0, \end{cases}$$

where $f = eE_E$.

One finds the transmission coefficient of tunneling to be

$$T(E) \cong e^{-(2/\hbar)\int_0^{(\Pi_0-E)/f} \sqrt{2m(\Pi_0-ax-E)}dx} = e^{-(4\sqrt{2m})/(3\hbar f)(\Pi_0-E)^{3/2}}.$$

Here, the x_f is found by taking note of $(\Pi_0 - fx) = E$. In particular, x_f and performing the integration we have the upper limit $x_f = (\Pi_0 - E)/f$.

Example 6.22

A proton of energy E is incident from the right on a nucleus of charge Ze. To estimate the transmission coefficient that provides one with the perception of how proton penetrates toward nucleus, one considers the repulsive Coulomb force of the nucleus. The radial Coulomb potential barrier is $\Pi(r) = -Z(r)e^2/(4\pi\varepsilon_0 r)$ or $\Pi(r) = -Z_{\text{eff}}e^2/(4\pi\varepsilon_0 r)$.

To simplify the resulting expression for $T(E)$, without loss of generality, let $\Pi(r) = Ze^2/r$.

Taking note of E, one finds $E = \Pi(r)|_{\text{at } b=r}$, and $b = Ze^2/E$. Thus, we have

$$T(E) \propto e^{-(2/\hbar)\int_b^0 \sqrt{2m((Ze^2/r)-E)}dr} = e^{-(2\sqrt{2mE}/\hbar)\int_{Ze^2/E}^0 \sqrt{(Ze^2/Er)-1}\,dr}.$$

By using a new variable $y = E/(Ze^2 r)$, one obtains

$$\frac{2\sqrt{2mE}}{\hbar}\int_{Ze^2/E}^{0}\sqrt{\frac{Ze^2}{Er}-1}\,dr = \frac{2Ze^2}{\hbar}\sqrt{\frac{2m}{E}}\int_0^1\sqrt{\frac{1}{y}-1}\,dy = \frac{Ze^2\pi}{\hbar}\sqrt{\frac{2m}{E}}$$

because

$$\int_0^1 \sqrt{\frac{1}{y}-1}\,dy = \frac{1}{2}\pi.$$

Hence, $T(E) \cong e^{-((\sqrt{2m}Ze^2\pi)/\hbar)(1/\sqrt{E})}$.

Example 6.23

The tunneling of a particle through the rectangular double barrier with the same potentials ($\Pi_0 = \Pi_{01} = \Pi_{02}$) is considered. Let $E < \Pi_0$. We denote the barrier width as $L(L = L_1 = L_2)$, and the barriers spacing as l.

By making use of the Schrödinger equation and having derived $\Psi_i(x)$, the analytic expression for the transmission coefficient is found to be

$$T(E) = \left| \frac{4A^2B^2}{C} \right|^2,$$

where

$$A = \sqrt{\frac{2m}{\hbar^2}(\Pi_0 - E)}, \quad B = \sqrt{\frac{2m}{\hbar^2}E}$$

and

$$C = e^{iB(l+2L)}[(e^{i2lB}(A^2 + B^2)^2 - A^4 - B^4)\sinh^2 LA + A^2B^2(1 + 3\cosh 2LA) + i2AB(A^2 - B^2)\sinh 2LA].$$

Let $m = m_e = 9.11 \times 10^{-31}$ kg, $\Pi_0 = 7$ eV $= 7 \times 1.6 \times 10^{-19}$ J $= 1.12 \times 10^{-18}$ J, and $L = 0.14$ nm. For two distinct l ($l = 4L = 0.56$ nm and $l = 6L = 0.84$ nm), the plots for $T(E)$ with three and four resonant states at different energies are shown in Figure 6.12. Significant changes of $T(E)$ are observed. The MATLAB statement to perform calculations and plotting is

```
m=9.11e-31;   h=1.055e-34;   U=7*1.6e-19;   N=4;
l=N*0.14e-9;   L=0.14e-9;   E=0:U/1e4:1*U;
A=sqrt(2*m*(U-E)./(h^2));   B=sqrt(2*m*E./(h^2));
C=exp(i*B.*(l+2*L)).*(((exp(i*2*B*l)).*((A.^2+...
B.^2).^2)-A.^4-B.^4).*sinh(A.*L).^2+(A.^2).*(B.^2).*...
   (1+3*cosh(2*A.*L))+i*2*A.*B.*(A.^2-...
   B.^2).*sinh(2*A.*L)); Kd=4.*(A.^2).*(B.^2); K=Kd./C;
T=(abs(K)).^2;
plot (E./U,T);
title('Transmission as a Function of Energy, \itT(E)','FontSize',14);
xlabel('Energy, \itE/\Pi\_0','FontSize',14);
ylabel('Transmission Coefficient, \itT','FontSize',14);
```

For the potential barriers in Figures 6.13a and b, one studies a tunneling problem examining the incident and reflected wave function amplitudes. As shown in Example 6.14, for $\Psi(x) = A\,e^{ikx} + B\,e^{-ikx}$, one has $J = (\hbar k/m) \times (|A|^2 - |B|^2)$, which can be defined as the difference between incident and reflected probability current densities, $J = J_I - J_R$. The reflection coefficient

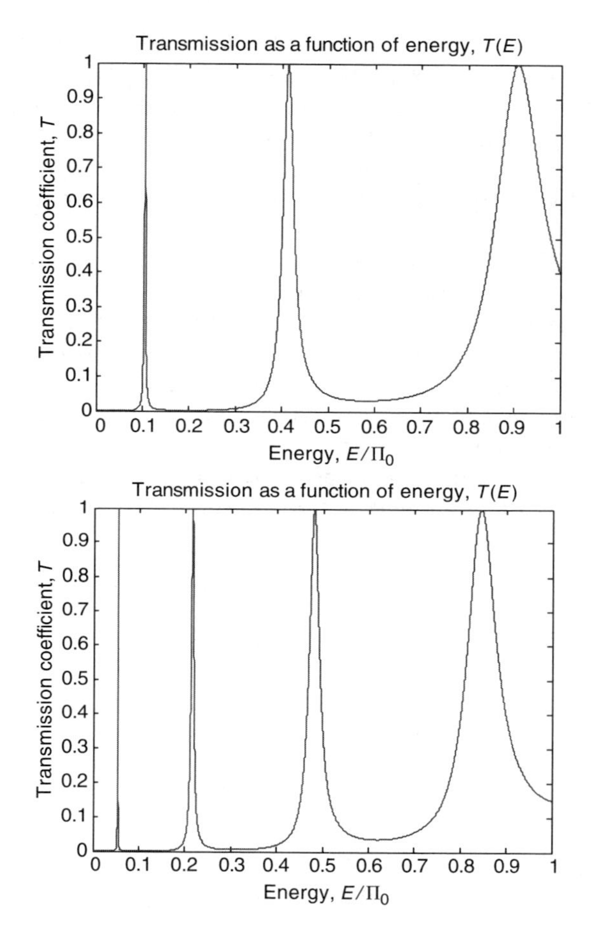

FIGURE 6.12
Tunneling as a function of energy for an electron in a rectangular double barrier with spacings $l = 4L$ and $l = 6L$.

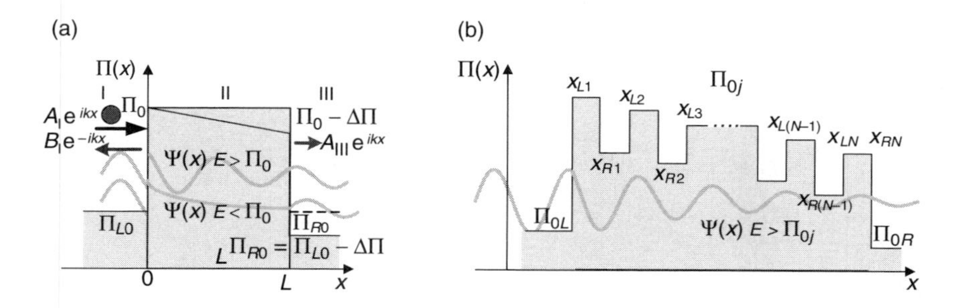

FIGURE 6.13
Electron tunneling through finite potential barriers: (a) single potential barrier; (b) multiple potential barriers.

is $R = J_R/J_I = |B|^2/|A|^2$. One may find the velocity and probability density of incoming, injected, and reflected electrons. The potential can vary as a function of the applied voltage (voltage bias is $\Delta V = V_L - V_R$), electric field, transitions, and so forth. Using the potential difference $\Delta\Pi$, the variation of a piecewise-continuous energy potential barrier $\Pi(x)$ is shown in Figure 6.13a. The analysis of wave function (if $E < \Pi$ or $E > \Pi$) is of particular interest. One may examine electrons that move from the region of negative values of coordinate x to the region of positive values of x. At x_{Lj} and x_{Rj} electrons encounter intermediate finite potentials Π_{0j} with width L_j; see Figure 6.13a and b. At the left and right (x_{L1} and x_{RN}) the finite potentials are denoted as Π_{0L} and Π_{0R}. There is a finite probability for transmission and reflection. The electrons on the left side that occupy the energy levels E_n can tunnel through the barrier to occupy empty energy levels E_n on the right side. The currents have contribution from all electrons.

Consider a finite multiple potential $\Pi(x)$ as illustrated in Figure 6.13b. In all regions, using the Schrödinger equation, one obtains a set of $(2N+2)$ second-order differential equations:

$$-\frac{\hbar^2}{2m}\frac{d^2\Psi_j}{dx^2} + \Pi_{0j}\Psi_j = E\Psi_j$$

or

$$\frac{d^2\Psi_j}{dx^2} + \kappa_{nj}^2\Psi_j = 0, \quad j = 0, 1, \ldots, 2N, 2N+1,$$

where κ_{nj} (κ_{1j} or κ_{2j}) depend on the particle energy E and potentials Π_{0L}, Π_{0j}, and Π_{0R}.

For $E > \Pi_{0L}$, $E > \Pi_{0j}$ and $E > \Pi_{0R}$, the general solutions are

$$\begin{aligned}
\Psi_{\mathrm{I}}(x) &= A_{\mathrm{I}}\, e^{i\kappa_{10}x} + B_{\mathrm{I}}\, e^{-i\kappa_{10}x}, \quad x < x_{L1}, \\
\Psi_{\mathrm{II}_j}(x) &= A_{\mathrm{II}_j}\, e^{i\kappa_{1_j}x} + B_{\mathrm{II}_j}\, e^{-i\kappa_{1_j}x}, \quad x_{L1} \le x < x_{R1}, \\
&\quad x_{R1} \le x < x_{L2}, \ldots, x_{R(N-1)} \le x < x_{LN}, \quad x_{LN} \le x \le x_{RN}, \\
&\quad j = 1, 2, \ldots, N-1, N, \\
\Psi_{\mathrm{III}}(x) &= A_{\mathrm{III}}\, e^{i\kappa_{1_{2N+1}}x} + B_{\mathrm{III}}\, e^{-i\kappa_{1_{2N+1}}x}, \quad x > x_{RN},
\end{aligned}$$

where

$$\kappa_{1_0}^2 = \frac{2m}{\hbar^2}(E - \Pi_{0L}), \quad \kappa_{1j}^2 = \frac{2m}{\hbar^2}(E - \Pi_{0j}) \quad \text{and} \quad \kappa_{1_{2N+1}}^2 = \frac{2m}{\hbar^2}(E - \Pi_{0R}).$$

If $E < \Pi_{0L}$, $E < \Pi_{0j}$ and $E < \Pi_{0R}$, we have

$$\Psi_{\mathrm{I}}(x) = A_{\mathrm{I}}\,\mathrm{e}^{-\kappa_{2_0}x} + B_{\mathrm{I}}\,\mathrm{e}^{\kappa_{2_0}x}, \quad x < x_{L1},$$

$$\Psi_{\mathrm{II}_j}(x) = A_{\mathrm{II}_j}\,\mathrm{e}^{-\kappa_{2_j}x} + B_{\mathrm{II}_j}\,\mathrm{e}^{\kappa_{2_j}x}, \quad x_{L1} \le x < x_{R1},$$

$$x_{R1} \le x < x_{L2}, \ldots, x_{R(N-1)} \le x < x_{LN}, \quad x_{LN} \le x \le x_{RN},$$

$$j = 1, 2, \ldots, N-1, N,$$

$$\Psi_{\mathrm{III}}(x) = A_{\mathrm{III}}\,\mathrm{e}^{-\kappa_{2_{2N+1}}x} + B_{\mathrm{III}}\,\mathrm{e}^{\kappa_{2_{2N+1}}x}, \quad x > x_{RN},$$

where

$$\kappa_{2_0}^2 = \frac{2m}{\hbar^2}(\Pi_{0L} - E), \quad \kappa_{2_j}^2 = \frac{2m}{\hbar^2}(\Pi_{0j} - E) \quad \text{and} \quad \kappa_{2_{2N+1}}^2 = \frac{2m}{\hbar^2}(\Pi_{0R} - E).$$

One may straightforwardly modify these solutions taking note of other possible relationships between potentials (Π_{0L}, Π_{0j}, and Π_{0R}) and E. The boundary and continuity conditions, as well as normalization, are used to obtain the wave functions and unknown A_{I}, $A_{\mathrm{II}j}$, A_{III}, B_{I}, $B_{\mathrm{II}j}$, and B_{III}. The interatomic bond lengths in various organic molecular aggregates are usually from 1 to 2 Å. For example, in fullerenes, the C—C, C—N, and C—B interatomic bond lengths are from 1.4 to 1.45 Å. Assuming that $L = (x_{Rj} - x_{Lj}) = \text{constant}$, the procedure for deriving $\Psi(x)$ an $T(E)$ can be simplified. However, a realistic $\Pi(\mathbf{r})$ usually does not impose any difficulties for differential equation solvers high-performance software toolboxes that are used to numerically solve a set of Schrödinger equations.

Molecular aggregates exhibit complex energy profile. The Schrödinger equation is

$$\left(\frac{\hbar^2}{2m_j}\frac{\mathrm{d}^2}{\mathrm{d}x^2} - \Pi_j(x) + E_j(x) + E_{aj}(x)\right)\Psi_j(x) = 0,$$

where $E_{aj}(x)$ is the applied external energy.

The boundary and continuity conditions to be used are

$$\Psi_j(x_j) = \Psi_{j+1}(x_j), \quad \left.\frac{1}{m_j}\frac{\partial \Psi_j(x)}{\partial x}\right|_{x=x_j} = \left.\frac{1}{m_{j+1}}\frac{\partial \Psi_{j+1}(x)}{\partial x}\right|_{x=x_j}.$$

The general solutions were reported for $E_{aj} = 0$ and $\Pi_j(x) = \text{constant}$. As was emphasized, a numerical solution can be found without any assumptions on $\Pi(\mathbf{r})$. For some problems, analytic solutions also can be derived providing the basic understanding.

For the energy profile, illustrated in Figure 6.14, the analytic solution is derived using Airy's functions Ai and Bi. In particular,

$$\Psi_j(x) = A_j \mathrm{Ai}[C_j(x)] + B_j \mathrm{Bi}[C_j(x)], \quad C_j(x) = \frac{\hbar^2 k_j^2 - 2m_j E_{aj} x}{(2m_j E_{aj})^{2/3}}.$$

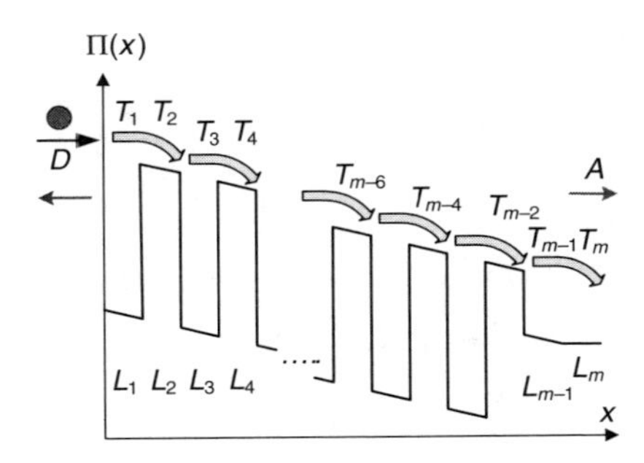

FIGURE 6.14
Energy profile.

For the scattering state, we have

$$M_j \begin{bmatrix} A_j \\ B_j \end{bmatrix} = M_{j+1} \begin{bmatrix} A_{j+1} \\ B_{j+1} \end{bmatrix}.$$

The transfer matrix is $M_{1\to m} = M_{1\to 2}M_{2\to 3}\ldots M_{(m-2)\to(m-1)}M_{(m-1)\to m}$.

The analytic solution of the Schrödinger equation has been emphasized. For practical problems, including electron transport, one departs from some assumptions and simplifications made in order to derive analytic solutions. Though the explicit expressions for wave function, incident/reflected/transmitted amplitudes and other quantities are of a significant interest, those analytic results are difficult to obtain for complex energy profiles $\Pi(\mathbf{r})$. Therefore, numerical solutions, computational algorithms, and numerical methods are emphasized.

Consider the Schrödinger equation

$$-\frac{\hbar^2}{2m}\frac{d^2\Psi(x)}{dx^2} + \Pi(x)\Psi(x) = E\Psi(x)$$

which is given as a second-order differential equation

$$\frac{d^2\Psi(x)}{dx^2} = -k^2(x)\Psi(x)$$

to be numerically solved. Here,

$$k^2(x) = \frac{2m}{\hbar^2}[E - \Pi(x)].$$

Many numerical methods and corresponding software are available, including MATLAB with a number of differential equations solvers. In order to illustrate the use of numerical methods, we apply the following concepts.

The Euler approximation is used to represent the first spatial derivative as a first difference:

$$\frac{\mathrm{d}\Psi(x)}{\mathrm{d}x} \approx \frac{\Psi_{n+1} - \Psi_n}{\Delta_h},$$

where Δ_h is the spatial discretization spacing.

Thus, the Schrödinger equation can be numerically solved through discretization by applying the built-in differential equations solvers within high-performance software or by developing (if needed) supporting application-specific software solutions.

Various discretization formulas and methods can be utilized. The Numerov three-point-difference expression is

$$\frac{\mathrm{d}^2\Psi(x)}{\mathrm{d}x^2} \approx \frac{\Psi_{n+1} - 2\Psi_n + \Psi_{n-1}}{\Delta_h^2}.$$

From $\mathrm{d}^2\Psi(x)/\mathrm{d}x^2 = -k^2(x)\Psi(x)$, one obtains a simple recursive equation

$$\Psi_{n+1} = \frac{2(1 - (5/12)k_n^2\Delta_h^2)\Psi_n - (1 + (1/12)k_{n-1}^2\Delta_h^2)\Psi_{n-1}}{1 + (1/12)k_{n+1}^2\Delta_h^2}.$$

Assigning the initial values for Ψ_{n-1} and Ψ_n (for example, Ψ_0 and Ψ_1), the value of Ψ_{n+1} is derived. The *forward* or *backward* calculations of Ψ_i are performed with the accuracy $0(\Delta_h^6)$. The initial values of Ψ_{n-1} and Ψ_n can be assigned using the boundary conditions. One assigns and refines a *trial* energy E_n guarantying stability and convergence of the solution. The application of other methods can provide one with more accurate and numerically robust results.

Using the Numerov three-point-difference expression, the Schrödinger equation is discretized as

$$\frac{\hbar^2}{2m}\left(\frac{(\Psi_{n+1} - \Psi_n) - (\Psi_n - \Psi_{n-1})}{\Delta_h^2}\right) - \Pi_n\Psi_n + E_n\Psi_n = 0.$$

Using the Hamiltonian matrix $\mathbf{H} \in \mathbb{R}^{(N+2)\times(N+2)}$, vector $\boldsymbol{\Psi} \in \mathbb{R}^{N+2}$ that contains Ψ_i, and the source vector $\mathbf{Q} \in \mathbb{R}^{N+2}$, the matrix equation

$$(E\mathbf{I} - \mathbf{H})\boldsymbol{\Psi} = \mathbf{Q}$$

should be solved. Here, $\mathbf{I} \in \mathbb{R}^{(N+2)\times(N+2)}$ is the identity matrix.

For a two-terminal ^{ME}device, the entities of the diagonal matrix $\mathbf{H}$ are $H_{n,n} = -(\hbar^2/2m\Delta_h^2) + \Pi_n$, except $H_{0,0}$ and $H_{(N+1)(N+1)}$, which depend on the self-energies that account for the interconnect interactions (chemical bonding, radiated/absorbed energetics, etc.). By taking note of notations used for the incoming wave function $\Psi(x) = A\ e^{ik_L x} + B\ e^{-ik_L x}$, which leads to $\Psi_{-1} = A\ e^{-ik_L \Delta_h} + B\ e^{ik_L \Delta_h} = A\ e^{-ik_L \Delta_h} + (\Psi_0 - A)e^{ik_L \Delta_h}$ and $\Psi_{N+2} = \Psi_{N+1}\ e^{ik_R \Delta_h}$, one has $H_{0,0} = -(\hbar^2/m\Delta_h^2)(1 + (1/2)\ e^{ik_L \Delta_h}) + \Pi_0$ and $H_{(N+1),(N+1)} = -(\hbar^2/m\Delta_h^2)(1+(1/2)\ e^{ik_R \Delta_h})+\Pi_{N+1}$. Hence, the solution of the Schrödinger equation is reduced to the solution of linear algebraic equation.

The probability current density is

$$J = (i\hbar/2m)(\Psi_n(\Psi_{n+1}^* - \Psi_n^*/\Delta_h) - \Psi_n^*(\Psi_{n+1} - \Psi_n/\Delta_h)).$$

6.6 Quantum Mechanics and Molecular Electronic Devices: Three-Dimensional Problem

The electron transport in 3D-topology ^{ME}devices must be examined in 3D applying quantum mechanics. The time-independent Schrödinger equation

$$-\frac{\hbar^2}{2m}\nabla^2\Psi(\mathbf{r}) + \Pi(\mathbf{r})\Psi(\mathbf{r}) = E(\mathbf{r})\Psi(\mathbf{r})$$

can be solved in different coordinate systems depending on the problem under the consideration.

In the Cartesian system we have

$$\nabla^2\Psi(\mathbf{r}) = \nabla^2\Psi(x,y,z) = \frac{\partial^2\Psi}{\partial x^2} + \frac{\partial^2\Psi}{\partial y^2} + \frac{\partial^2\Psi}{\partial z^2},$$

while in the cylindrical and spherical systems one has

$$\nabla^2\Psi(\mathbf{r}) = \nabla^2\Psi(r,\phi,z) = \frac{1}{r}\frac{\partial}{\partial r}\left(r\frac{\partial\Psi}{\partial r}\right) + \frac{1}{r^2}\frac{\partial^2\Psi}{\partial\phi^2} + \frac{\partial^2\Psi}{\partial z^2}$$

and

$$\nabla^2\Psi(\mathbf{r}) = \nabla^2\Psi(r,\theta,\phi) = \frac{1}{r^2}\frac{\partial}{\partial r}\left(r^2\frac{\partial\Psi}{\partial r}\right) + \frac{1}{r^2\sin\theta}\frac{\partial}{\partial\theta}\left(\sin\theta\frac{\partial\Psi}{\partial\theta}\right) + \frac{1}{r^2\sin^2\theta}\frac{\partial^2\Psi}{\partial\phi^2}.$$

The solution of the Schrödinger equation is obtained by using different analytical and numerical methods. The analytical solution can be found by using the separation of variables. For example, if the potential is

$$\Pi(x,y,z) = \Pi_x(x) + \Pi_y(y) + \Pi_z(z),$$

one has

$$[H_x(x) + H_y(y) + H_z(z)]\Psi(x,y,z) = E\Psi(x,y,z),$$

where the Hamiltonians are

$$H_x(x) = -\frac{\hbar^2}{2m}\frac{\partial^2}{\partial x^2} + \Pi_x(x), \quad H_y(y) = -\frac{\hbar^2}{2m}\frac{\partial^2}{\partial y^2} + \Pi_y(y), \quad \text{and}$$

$$H_z(z) = -\frac{\hbar^2}{2m}\frac{\partial^2}{\partial z^2} + \Pi_z(z).$$

The wave function is given as a product of three functions

$$\Psi(x, y, z) = X(x)Y(y)Z(z).$$

This results in

$$\left[-\frac{\hbar^2}{2m}\frac{1}{X(x)}\frac{\mathrm{d}^2X(x)}{\mathrm{d}x^2} + \Pi_x(x)\right] + \left[-\frac{\hbar^2}{2m}\frac{1}{Y(y)}\frac{\mathrm{d}^2Y(y)}{\mathrm{d}y^2} + \Pi_y(y)\right] + \left[-\frac{\hbar^2}{2m}\frac{1}{Z(z)}\frac{\mathrm{d}^2Z(z)}{\mathrm{d}z^2} + \Pi_z(z)\right] = E,$$

where the constant total energy is $E = E_x + E_y + E_z$.

The separation of variables technique results in reduction of 3D Schrödinger equation to three independent one-dimensional equations:

$$\left[-\frac{\hbar^2}{2m}\frac{\mathrm{d}^2}{\mathrm{d}x^2} + \Pi_x(x)\right]X(x) = E_xX(x),$$

$$\left[-\frac{\hbar^2}{2m}\frac{\mathrm{d}^2}{\mathrm{d}y^2} + \Pi_y(y)\right]Y(y) = E_yY(y)$$

and

$$\left[-\frac{\hbar^2}{2m}\frac{\mathrm{d}^2}{\mathrm{d}z^2} + \Pi_z(z)\right]Z(z) = E_zZ(z).$$

The cylindrical and spherical systems can be effectively used to reduce the complexity and make the problem tractable. In the spherical system, one uses $\Psi(r, \theta, \phi) = R(r)Y(\theta, \phi)$. The Schrödinger differential equation is solved using the continuity and boundary conditions, and the wave function is normalized as $\int_V \Psi^*(\mathbf{r})\Psi(\mathbf{r})\mathrm{d}V = 1$.

Example 6.24

Consider a *microscopic* particle that moves in the x and y directions (two-dimensional problem); see Figure 6.15. We study a particle confined in a rectangular area where the potential is zero, and the potential is infinite outside the box.

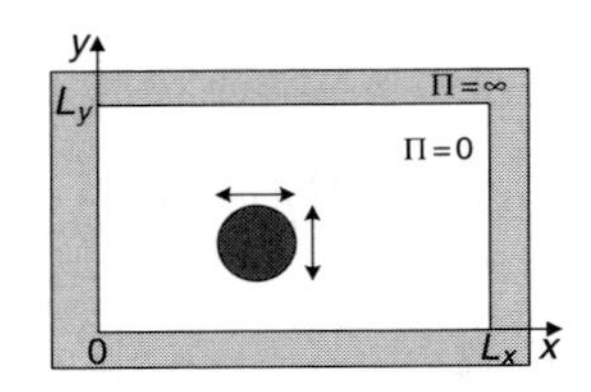

FIGURE 6.15
Particle in two-dimensional potential well.

The Schrödinger equation becomes

$$-\frac{\hbar^2}{2m}\left(\frac{\partial^2\Psi(x,y)}{\partial x^2}+\frac{\partial^2\Psi(x,y)}{\partial y^2}\right)+\Pi(x,y)\Psi(x,y)=E(x,y)\Psi(x,y).$$

This equation can be solved by using the separation of variables technique. We let

$$\Psi(x,y)=X(x)Y(y),$$

where $X(x)=C_x\sin k_x x+D_x\cos k_x x$ and $Y(y)=C_y\sin k_y y+D_y\cos k_y y$.

Substitution of these $X(x)$ and $Y(y)$ in the Schrödinger equations results in

$$-\frac{\hbar^2}{2m}\left(Y(y)\frac{\mathrm{d}^2X(x)}{\mathrm{d}x^2}+X(x)\frac{\mathrm{d}^2Y(y)}{\mathrm{d}y^2}\right)=E(x,y)X(x)Y(y).$$

The division by $-(2m/\hbar^2)(1/X(x)Y(y))$ gives

$$\frac{1}{X(x)}\frac{\mathrm{d}^2X(x)}{\mathrm{d}x^2}+\frac{1}{Y(y)}\frac{\mathrm{d}^2Y(y)}{\mathrm{d}y^2}=-\frac{2m}{\hbar^2}E(x,y).$$

Assuming $E(x,y)=E_x(x)+E_y(y)$, we have a set of two equations

$$\frac{1}{X(x)}\frac{\mathrm{d}^2X(x)}{\mathrm{d}x^2}=-\frac{2m}{\hbar^2}E_x(x),$$

or,

$$\frac{\mathrm{d}^2X(x)}{\mathrm{d}x^2}=-\frac{2m}{\hbar^2}E_x(x)X(x),$$

and

$$\frac{1}{Y(y)}\frac{\mathrm{d}^2Y(y)}{\mathrm{d}y^2}=-\frac{2m}{\hbar^2}E_y(y)$$

or,

$$\frac{\mathrm{d}^2Y(y)}{\mathrm{d}y^2} = -\frac{2m}{\hbar^2}E_y(y)Y(y).$$

The derived equations are similar to the one-dimensional case considered in the Section 6.5 examples. By normalizing the wave function as $\int_{-\infty}^{+\infty}\int_{-\infty}^{+\infty} \Psi_n^2(x,y)\mathrm{d}x\,\mathrm{d}y = 1$, the resulting wave function and energy are

$$\Psi_{n_x n_y} = X(x)Y(y) = \sqrt{\frac{2}{L_x}}\sin\left(\frac{n_x\pi}{L_x}x\right)\sqrt{\frac{2}{L_y}}\sin\left(\frac{n_y\pi}{L_y}y\right),$$

$$E_{n_x n_y} = E_{n_x} + E_{n_y} = \frac{\hbar^2\pi^2}{2m}\left(\frac{n_x^2}{L_x^2} + \frac{n_y^2}{L_y^2}\right).$$

Compared with the one-dimensional problem, two states now can have equal energies (states are degenerate) if $L_x = L_y$.

Example 6.25

For an infinite spherical potential well, let

$$\Pi(r) = \begin{cases} 0 & \text{for } r \le a, \\ \infty & \text{for } r > a. \end{cases}$$

For a particle in $\Pi(r)$ as sown in Figure 6.16, the Schrödinger equation is

$$-\frac{\hbar^2}{2m}\left[\frac{1}{r^2}\frac{\partial}{\partial r}\left(r^2\frac{\partial\Psi}{\partial r}\right) + \frac{1}{r^2\sin\theta}\frac{\partial}{\partial\theta}\left(\sin\theta\frac{\partial\Psi}{\partial\theta}\right) + \frac{1}{r^2\sin^2\theta}\frac{\partial^2\Psi}{\partial\phi^2}\right]$$
$$+\,\Pi(r,\theta,\phi)\Psi(r,\theta,\phi) = E\Psi(r,\theta,\phi).$$

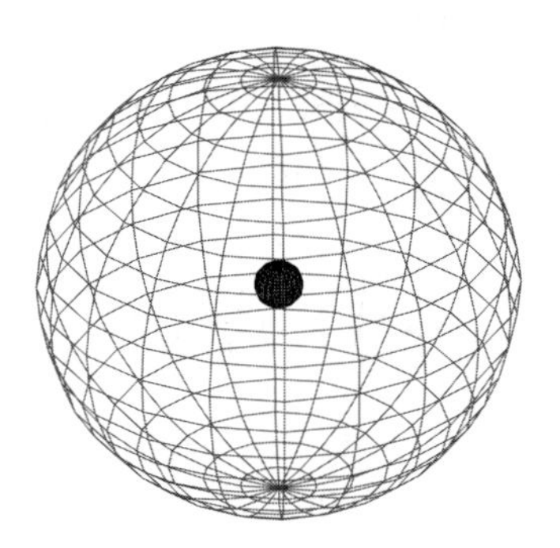

FIGURE 6.16
Particle in an infinite spherical potential well $\Pi(r)$.

We apply the separation of variables concept. The wave function is given as

$$\Psi(r,\theta,\phi) = R(r)Y(\theta,\phi).$$

Outside the well, when $r > a$, the wave function is zero. The stationary states are labeled using three quantum numbers n, l, and m_l. Our goal is to derive the expression for $\Psi_{nlm_l}(r,\theta,\phi)$.

The energy depends only on n and l, so we need to find E_{nl}.

In general,

$$\Psi_{nlm_l}(r,\theta,\phi) = A_{nl}S_{Bl}(s_{nl}r/a)Y_l^{m_l}(\theta,\phi),$$

where A_{nl} is the constant that must be found through the normalization of wave function; S_{Bl} is the spherical Bessel function of order l, $S_{Bl}(x) = (-x)^l((1/x)(\mathrm{d}/\mathrm{d}x))^l(\sin x/x)$, and for $l = 0$ and $l = 1$, we have $S_{B0} = \sin x/x$ and $S_{B1} = \sin x/x^2 - \cos x/x$; s_{nl} is the nth zero of the lth spherical Bessel function.

Inside the well, the radial equation is

$$\frac{\mathrm{d}^2u}{\mathrm{d}r^2} = \left(\frac{l(l+1)}{r^2} - k^2\right)u, \quad k^2 = \frac{2mE}{\hbar^2}.$$

The general solution of this equation for an arbitrary integer l is

$$u(r) = ArS_{Bl}(kr) + BrS_{Nl}(kr),$$

where S_N is the spherical Neumann function of order l, $S_{Nl}(x) = -(-x)^l \times ((1/x)(\mathrm{d}/\mathrm{d}x))^l(\cos x/x)$, and for $l = 0$ and $l = 1$, one finds $S_{N0} = -\cos x/x$ and $S_{N1} = -\cos x/x^2 - \sin x/x$.

The radial wave function is $R(r) = u(r)/r$.

We use the boundary condition $u(a) = 0$.

For $l = 0$, from $(\mathrm{d}^2u/\partial r^2) = -k^2u$, we have

$$u(r) = A\sin kr + B\cos kr, \text{ where } B = 0.$$

Taking note of the boundary condition, from $\sin ka = 0$, one obtain $ka = n\pi$. The normalization of $u(r)$ gives $A = \sqrt{2/a}$.

The angular equation is

$$\sin\theta\frac{\partial}{\partial\theta}\left(\sin\theta\frac{\partial Y}{\partial\theta}\right) + \frac{\partial^2 Y}{\partial\phi^2} = -l(l+1)\sin^2\theta Y.$$

By applying $Y(\theta,\phi) = \Theta(\theta)\Phi(\phi)$, the normalized angular wave function (spherical harmonics) is known to be

$$Y_l^{m_l}(\theta,\phi) = \gamma\sqrt{\frac{2l+1}{4\pi}\frac{(l-|m_l|)!}{(l+|m_l|)!}}e^{im_l\phi}L_l^{m_l}(\cos\theta),$$

where $\gamma = (-1)^{m_l}$ for $m_l \geq 0$ and $\gamma = 1$ for $m_l \leq 0$; $L_l^{m_l}(x)$ is the Legendre function, $L_l^{m_l}(x) = (1 - x^2)^{(1/2)|m_l|}(\mathrm{d}/\mathrm{d}x)^{|m_l|}L_l(x)$; $L_l(x)$ is the lth Legendre polynomial, $L_l(x) = (1/(2^l l!))(\mathrm{d}/\mathrm{d}x)^l(x^2 - 1)^l$.

Thus, the angular component of the wave function for $l = 0$ and $m_l = 0$ is

$$Y_0^0(\theta, \phi) = 1/\sqrt{4\pi}.$$

Hence,

$$\Psi_{n00} = \frac{1}{\sqrt{2\pi a}} \frac{1}{r} \sin \frac{n\pi r}{a},$$

and the allowed energies are

$$E_{n0} = \frac{\pi^2\hbar^2}{2ma^2} n^2, \quad n = 1, 2, 3, \ldots.$$

Using the nth order of the lth spherical Bessel function S_{Bnl}, the allowed energies are

$$E_{nl} = \frac{\pi^2\hbar^2}{2ma^2} S_{Bnl}^2.$$

The Schrödinger differential equation is numerically solved in all regions for the specified potentials, energies, potential widths, boundaries, and so forth. For 3D-topology ^{ME}devices, using potentials, tunneling paths, interatomic bond lengths, and other data, having found $\Psi(t, \mathbf{r})$, one obtains, $P(t, \mathbf{r})$, $T(t, E)$, expected values of variables, and other quantities of interest. For example, having found the velocity (or momentum) of a charged particle as a function of control variables (time-varying external electric or magnetic field) and parameters (mass, interatomic lengths, permittivity, etc.), the electric current is derived. As documented, the particle momentum, velocity, transmission coefficient, traversal time, and other variables change as functions of the time-varying external electromagnetic field. Therefore, depending on the device physics varying, for example, $\mathbf{E}_E(t, \mathbf{r})$ or $\mathbf{B}(t, \mathbf{r})$, one controls electron transport. Different dynamic and steady-state characteristics are examined. For example, the steady-state I–V and G–V characteristics can be derived using theoretical fundamentals reported.

For the planar solid-state semiconductor devices, to derive the transmission coefficient $T(E)$, the Green function $G(E)$ has been used. In particular, we have

$$T(E) = \mathrm{tr}[E_{\mathrm{BL}} G(E) E_{\mathrm{BR}} G^*(E)].$$

To obtain the I–V characteristics, one self-consistently solves the coupled transport and Poisson's equations [1–5]. The Poisson equation

$$\nabla \cdot (\varepsilon(\mathbf{r}) \nabla V(\mathbf{r})) = -\rho(\mathbf{r})$$

is solved to find the electric field intensity and electrostatic potential. Here, $\rho(\mathbf{r})$ is the charge density; $\varepsilon(\mathbf{r})$ is the dielectric tensor.

For example, letting $\rho_x = \rho_0 \operatorname{sech}(x/L)\tanh(x/L)$, we solve $\nabla^2 V_x = -\rho_x/\varepsilon$ obtaining the following expressions

$$E_x = -\rho_0/\varepsilon L \operatorname{sech}\frac{x}{L} \text{ and } V_x = 2\frac{\rho_0}{\varepsilon}L^2\left(\tan^{-1}e^{x/L} - \frac{1}{4}\pi\right).$$

Example 6.26

Examine the electric field inside and outside a spherical cloud of electrons in the carbon nanostructure assuming a uniform charge density $\rho = -\rho_0$ for $0 \le r < R$ and $\rho = 0$ for $r \ge R$.

The solution is found by applying the Poisson and Laplace equations for the electric potential that is defined as $\mathbf{E} = -\nabla V$. In linear isotropic media, $\nabla \cdot \varepsilon\mathbf{E} = \rho$. In general, ε is a function of position, but in the homogeneous media, ε is constant. Thus, we obtain

$$\nabla^2 V = -\frac{\rho}{\varepsilon},$$

where the Laplacian operator stands for the divergence of the gradient.

To solve Poisson's equation one must use the boundary conditions. Inside the electron cloud, we solve the Poisson equation. In the spherical coordinate system, we have

$$\nabla^2 V(r) = \frac{1}{r^2}\frac{d}{dr}\left(r^2\frac{dV(r)}{dr}\right) = \frac{\rho_0}{\varepsilon_0} \quad \text{or} \quad \frac{d}{dr}\left(r^2\frac{dV(r)}{dr}\right) = r^2\frac{\rho_0}{\varepsilon_0}.$$

Integration results in the following expression:

$$\frac{dV(r)}{dr} = \frac{\rho_0}{3\varepsilon_0}r + \frac{c_1}{r^2}.$$

The electric field intensity is found to be

$$\mathbf{E} = -\nabla V = -\mathbf{a}_r\frac{dV(r)}{dr} = -\mathbf{a}_r\left(\frac{\rho_0}{3\varepsilon_0}r + \frac{c_1}{r^2}\right).$$

We need to find the integration constant c_1. The electric field cannot be infinite at $r = 0$. Thus, $c_1 = 0$. Therefore, $\mathbf{E} = -\mathbf{a}_r(\rho_0/3\varepsilon_0)r$ for $0 \le r < R$.

Outside the cloud, $\rho = 0$ for $r \ge R$. If $\rho = 0$, the Poisson equation becomes the Laplace equation, $\nabla^2 V = 0$. Correspondingly, one solves the Laplace equation

$$\nabla^2 V = 0 \quad \text{or} \quad \frac{1}{r^2}\frac{d}{dr}\left(r^2\frac{dV(r)}{dr}\right) = 0.$$

Integration gives $dV(r)/dr = c_2/r^2$. Thus,

$$\mathbf{E} = -\nabla V = -\mathbf{a}_r\frac{dV(r)}{dr} = -\mathbf{a}_r\frac{c_2}{r^2}.$$

The integration constant c_2 is found using the boundary condition at $r = R$. In particular, from $c_2/R^2 = (\rho_0/3\varepsilon_0)R$, one has $c_2 = (\rho_0/3\varepsilon_0)R^3$.
Thus, we obtain $\mathbf{E} = -\mathbf{a}_r(\rho_0/3\varepsilon_0)(R^3/r^2)$ for $r \geq R$.

For 3D-topology ${}^{\text{ME}}$devices, the Poisson equation is of a great importance to find self-consistent solution. The Schrödinger and Poisson equations are solved utilizing robust numerical methods using the difference expressions for the Laplacian, integration–differentiation concepts, and so forth. It is possible to solve differential equations in 3D using a finite-difference method that gives lattices. Generalizing the results reported for the one-dimensional problem, for the Laplace equation one has

$$\frac{\partial^2 V(i,j,k)}{\partial^2 r} = \frac{V(i+1,j,k) - 2V(i,j,k) + V(i-1,j,k)}{\Delta_h^2},$$

where (i, j, k) gives a grid point; Δ_h is the spatial discretization spacing in the x, y, or z directions.
For Poisson's equation, we have

$$\begin{aligned}\nabla \cdot (\varepsilon(\mathbf{r})\nabla V(\mathbf{r})) = {} & \frac{C_{i,j,k}^{i+1,j,k}(V_{i+1,j,k} - V_{i,j,k}) - C_{i-1,j,k}^{i,j,k}(V_{i,j,k} - V_{i-1,j,k})}{\Delta_x^2} \\ & + \frac{C_{i,j,k}^{i,j+1,k}(V_{i,j+1,k} - V_{i,j,k}) - C_{i,j-1,k}^{i,j,k}(V_{i,j,k} - V_{i,j-1,k})}{\Delta_y^2} \\ & + \frac{C_{i,j,k}^{i,j,k+1}(V_{i,j,k+1} - V_{i,j,k}) - C_{i,j,k-1}^{i,j,k}(V_{i,j,k} - V_{i,j,k-1})}{\Delta_z^2},\end{aligned}$$

$$C_{l,m,n}^{i,j,k} = \frac{2\varepsilon_{i,j,k}\varepsilon_{l,m,n}}{\varepsilon_{i,j,k} + \varepsilon_{l,m,n}}.$$

Thus, using the number of grid points, the equation $\nabla \cdot (\varepsilon(\mathbf{r})\nabla V(\mathbf{r})) = -\rho(\mathbf{r})$ is represented and solved as

$$\mathbf{A}V = \mathbf{B},$$

where $\mathbf{A} \in \mathbb{R}^{N \times N}$ is the matrix; $\mathbf{B} \in \mathbb{R}^N$ is the vector of the boundary conditions.
The self-consistent problem that integrates the solution of the Schrödinger (gives wave function, energy, etc.) and Poisson (provides the potential) equations is solved updating the potentials and other variables obtained through iterations. Convergence is enforced and specified accuracy is guaranteed by robust numerical methods.

6.7 Electromagnetic Field and Control of Particles in Molecular Electronic Devices

6.7.1 Microscopic Particle in an Electromagnetic Field

For a free particle in the Cartesian coordinate system,

$$E(\mathbf{r}) = \frac{\mathbf{p}^2}{2m}, \quad \mathbf{p}^2 = p_x^2 + p_y^2 + p_z^2.$$

Taking into account a potential, one uses

$$E(\mathbf{r}) = \frac{\mathbf{p}^2}{2m} + \Pi(\mathbf{r}).$$

In a magnetic field, the interaction of a magnetic moment $\boldsymbol{\mu}$ with a magnetic field $\mathbf{B}$ changes the energy of the particle by $-\boldsymbol{\mu} \cdot \mathbf{B}$.

Consider a particle with a charge q and mass m in a one-dimensional potential $\Pi(x)$. Let a particle propagates under an external time-varying electric field $E_E(t,x)$. The particle Hamiltonian is

$$H = \frac{1}{2m}p^2 + \Pi(x) + qE_E(t,x)x.$$

For example, $E_E(t) = E_{E0} \sin \omega t$, where E_{E0} is the amplitude of the electrostatic field.

The operators are commonly used deriving the expressions for the Hamiltonian, which can be time-invariant or time-dependent. The external electromagnetic field, which can be controlled, affects the Hamiltonian. In general, for a particle with a charge q in an uniform magnetic field $\mathbf{B}$, one has

$$H = \frac{1}{2\mu}\mathbf{p}^2 + \Pi(\mathbf{r}) - \frac{q}{2\mu c}\mathbf{B} \cdot \mathbf{L} + \frac{q^2}{8\mu c^2}[B^2 r^2 - (\mathbf{B} \cdot \mathbf{r})^2],$$

where μ is the angular momentum; $\mathbf{L}$ is the orbital angular momentum.

In H, the term $-(q/2\mu c)\mathbf{B} \cdot \mathbf{L} = -\boldsymbol{\mu}_L \cdot \mathbf{B}$ represents the energy resulting from the interaction between the particle orbital magnetic moment $\boldsymbol{\mu}_L = q\mathbf{L}/(2\mu c)$ and the magnetic field $\mathbf{B}$.

If the charge q has an intrinsic spin $\mathbf{S}$, the spinning motion results in the magnetic dipole moment $\boldsymbol{\mu}_S = q\mathbf{S}/(2\mu c)$, which interacts with an external magnetic field generating the energy $-\boldsymbol{\mu}_S \cdot \mathbf{B}$. Thus, we have

$$H = \frac{1}{2\mu}\mathbf{p}^2 + \Pi(\mathbf{r}) - \boldsymbol{\mu}_L \cdot \mathbf{B} - \boldsymbol{\mu}_S \cdot \mathbf{B} + \frac{q^2}{8\mu c^2}[B^2 r^2 - (\mathbf{B} \cdot \mathbf{r})^2].$$

Consider the hydrogen atom under an external uniform magnetic field **B**. The atom energy levels are shifted, and this energy shift is known as the Zeeman effect. In the normal Zeeman effect, neglecting the electron spin (the anomalous Zeeman effect takes into the consideration the spin of the electron utilizing the perturbation theory), we assume that $\mathbf{B} = B_z\mathbf{z}$, that is, $\mathbf{B} = [0, 0, B_z]$. The Hamiltonian is

$$H = \frac{1}{2\mu}\mathbf{p}^2 - \frac{e^2}{4\pi\varepsilon_0 r} + \frac{e}{2\mu c}B_z L_z + \frac{e^2 B_z^2}{8\mu c^2}(x^2 + y^2),$$

where $H_0 = (1/2\mu)\mathbf{p}^2 - (e^2/4\pi\varepsilon_0 r)$ is the atomic Hamiltonian in the absence of the magnetic field; L_z is the orbital angular momentum.

The third term of H is usually rewritten as $(e/2\mu c)B_z L_z = (\mu_{\mathrm{B}}/\hbar)B_z L_z$, where μ_B is the Bohr magneton,

$$\mu_{\mathrm{B}} = \frac{e\hbar}{2\mu c} = \frac{e\hbar}{2m_e} = 9.274 \times 10^{-24}\ \mathrm{J/T} = 5.7884 \times 10^{-5}\ \mathrm{eV/T}.$$

The electron's orbital magnetic dipole moment, resulting from the orbiting motion of the electron about the proton, is $\boldsymbol{\mu}_L = -e\mathbf{B}/(2\mu c)$.

The term $(e^2 B_z^2/8\mu c^2)(x^2 + y^2)$ may be negligibly small. The spherical and Cartesian coordinates are related as $x = r\sin\theta\cos\varphi$, $y = r\sin\theta\sin\varphi$, and $z = r\cos \mathrm{f}\theta$.

One concludes that the propagation of electrons can be effectively controlled by changing the electromagnetic field in ^{ME}devices. The control variables (one can effectively vary **E** and **B**) are time-varying. One examines a time-dependent Schrödinger equation

$$H(t, \mathbf{r})\Psi(t, \mathbf{r}) = i\hbar\frac{\partial\Psi(t, \mathbf{r})}{\partial t}.$$

Consider a time-varying one-dimensional potential $\Pi(t, x)$ as given by

$$\Pi(t, x) = \Pi_t(t, x) + \Pi_0(x).$$

If $\Pi(t, x) = \Pi_0(x)$, the solution of the Schrödinger equation is

$$\Psi_n(t, x) = \Psi_n(t)\Psi_n(x) = e^{-(iE_n/\hbar)t}\Psi_n(x),$$

where E_n and $\Psi_n(x)$ are the unperturbed eigenvalues and eigenfunctions.

Taking note of a time-varying $\Pi(t, x)$, the solution is

$$\Psi(t, x) = \sum_n a_n(t)\Psi_n(t, x),$$

where $a_n(t)$ is the time-varying function found by solving a set of differential equations depending on the problem under the consideration.

The transition probability is related to $a_n(t)$ as $P_m = \sum_{n,n\neq m} a_n^*(t)a_n(t)$.

Our goal is to study how the quantum state, given by $\Psi(t)$, evolves in time. In particular, for a given initial state $\Psi(t_0)$, the system's dynamic behavior governed by the Schrödinger equation to the following (intermediate or final) state with $\Psi(t_f)$, is of our interest. We have

$$\Psi(t) = U(t_0, t)\Psi(t_0), \quad t > t_0,$$

where $U(t_0, t)$ is the unitary operator that gives the finite time transition.

To find the *time-evolution* operator $U(t_0, t)$, one substitutes $\Psi(t) = U(t_0, t)\Psi(t_0)$ in the time-dependent Schrödinger equation yielding

$$\frac{\partial U(t_0, t)}{\partial t} = -\frac{i}{\hbar} HU(t_0, t).$$

If the Hamiltonian H is not a function of time, using the unit initial condition $U(t_0, t_0) = I$, we have

$$U(t_0, t) = \mathrm{e}^{-iH(t-t_0)/\hbar} \quad \text{and} \quad \Psi(t) = \Psi(t_0)\mathrm{e}^{-iH(t-t_0)/\hbar}.$$

One can find a solution for a time-varying potential

$$\Pi(t, x) = \Pi_t(t, x) + \Pi_0(x).$$

Let $\Pi_0(x) \gg \Pi_t(t, x)$ and assume

$$\Pi(t) = \begin{cases} \Pi(t) & \text{for } 0 \leq t \leq \tau, \\ 0 & \text{for } t < 0,\ t > \tau. \end{cases}$$

The solution of the Schrödinger equation in $0 \leq t \leq \tau$ gives

$$\Psi(t) = U_H(t_0, t)\Psi(t_0),$$

where $U_H(t_0, t) = \mathrm{e}^{(iH_0/\hbar)t} U(t_0, t)\mathrm{e}^{-(iH_0/\hbar)t}$; H_0 is the time-independent part of the Hamiltonian, $H_0 > \Pi(t)$.

From the time-dependent Schrödinger equation, one obtains

$$i\hbar\frac{\partial U_H(t_0, t)}{\partial t} = \mathrm{e}^{(iH_0/\hbar)t}\Pi(t)\mathrm{e}^{-(iH_0/\hbar)t} U_H(t_0, t).$$

The solution of this equation is

$$U_H(t_0, t) = I - \frac{i}{\hbar}\int_{t_0}^{t} \mathrm{e}^{(iH_0/\hbar)t}\Pi(t)\mathrm{e}^{-(iH_0/\hbar)t} U_H(t_0, t)\, \mathrm{d}t.$$

The time-dependant perturbation theory provides the first-, second-, third-, and other high-order approximations. The first-order approximation is derived substituting $U_H(t_0, t) = I$. That is,

$$U_H^{(1)}(t_0, t) = I - \frac{i}{\hbar} \int_{t_0}^{t} e^{(iH_0/\hbar)t} \Pi(t) e^{-(iH_0/\hbar)t} dt.$$

Having found the initial and final states defined by Ψ_i and Ψ_f, the transition probability is $P_{if}(t) = |\Psi_f U_H(t_0, t)\Psi_i|^2$, and the second-order approximation is

$$P_{if}(t) = \left| -\frac{i}{\hbar} \int_0^t \Psi_f \Pi(t') \Psi_i \, e^{i\omega_t t'} \, dt' \right|^2,$$

where ω_f is the transition frequency between the initial and final system's states,

$$\omega_t = \frac{E_f - E_i}{\hbar} = \frac{\Psi_f H_0 \Psi_f - \Psi_i H_0 \Psi_i}{\hbar}.$$

For practical engineering problems, the time-dependent problem can be solved numerically. In general, numerical formulation and solutions relax the complexity of analytic results, which are usually based on a number of assumptions and approximations of the time-dependent perturbation theory.

6.7.2 Dynamics and Control of Particles in Molecular Electronic Devices

Consider a *microscopic* particle that moves in an external time-dependent electromagnetic field. The vector and scalar potentials of the electromagnetic field are $\mathbf{A}(t, \mathbf{r})$ and $V(t, \mathbf{r})$, respectively. One obtains the Hamiltonian as

$$H(t, \mathbf{r}, \mathbf{p}) = \frac{1}{2m} \left(\mathbf{p} - \frac{q}{c} \mathbf{A} \right)^2 + qV.$$

For a *microscopic* particle, the time rate of change of the expectation value of $\mathbf{r}$ is

$$\frac{d}{dt} \langle \mathbf{r} \rangle = \frac{1}{i\hbar} \langle [\mathbf{r}, H] \rangle = \frac{1}{m} \left\langle \mathbf{p} - \frac{q}{c} \mathbf{A} \right\rangle$$

which defines the velocity operator as $\mathbf{v} = (1/m)(\mathbf{p} - (q/c)\mathbf{A})$.

Newton's second law in the quantum mechanical form is written as

$$\frac{d}{dt} \langle \mathbf{v} \rangle = \frac{1}{i\hbar} \langle [\mathbf{v}, H] \rangle + \left\langle \frac{\partial \mathbf{v}}{\partial t} \right\rangle = \frac{1}{i\hbar} \left\langle \left[\mathbf{v}, \frac{1}{2} m \mathbf{v} \cdot \mathbf{v} \right] \right\rangle + \frac{1}{i\hbar} \langle [\mathbf{v}, qV] \rangle - \frac{q}{mc} \left\langle \frac{\partial \mathbf{A}}{\partial t} \right\rangle.$$

Hence,

$$\frac{\mathrm{d}}{\mathrm{d}t}\langle\mathbf{v}\rangle = \frac{q}{2mc}\langle\mathbf{v}\times\mathbf{B}-\mathbf{B}\times\mathbf{v}\rangle - \frac{q}{m}\langle\nabla V\rangle - \frac{q}{mc}\left\langle\frac{\partial\mathbf{A}}{\partial t}\right\rangle.$$

By taking note of the electric field equation $\mathbf{E} = -\nabla V - (1/c)(\partial\mathbf{A}/\partial t)$, the acceleration of the *microscopic* particle in terms of the Lorenz force is given as

$$\frac{\mathrm{d}}{\mathrm{d}t}\langle\mathbf{v}\rangle = \frac{q}{2mc}\langle\mathbf{v}\times\mathbf{B}-\mathbf{B}\times\mathbf{v}\rangle + \frac{q}{m}\langle\mathbf{E}\rangle.$$

For an electron with a charge $-e$, by making use of

$$H(t,\mathbf{r},\mathbf{p}) = \frac{1}{2m}\left(\mathbf{p}+\frac{e}{c}\mathbf{A}\right)^2 - eV,$$

one finds the Hamiltonian operator to be

$$H = -\frac{\hbar^2}{2m}\nabla^2 - eV(t,\mathbf{r}) + \frac{e}{mc}\mathbf{A}(t,\mathbf{r})\cdot\frac{\hbar}{i}\nabla - \frac{ie\hbar}{2mc}[\nabla\cdot\mathbf{A}(t,\mathbf{r})] + \frac{e^2}{2mc^2}[\mathbf{A}(t,\mathbf{r})]^2.$$

Using the Maxwell's theory (*Coulomb gauge*), the Hamiltonian is simplified to

$$H = \frac{1}{2m}\mathbf{p}^2 + \Pi(\mathbf{r}) + \frac{e}{mc}\mathbf{A}(t,\mathbf{r})\cdot\mathbf{p},$$

where $(e/mc)\mathbf{A}(t,\mathbf{r})\cdot\mathbf{p}$ represents the external perturbations.

From the relativistic electron theory, using the one-electron Dirac theory, the relativistic wave equation is

$$i\hbar\frac{\partial\Psi}{\partial t} = \left[c\mathbf{\Lambda}\cdot\left(\frac{\hbar}{i}\nabla+\frac{e}{c}\mathbf{A}\right) - eV + \mathbf{M}mc^2\right]\Psi,$$

where the matrices $\mathbf{\Lambda}$ and $\mathbf{M}$ are the dynamic variable, and $\mathbf{\Lambda}$ evolves as $i\hbar(\mathrm{d}\mathbf{\Lambda}/\mathrm{d}t) = [\mathbf{\Lambda}(t), H(t)]$.

For a *microscopic* particle (electron), the Hamiltonian and relativistic equation of motion are

$$H = c\mathbf{\Lambda}\cdot\left(\mathbf{p}+\frac{e}{c}\mathbf{A}\right) - eV + \mathbf{M}mc^2,$$

$$\mathbf{v} = \frac{\mathrm{d}\mathbf{r}}{\mathrm{d}t} = \frac{1}{i\hbar}[\mathbf{r},H] = c\mathbf{\Lambda},$$

$$\frac{\mathrm{d}}{\mathrm{d}t}\left(\mathbf{p}+\frac{e}{c}\mathbf{A}\right) = -e\mathbf{E} - e\mathbf{\Lambda}\times\mathbf{B} = -e\mathbf{E} - e\frac{\mathbf{v}}{c}\times\mathbf{B}.$$

If $\mathbf{A} = 0$ and $V = 0$, one finds $\mathrm{d}\mathbf{p}/\mathrm{d}t = 0$ and $\mathbf{p} = \text{const}$.

The free particle Hamiltonian is $H = c\mathbf{\Lambda} \cdot \mathbf{p} + \mathbf{M}mc^2$.
The operator $\mathbf{\Lambda}$ evolves as

$$\frac{\mathrm{d}\mathbf{\Lambda}}{\mathrm{d}t} = \frac{1}{i\hbar}[\mathbf{\Lambda}(t), H(t)] = \frac{2}{i\hbar}(c\mathbf{p} - H\mathbf{\Lambda}),$$

while $\mathbf{M}H + H\mathbf{M} = 2mc^2$, and in a state of energy E $\langle \mathbf{M} \rangle = (mc^2/E) = \pm\sqrt{1 - c^2p^2/E^2}$ approaching ± 1 in the nonrelativistic approximation, and 0 at c.

Taking note of $H =$ constant, one finds

$$\mathbf{v}(t) = c\mathbf{\Lambda}(t) = c^2H^{-1}\mathbf{p} + c\mathrm{e}^{2(i/\hbar)Ht}[\mathbf{\Lambda}_0 - cH^{-1}\mathbf{p}],$$

which gives

$$\mathbf{r}(t) = \mathbf{r}_0 + c^2H^{-1}\mathbf{p}t + \tfrac{1}{2}i\hbar cH^{-1}(1 - \mathrm{e}^{2(i/\hbar)Ht})[\mathbf{\Lambda}_0 - cH^{-1}\mathbf{p}].$$

Here, the first two terms describe the uniform motion of a free microscopic particle, while

$$\tfrac{1}{2}i\hbar cH^{-1}(1 - \mathrm{e}^{2(i/\hbar)Ht})[\mathbf{\Lambda}_0 - cH^{-1}\mathbf{p}]$$

gives the high-frequency vibrations with amplitude $\hbar/(mc)$ and frequency $\sim mc^2/\hbar$.

Example 6.27

By making use of

$$\frac{\mathrm{d}}{\mathrm{d}t}\langle \mathbf{v} \rangle = \frac{q}{2mc}\langle \mathbf{v} \times \mathbf{B} - \mathbf{B} \times \mathbf{v} \rangle - \frac{q}{m}\langle \nabla V \rangle - \frac{q}{mc}\left\langle \frac{\partial \mathbf{A}}{\partial t} \right\rangle,$$

taking into account the noise, the dynamics of a microscopic particle in a ^{ME}devices is described by stochastic nonlinear differential equations

$$\frac{\mathrm{d}\mathbf{v}}{\mathrm{d}t} = -A\mathbf{v} + \frac{q}{mc}\mathbf{v} \times \mathbf{B} + D\mathbf{w}.$$

We consider a ^{ME}device in which the *microscopic* particles must be controlled. For the uncontrolled particle, assuming

$$\mathbf{B} = \underbrace{B_0\mathbf{a}_z}_{B_{\text{uniform}}} + \underbrace{B_1\left[\cos\left(\frac{2\pi}{a}x\right) + \cos\left(\frac{2\pi}{a}y\right)\right]}_{B_{\text{varying}}}\mathbf{a}_z,$$

the stochastic nonlinear differential equations are

$$\frac{d\mathbf{v}}{dt} = \begin{bmatrix} -\frac{1}{T} & 0 \\ 0 & -\frac{1}{T} \end{bmatrix} \begin{bmatrix} v_x \\ v_y \end{bmatrix} + \begin{bmatrix} \frac{q(B_0 + B_1)}{mc} \left(\cos\left(\frac{2\pi}{a}x\right) + \cos\left(\frac{2\pi}{a}y\right) \right) v_y \\ -\frac{q(B_0 + B_1)}{mc} \left(\cos\left(\frac{2\pi}{a}x\right) + \cos\left(\frac{2\pi}{a}y\right) \right) v_x \end{bmatrix}$$
$$+ \begin{bmatrix} D_x & 0 \\ 0 & D_y \end{bmatrix} \begin{bmatrix} w_x \\ w_y \end{bmatrix},$$
$$\frac{dx}{dt} = v_x, \quad \frac{dy}{dt} = v_y.$$

The analytic and numerical solutions are obtained assigning $q = 1.6 \times 10^{-19}$ C, $m = 9.1 \times 10^{-31}$ kg, $a = 1 \times 10^{-9}$ m, $B_0 = 1$ T, and $T = 5 \times 10^{-15}$ sec. Using the SIMULINK environment, the model to simulate and analyze stochastic dynamics, is developed and illustrated in Figure 6.17.

The uncontrolled system evolutions for v_x and v_y are shown in Figure 6.18 for initial conditions $v_{x0} = 100$ and $v_{y0} = 100$ m/sec.

With the ultimate objective to control a *microscopic* particle, we solve the stochastic control problem. One considers the following system to be controlled

$$\frac{d\mathbf{v}}{dt} = \begin{bmatrix} -\frac{1}{T} & 0 \\ 0 & -\frac{1}{T} \end{bmatrix} \begin{bmatrix} v_x \\ v_y \end{bmatrix} + \begin{bmatrix} \frac{q(B_0 + B_1)}{mc} \left(\cos\left(\frac{2\pi}{a}x\right) + \cos\left(\frac{2\pi}{a}y\right) \right) v_y \\ -\frac{q(B_0 + B_1)}{mc} \left(\cos\left(\frac{2\pi}{a}x\right) + \cos\left(\frac{2\pi}{a}y\right) \right) v_x \end{bmatrix}$$
$$+ \begin{bmatrix} \frac{q(B_0 + B_1)}{mc} \left(\cos\left(\frac{2\pi}{a}x\right) + \cos\left(\frac{2\pi}{a}y\right) \right) v_y \\ -\frac{q(B_0 + B_1)}{mc} \left(\cos\left(\frac{2\pi}{a}x\right) + \cos\left(\frac{2\pi}{a}y\right) \right) v_x \end{bmatrix} \begin{bmatrix} u_{B_x} \\ u_{B_y} \end{bmatrix}$$
$$+ \begin{bmatrix} D_x & 0 \\ 0 & D_y \end{bmatrix} \begin{bmatrix} w_x \\ w_y \end{bmatrix},$$
$$\frac{dx}{dt} = v_x, \quad \frac{dy}{dt} = v_y.$$

The controller is found to be

$$u_{B_x} = 0.95 \times 10^8 e_{v_x} \quad \text{and} \quad u_{B_y} = 0.95 \times 10^8 e_{v_y},$$

where the tracking errors are $e_{v_x} = r_{v_x} - v_x$ and $e_{v_y} = r_{v_y} - v_y$.

The transient dynamics is analyzed using the SIMULINK model, which is illustrated in Figure 6.19. The closed-loop system evolutions are shown in Figure 6.20. The desired velocities are assigned to be 2×10^5 and 2×10^5 m/sec.

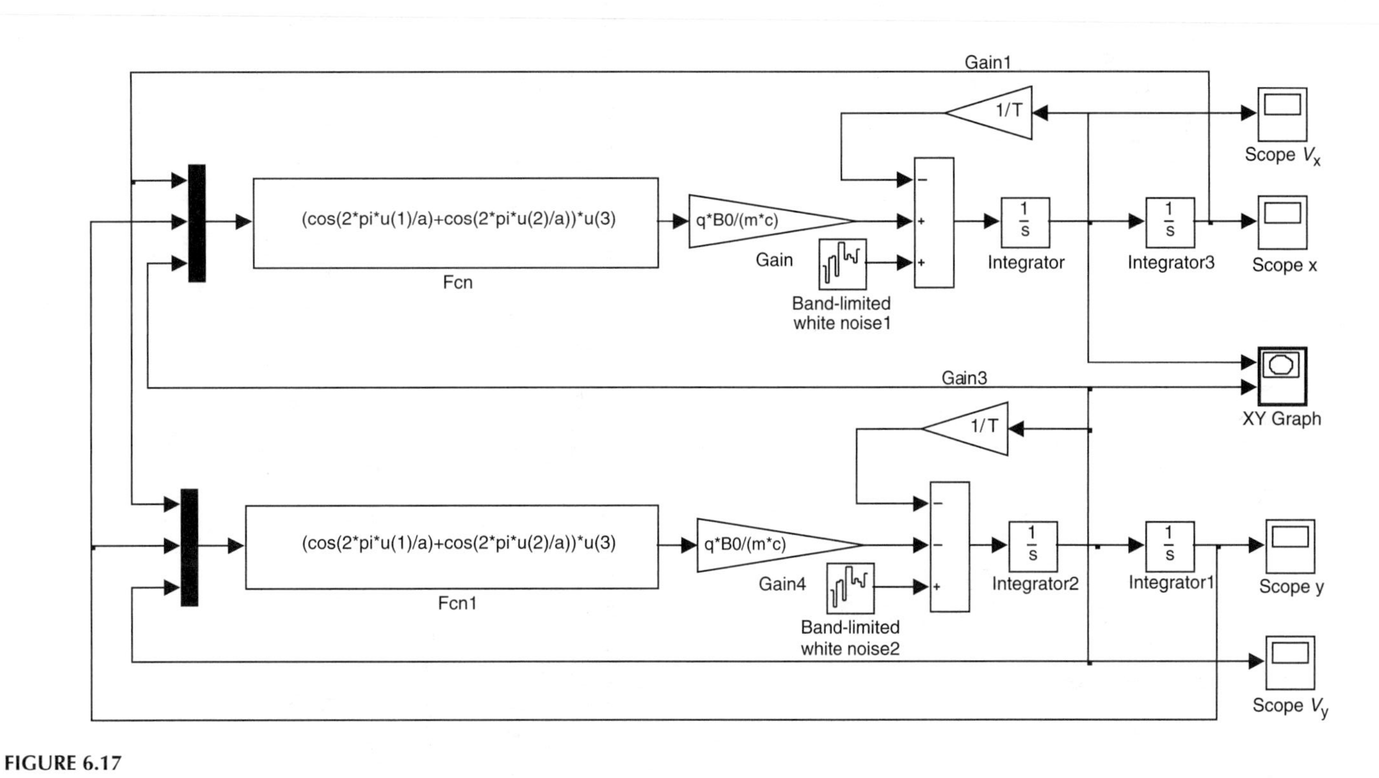

FIGURE 6.17
SIMULINK model to simulate the system dynamic evolutions.

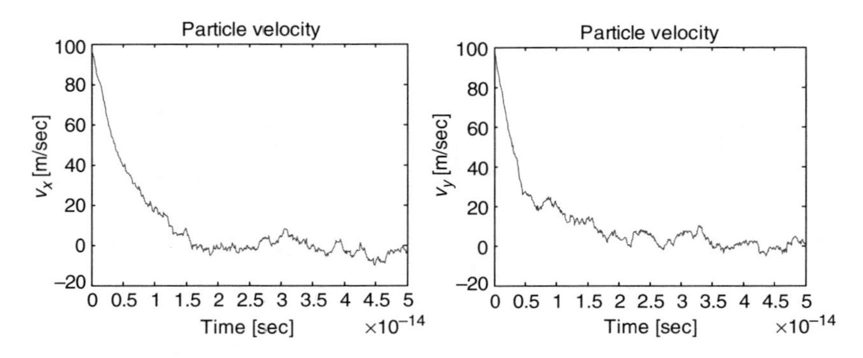

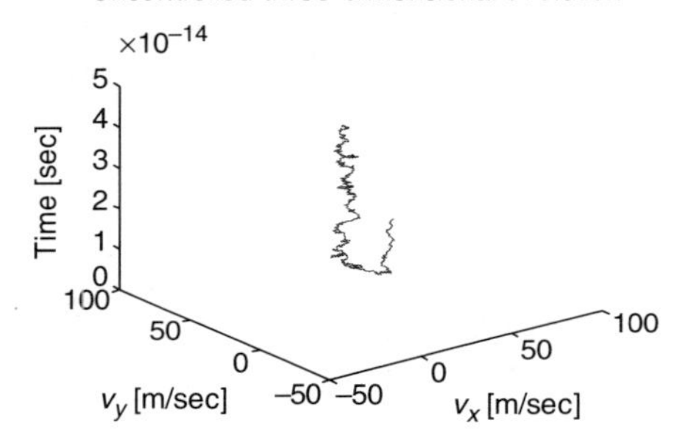

FIGURE 6.18
Uncontrolled stochastic dynamics of the particle.

It is evident, that the tracking is achieved guarantying the stochastic stability. The settling time is 0.6 fsec. The velocities reach the desired values. The steady-state tracking error is due to the use of the proportional controller. This error can be eliminated by using the proportional-integral-derivative controllers.

6.8 Green Function Formalism

Electronic devices can be modeled using the Green function method [1–5]. The time-independent Schrödinger equation $-(\hbar^2/2m)\nabla^2\Psi(\mathbf{r}) + \Pi(\mathbf{r})\Psi(\mathbf{r}) = E(\mathbf{r})\Psi(\mathbf{r})$ is written as the Helmholtz equation by using the inhomogeneous term $Q(\Psi)$. We have

$$(\nabla^2 + k^2)\Psi = Q,$$

where

$$k^2 = \frac{2mE}{\hbar^2} \quad \text{and} \quad Q = \frac{2m}{\hbar^2}\Pi\Psi.$$

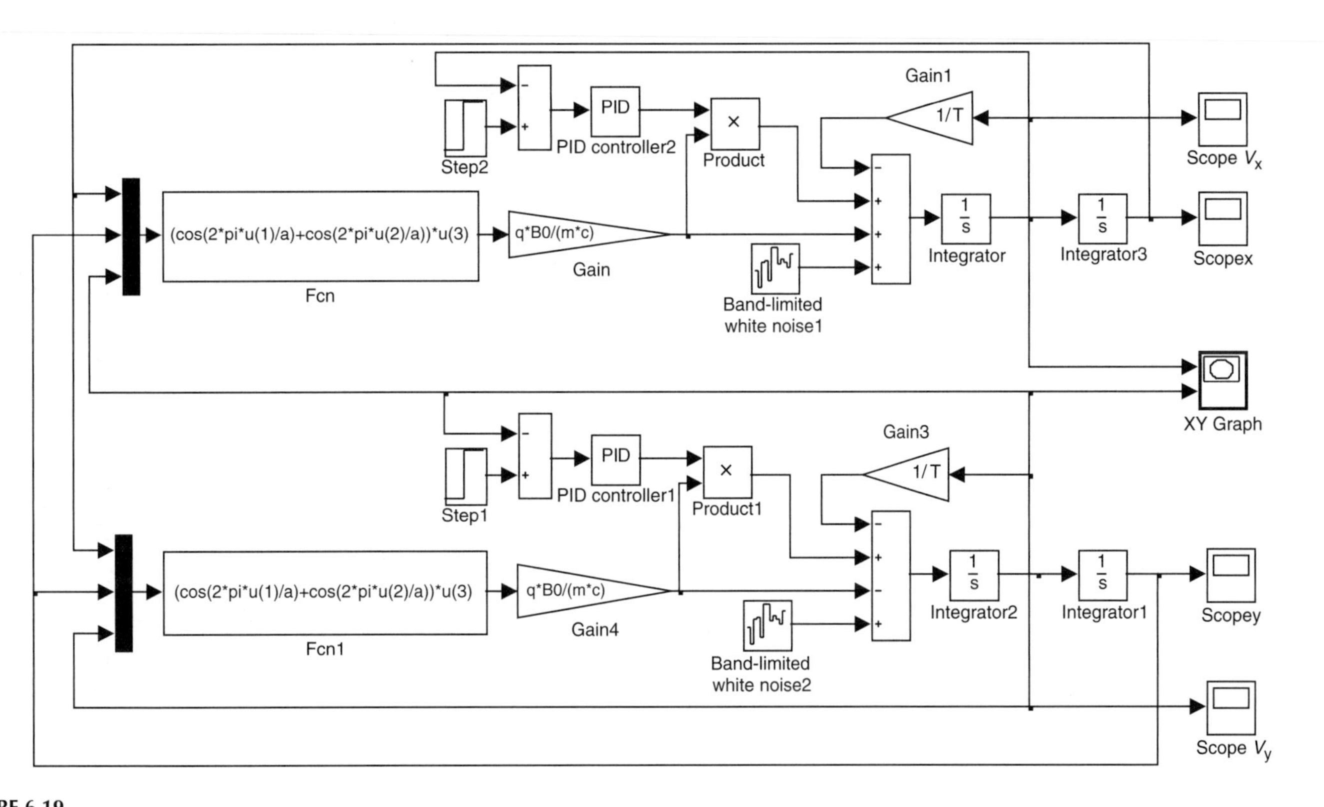

FIGURE 6.19
SIMULINK model to simulate and analyze the closed-loop dynamics.

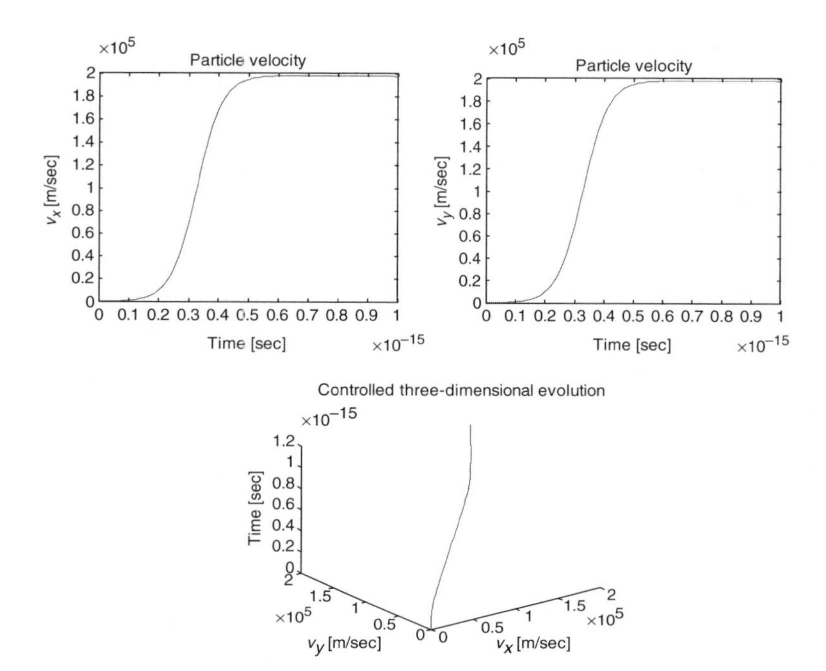

FIGURE 6.20
Controlled particle evolution.

Our goal is to find a function $G(\mathbf{r})$, called the Green function, which solves the Helmholtz equation with a delta-function *source*, which is given as

$$(\nabla^2 + k^2)G(\mathbf{r}) = \delta^3(\mathbf{r}).$$

The wave function

$$\Psi(\mathbf{r}) = \int G(\mathbf{r} - \mathbf{r}_0)Q(\mathbf{r}_0)d^3\mathbf{r}_0$$

satisfies the Schrödinger equation

$$(\nabla^2 + k^2)\Psi(\mathbf{r}) = \int [(\nabla^2 + k^2)G(\mathbf{r} - \mathbf{r}_0)]Q(\mathbf{r}_0)d^3\mathbf{r}_0$$
$$= \int \delta^3(\mathbf{r} - \mathbf{r}_0)Q(\mathbf{r}_0)d^3\mathbf{r}_0 = Q(\mathbf{r}).$$

The general solution of the Schrödinger equation is

$$\Psi(\mathbf{r}) = \Psi_0(\mathbf{r}) - \frac{m}{2\pi\hbar^2}\int \frac{e^{ik|\mathbf{r}-\mathbf{r}_0|}}{|\mathbf{r}-\mathbf{r}_0|}\Pi(\mathbf{r}_0)\Psi(\mathbf{r}_0)d^3\mathbf{r}_0,$$

where $\Psi_0(\mathbf{r})$ satisfies the homogeneous equation $(\nabla^2 + k^2)\Psi_0 = 0$.

To solve the integral Schrödinger equation derived, one must know the solution because $\Psi(\mathbf{r}_0)$ is under the integration sign.

Using the Hamiltonian H, one obtains the equation

$$(E - H)G(\mathbf{r}, \mathbf{r}_E) = \delta(\mathbf{r} - \mathbf{r}_E).$$

Studying electron–electron interactions in the π-conjugated molecules, one may apply the semi-empirical Hamiltonian [11–13]. For a molecule, one has

$$H_\mathrm{M} = \sum_{i,\sigma} E_i a_{i\sigma}^+ a_{i\sigma} - \sum_{\langle ij\rangle,\sigma} (t_{ij} a_{i\sigma}^+ a_{j\sigma} + t_{ij}^* a_{j\sigma}^+ a_{i\sigma}) + U \sum_i n_{i,\uparrow} n_{i,\downarrow}$$
$$+ \frac{1}{2} \sum_{i,j,i\neq j} U_{ij} \Big(\sum_\sigma a_{i\sigma}^+ a_{i\sigma} - 1\Big)\Big(\sum_\sigma a_{j\sigma}^+ a_{j\sigma} - 1\Big),$$

where E_i are the orbital energies; $a_{i\sigma}^+$ and $a_{i\sigma}$ are the creation and annihilation operators for the π-electron of ith atom with spin $\sigma(\uparrow, \downarrow)$; t_{ij} are the tight-binding hopping matrix entities for p_z orbitals of nearest neighbor atoms; $\langle ij\rangle$ denotes the sum of the nearest neighbor sites i and j; U is the on-site Coulomb repulsion between two electrons occupying the same atom p_z orbital; n_i is the total number of π-electrons on site i, $n_i = \sum_\sigma a_{i\sigma}^+ a_{i\sigma}$; U_{ij} is the intersite Coulomb interaction.

In H_M, the third and fourth terms represent the electron–electron interactions. The electron–electron interactions depend on the distance. For the interaction energies, the parametrization equation is

$$U_{ij} = \frac{U}{\sqrt{1 + k_r r_{ij}^2}},$$

where r_{ij} is the distance between sites i and j; k_r is the screening constant. The parameters can be obtained. In particular, t_{ij} varies from 2 to 3 eV for orbitals depending on atom placement, bonds (single, double, or triple), while $U \sim 10$ eV and $k_r \sim 50$ (for r_{ij} given in nm).

The Hamiltonian for the molecular complex is constructed by using molecule H_M, tunneling H_T, terminal H_C, and external H_E Hamiltonians. That is, we have

$$H = H_\mathrm{M} + H_\mathrm{T} + H_\mathrm{C} + H_\mathrm{E}.$$

For the oscillator with mass m, momentum p, and resonant angular frequency of radial vibrations ω_0, one obtains $H_0 = (p^2/2m) + \frac{1}{2} m\omega_0^2 x^2$. However, H_M integrates the core energy-based single-electron Hamiltonian and electron–electron interaction terms resulting in:

$$H_\mathrm{M} = \sum_i \left(\frac{1}{2m}\mathbf{p}_i^2 + \frac{1}{2} m\omega_{0i}^2 \mathbf{r}_i^2\right) + \sum_{i,j} B_{ij} a_i^+ a_j,$$

where $B_{ij}(t)$ are the time-varying amplitudes; a and a^+ are the electron annihilation and creation operators, and the steady-state number of electrons is $N = \sum_{i,j} \langle a_i^+ \quad a_j \rangle$; i and j are indices that run over the molecule.

Using the molecular orbital indices n and l, the last term in equation for H_M can be expressed as $\sum_{(in)(jl)} B_{(in)(jl)} a_{(in)}^+ a_{(jl)}$. The current can be estimated as $I = -e(\mathrm{d}/\mathrm{d}t) \sum_{i,j} \langle a_i^+(t) \quad a_j(t) \rangle$, where the equations of motion for $a(t)$ and $a^+(t)$ are derived by using the Hamiltonian.

The tunneling Hamiltonian that describes the electrons transport to and from molecule is

$$H_T = -\sum_{i,j \in \text{Terminals}} \mathrm{e}^{-\mathbf{r}/\lambda_j} \left(T_{ij} a_i^+ b_j + T_{ij}^* a_i^+ b_j \right),$$

where $T_{ij}(t)$ are the time-varying tunneling amplitudes; b and b^+ are the electron annihilation and creation operators at the input (L) and output (R) terminals; λ_j are the tunneling lengths between molecule conducting and terminal atoms; T_{ij_L} and T_{ij_R} are the amplitudes of the electron transfer, for example, from the j_Lth occupied orbital to the molecule's lowest unoccupied molecular orbital $|i_{\mathrm{LUMO}}\rangle$, and to the j_Rth unoccupied orbital; $|j_L\rangle$ and $|j_R\rangle$ are the contacts orbitals; $|i_{\mathrm{LUMO}}\rangle$ is the molecule's lowest unoccupied molecular orbital.

The terminal Hamiltonian is expressed as

$$H_C = \sum_{j \in L,R} C_j b_j^+ b_j,$$

where $C_j(t)$ are the time-varying energy amplitudes.

The external Hamiltonian depends on the device physics as was documented in Section 6.7. For example,

$$H_E = -\sum_{i,j} E_{ij} \cdot \mathbf{m}_{ij} a_i^+ a_j,$$

where $E_{ij}(t)$ is the function that is controlled by varying the electrostatic potential; $\mathbf{m}_{ij}$ is the electron dipole moment vector.

Taking note of the Hamiltonians derived, one finds a total Hamiltonian H. To examine the functionality and characteristics of ^{ME}devices (input/control bonds–molecule–output bonds), the wave function should be derived by solving the Schrödinger equation. Alternatively, the Keldysh nonequilibrium Green function concept can be applied. Green's function is a wave function of energies at $\mathbf{r}$ resulting from an excitation applied at $\mathbf{r}_E$. We study the retarded Green's function G that represents the behavior of the aggregated molecule. The equation

$$(E - H)G(\mathbf{r}, \mathbf{r}_E) = \delta(\mathbf{r} - \mathbf{r}_E)$$

is used. The boundary conditions must be satisfied for the transport and Poisson equations.

To examine a finite molecular complex, the molecular Hamiltonian of the isolated system and the complex self-energy functions are used instead of the single-energy potential and broadening energies. In the matrix notations, using the overlap matrix S, one has

$$[G(E)] = \left(E[S] - [H] - [V_{SC}] - \sum_i [E_i]\right)^{-1},$$

where $[E_i] = [S_i][G_i][S_i^*]$; S_i is the geometry-dependent terminal coupling matrix between the molecular terminals.

The imaginary non-Hermitian self-energy functions of the input and output electron reservoirs E_L and E_R are

$$[E_L] = [S_L][G_L][S_L^*] \quad \text{and} \quad [E_R] = [S_R][G_R][S_R^*],$$

where G_L and G_R are the surface Green's functions that are found applying the recursive methods.

By taking note of the Green's function, the density of state $D(E)$ is found as

$$D(E) = -\frac{1}{\pi}\text{Im}\{G(E)\}.$$

The spectral function $A(E)$ is the anti-Hermitian term of the Green's function:

$$A(E) = i[G(E) - G^*(E)] = -2\,\text{Im}[G(E)].$$

One obtains

$$D(E) = \frac{1}{2\pi}\text{tr}[A(E)S],$$

where tr is the trace operator; S is the identity overlap matrix for orthogonal basis functions.

Using the broadening energy functions E_{BL} and E_{BR}, one obtains the spectral functions

$$[A_L(E)] = [G(E)][E_{BL}(E)][G^*(E)] \quad \text{and} \quad [A_R(E)] = [G(E)][E_{BR}(E)][G^*(E)].$$

Multiterminal ^{ME}devices attain equilibrium at the Fermi level, and the nonequilibrium charge density matrix is

$$[\rho(E)] = \frac{1}{2\pi}\int_{-\infty}^{\infty} \sum_{k,i\in L,j\in R} f(E_{Vk}, V_{Fi,j})[A_{i,j}(E)]\,\mathrm{d}E$$

$$= \frac{1}{2\pi}\int_{-\infty}^{\infty} \sum_{k,i\in L,j\in R} f(E_{Vk}, V_{Fi,j})[G(E)][E_{Bi,j}(E)][G^*(E)]\,\mathrm{d}E,$$

where $V_{Fi,j}$ are the potentials; $f(E_V, V_{Fi,j})$ are the distribution functions.

Utilizing the transmission matrix

$$T(E) = \text{tr}[E_{BL}G(E)E_{BR}G^*(E)]$$

and taking note of the broadening, the current between terminals is found as

$$I_k = \frac{2e}{h}\int_{-\infty}^{+\infty} \text{tr}[E_{BL}G(E)E_{BR}G^*(E)] \sum_{k,i\in L,j\in R} f(E_{Vk}, V_{Fi,j})\, \text{d}E.$$

For a two-terminal ^{ME}device,

$$\rho(E) = \frac{1}{2\pi}\int_{-\infty}^{\infty} [f(E_V, V_{FL})G(E)E_{BL}G^*(E) + f(E_V, V_{FR})G(E)E_{BR}G^*(E)]\text{d}E$$

and

$$I = \frac{2e}{h}\int_{-\infty}^{+\infty} \text{tr}[E_{BL}G(E)E_{BR}G^*(E)]\,[f(E_V, V_{FL}) - f(E_V, V_{FR})]\, \text{d}E.$$

As emphasized in Sections 2.1.2 and 6.2, one may apply these equations using the applicable distribution functions after examining the assumptions of the statistical mechanics and their validity for the specific electronic device under the consideration.

6.9 Case Study: Multiterminal Quantum Effect Molecular Electronic Device

Solid-Sate Devices: Reference [14] thoroughly reports the device physics and application of the basic laws to straightforwardly obtain and examine the steady-state and dynamic characteristics of FETs, BJTs, and other solid-state electronic devices. The deviations are straightforward, and some well-known basics are briefly reported here. For FETs, one may find the total charge in the channel Q and the *transit* time t, which gives the time that it takes an electron to pass between source and drain. The drain-to-source current is $I_{\text{DS}} = Q/t$.

The electron velocity is $\mathbf{v} = -\mu_n \mathbf{E}_E$, where μ_n is the electron mobility; $\mathbf{E}_E$ is the electric field intensity. One also has $\mathbf{v} = \mu_p \mathbf{E}_E$, where μ_p is the hole mobility. At room temperature, for intrinsic silicon, μ_n and μ_p reach $\sim$1400 cm^2/V sec and $\sim$450 cm^2/V sec, respectively. It should be emphasized that μ_n and μ_p are functions of the field intensity, voltage, temperature, and other quantities. Therefore, the *effective* $\mu_{n\text{eff}}$ and $\mu_{p\text{eff}}$ are used.

Using the x component of the electric field, we have $E_{Ex} = -V_{\text{DS}}/L$, where L is the channel length.

Thus, $v_x = -\mu_n E_{Ex}$, and $t = L/v_x = L^2/\mu_n V_{\text{DS}}$.

The channel and the gate form a parallel capacitor with plates separated by an insulator (gate oxide). From $Q = CV$, taking the note that the charge appears when the voltage between the gate and the channel V_{GC} exceeds the n-channel threshold voltage V_t, one has $Q = C(V_{GC} - V_t)$.

Using the equation for a parallel-plate capacitors with length L, width W, and plate separation equal to the gate-oxide thickness T_{ox}, the gate capacitance is $C = WL\varepsilon_{ox}/T_{ox}$, where ε_{ox} is the gate-oxide dielectric permittivity, and for silicon dioxide SiO_2, ε_{ox} is $\sim 3.5 \times 10^{-11}$ F/m.

We briefly reported the baseline equations in deriving the size-dependant quantities, such as current, capacitance, velocity, *transit* time. Furthermore, the analytic equations for the I–V characteristics for FETs and BJTs are straightforwardly obtained and reported in [14]. The derived expressions for the so-called Level 1 model of nFETs in the *linear* and *saturation* regions are

$$I_D = \mu_n \frac{\varepsilon_{ox}}{T_{ox}} \frac{W_c}{L_c - 2L_{GD}} \left[(V_{GS} - V_t)V_{DS} - \frac{1}{2}V_{DS}^2 \right] (1 + \lambda V_{DS})$$
$$\text{for } V_{GS} \geq V_t,\ V_{DS} < V_{GS} - V_t$$

and

$$I_D = \frac{1}{2}\mu_n \frac{\varepsilon_{ox}}{T_{ox}} \frac{W_c}{L_c - 2L_{GD}} (V_{GS} - V_t)^2 (1 + \lambda V_{DS})$$
$$\text{for } V_{GS} \geq V_t,\ \ V_{DS} \geq V_{GS} - V_t,$$

where I_D is the drain current; V_{GS} and V_{DS} are the gate source and drain source voltages; L_c and W_c are the channel length and width; L_{GD} is the gate–drain overlap; L_{GD} is the channel length modulation coefficient.

For pFETs, in the equations for I_D one uses μ_p. The coefficients and parameters used to calculate the characteristics of nFETs and pFETs are different. Owing to distinct device physics and the effects exhibited and utilized, the foundations of semiconductor devices are not applicable to ^{ME}devices. For example, the electron velocity and I–V characteristics can be found using $\Psi(t, \mathbf{r})$, which depends on three-dimensional $\mathbf{E}_E(\mathbf{r})$ as documented in Section 6.7.

Quantum-well resonant tunneling diodes and FETs, Schottky-gated resonant tunneling, heterojunction bipolar, resonant tunneling bipolar and other transistors have been introduced to enhance the microelectronic device performance. The tunneling barriers are formed using AlAs, AlGaAs, AlInAs, AlSb, GaAs, GaSb, GaAsSb, GaInAs, InP, InAs, InGaP, and other composites and spacers with the thickness in the range from ~one to tens of nanometers. CMOS-technology high-speed double-heterojunction bipolar transistors ensure a cut-off frequency of ~300 GHz, breakdown voltage of ~5 V, and current density of $\sim 1 \times 10^5$ A/cm^2. The one-dimensional potential energy profile, shown in Figure 6.21, schematically depicts the first barrier (L_1, L_2), the well region (L_2, L_3) and the second barrier (L_3, L_4) with the quasi-Fermi

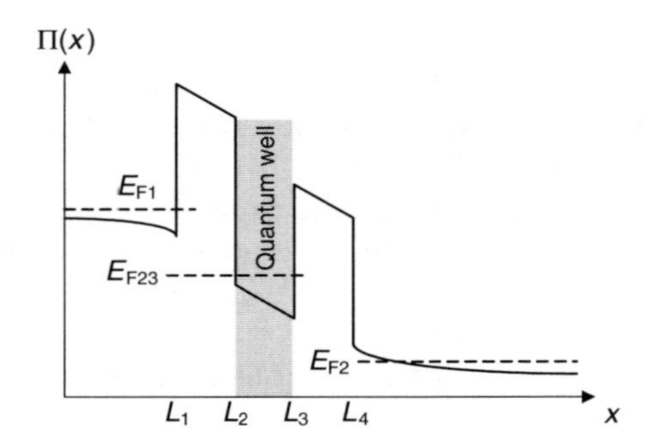

FIGURE 6.21
One-dimensional potential energy profile and quasi-Fermi levels in the double-barrier single-well heterojunction transistors.

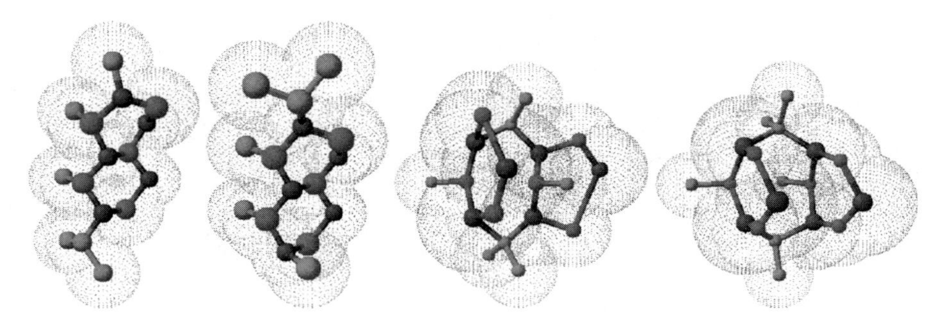

FIGURE 6.22
^{M}AND, ^{M}NAND, ^{M}OR, and ^{M}NOR gates comprised of cyclic molecules.

levels E_{F1}, E_{F23}, and E_{F2}. The device physics of these transistors is reported [14], and the electron transport in double-barrier single-quantum-well is straightforwardly examined by applying a self-consistent approach and numerically solving the one- or two-dimensional Schrödinger and Poisson equations.

The ^{M}AND gate was shown in Figure 4.6. Different ^{M}gates can be implemented using cyclic molecules as multiterminal ^{ME}devices if the electronic characteristics and acceptable performance are achieved. Figure 6.22 illustrates the overlapping molecular orbitals for monocyclic molecules used to implement ^{ME}devices that form ^{M}AND, ^{M}NAND, ^{M}OR, and ^{M}NOR gates.

In Chapter 5, we reported 3D-topology multiterminal ^{ME}devices formed using cyclic molecules with a carbon interconnecting framework; see Figure 5.9. In this section, we consider a three-terminal ^{ME}device with the *input*, *control*, and *output* terminals as shown in Figure 6.23. The device physics of the proposed ^{ME}device is based on the quantum interaction and controlled electron transport. The applied $V_{\text{control}}(t)$ changes the charge distribution $\rho(t, \mathbf{r})$ and $\mathbf{E}_E(t, \mathbf{r})$ affecting the electron transport. This ^{ME}device operates

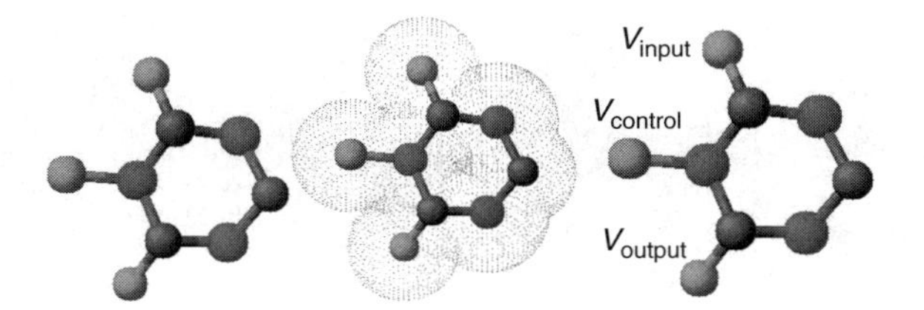

FIGURE 6.23
Three-terminal ^{ME}device comprised of a monocyclic molecule with a carbon interconnecting framework.

in the controlled electron-exchangeable environment exhibiting quantum transitions and interactions. The controlled super-fast potential-assisted electron transport is achieved. The electron-exchangeable environment interactions qualitatively and quantitatively modify the device behavior and its characteristics. Consider the electron in the time- and spatial-varying metastable potential $\Pi(t, \mathbf{r})$. From the quantum theory viewpoints, it is evident that the changes in the Hamiltonian result in

- Changes of tunneling $T(E)$
- Quantum interactions due to variations of $\rho(t, \mathbf{r})$, $\mathbf{E}_E(t, \mathbf{r})$, and $\Pi(t, \mathbf{r})$

One achieves the controlled electron transport between the *input* and *output* terminals. The device controllability is ensured by varying $V_{\text{control}}(t)$ that affects $\rho(t, \mathbf{r})$, $\mathbf{E}_E(t, \mathbf{r})$, and $\Pi(t, \mathbf{r})$, leading to variations of $T(E)$. Hence, the device switching, I–V, and other characteristics are controlled.

We perform high-fidelity modeling. The data-intensive analysis for the studied ^{ME}device should be studied by performing heterogeneous simulations. For heterojunction microelectronic devices, one usually solves the one-dimensional Schrödinger and Poisson equations applying the Fermi–Dirac distribution function. In contrast, for the devised ^{ME}devices, a 3D problem arises that cannot be simplified. Furthermore, the solid-state-device-centered distribution functions and statistical mechanics postulates may not be applied.

We consider nine atoms with motionless protons with charges q_i. The radial Coulomb potentials are $\Pi_i(r) = -(Z_{\text{eff}\,i} q_i^2 / 4\pi\varepsilon_0 r)$. For example, for carbon $Z_{\text{eff C}} = 3.14$. Using the spherical coordinate system, the Schrödinger equation

$$-\frac{\hbar^2}{2m}\left[\frac{1}{r^2}\frac{\partial}{\partial r}\left(r^2\frac{\partial\Psi}{\partial r}\right) + \frac{1}{r^2\sin\theta}\frac{\partial}{\partial\theta}\left(\sin\theta\frac{\partial\Psi}{\partial\theta}\right) + \frac{1}{r^2\sin^2\theta}\frac{\partial^2\Psi}{\partial\phi^2}\right]$$
$$+\,\Pi(r,\theta,\phi)\Psi(r,\theta,\phi) = E\Psi(r,\theta,\phi)$$

should be solved. For the problem under our consideration, it is virtually impractical to find the analytic solution as obtained in Example 6.25 by using

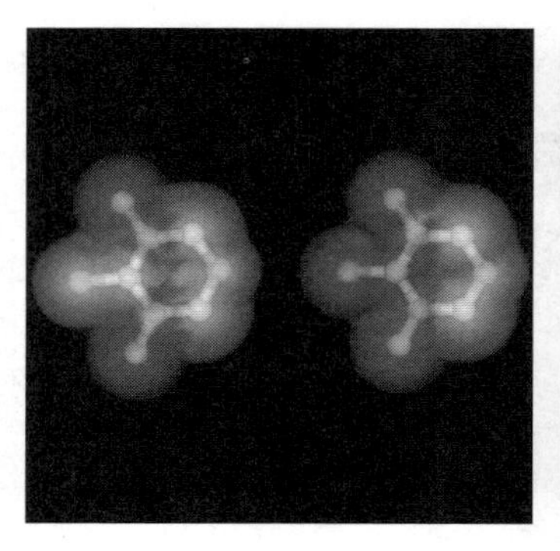 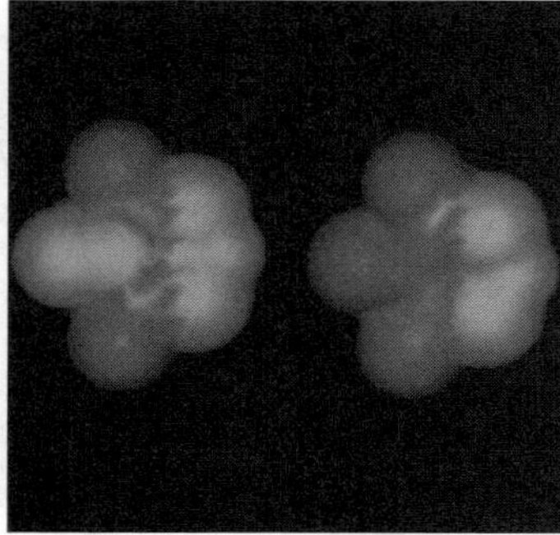

FIGURE 6.24
Charge distribution $\rho(\mathbf{r})$.

the separation of variables concept. We represented the wave function as $\Psi(r,\theta,\phi) = R(r)Y(\theta,\phi)$ in order to derive and analytically solve the radial and angular equations. In contrast, we discretize the Schrödinger and Poisson equations to numerically solve these differential equations. The magnitude of the time-varying potential applied to the *control* terminal is bounded due to the thermal stability of the molecule, that is, $|V_{\text{control}}| \leq V_{\text{control max}}$. In particular, we let $|V_{\text{control}}| \leq 0.25$ V. The charge distribution is of our particular interest. Figure 6.24 documents a 3D charge distribution in the molecule if $V_{\text{control}} = 0.1$ V and $V_{\text{control}} = 0.2$ V. The total molecular charge distribution is found by summing the individual orbital densities.

The Schrödinger and Poisson equations are solved using a self-consistent algorithm in order to verify the device physics soundness and examine the baseline performance characteristics. To obtain the current density $\mathbf{j}$ and current in the ^{ME}device, the velocity and momentum of the electrons are obtained by making use of $\langle p\rangle = \int_{-\infty}^{\infty} \Psi^*(t,\mathbf{r})(-i\hbar(\partial/\partial\mathbf{r}))\Psi(t,\mathbf{r})d\mathbf{r}$. The wave function $\Psi(t,\mathbf{r})$ is numerically derived for distinct values of V_{control}. The I–V characteristics of the studied ^{ME}device for two different control currents (0.1 and 0.2 nA) are shown in Figure 6.25. The results documented imply that the proposed ^{ME}device may be effectively used as a multiple-valued primitive in order to design enabling multiple-valued logics and memories.

The traversal time of electron transport is derived from the expression $\tau(E) = \int_{\mathbf{r}_0}^{\mathbf{r}_f} \sqrt{(m/2[\Pi(\mathbf{r}) - E])}d\mathbf{r}$. It is found that τ varies from 2.4×10^{-15} to 5×10^{-15} sec. Hence, the proposed ^{ME}device ensures super-fast switching.

The reported monocyclic molecule can be used as a six-terminal ^{ME}device as illustrated in Figure 6.26. The proposed carbon-centered molecular hardware solution, in general,

- Ensures a sound *bottom-up* synthesis at the device, gate and module levels
- Guarantees aggregability to form complex ^{M}ICs
- Results in the experimentally characterizable ^{ME}devices and ^{M}gates

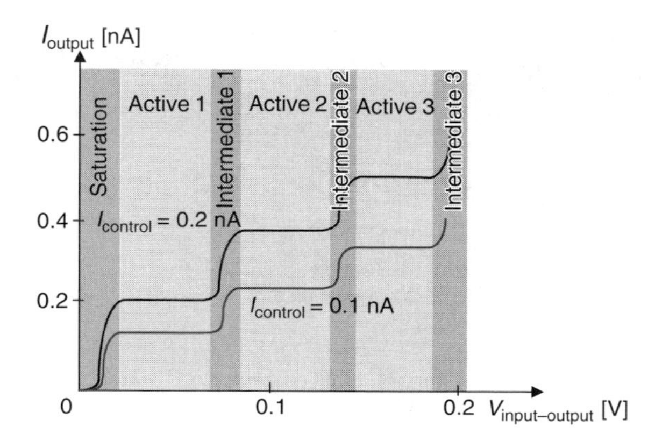

FIGURE 6.25
Multiple-valued I–V characteristics.

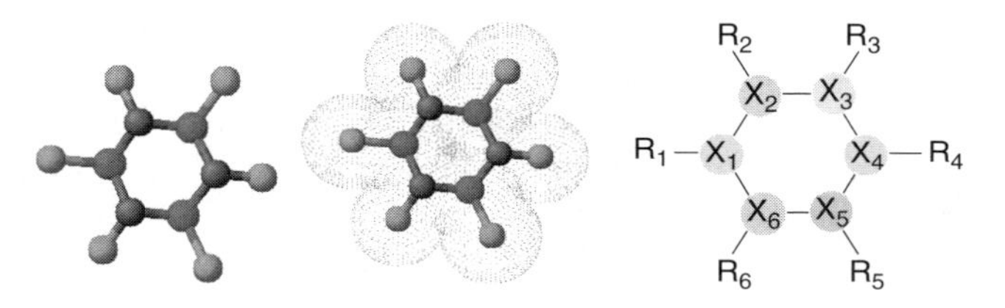

FIGURE 6.26
Six-terminal ^{ME}devices.

The use of the side groups R_i, shown in Figure 6.26, ensures the variations of the energy barriers and wells potential surfaces $\Pi(t, \mathbf{r})$. This results in

- Controlled quantum transitions due quantum effects and phenomena exhibited
- Controllable electron transport, tunneling, scattering, hopping, and so forth.
- Varying quantum interaction ensuring device functionality

As reported, ^{ME}devices can be utilized in combinational and memory ^{M}ICs. In addition, those devices can be used as routers. In particular, one achieves a reconfigurable networking-processing-and-memory as covered in Section 3.6 and 3.7 for fluidic platforms. The proposed ^{ME}device can be used as a *switch* or transmission device allowing one to design the neuromorphological reconfigurable solid ^{M}PPs.

We briefly report a *generic* modeling concept, as applied to control, evaluation, and other tasks. An ^{M}device may have two or more terminals. The interconnected ^{M}devices are well defined in the sense of their time-varying variables, for example, input $\mathbf{r}(t)$, control $\mathbf{u}(t)$, output $\mathbf{y}(t)$, state $\mathbf{x}(t)$, disturbance $\mathbf{d}(t)$, and noise $\boldsymbol{\xi}(t)$ vectors. The potential and electric or magnetic

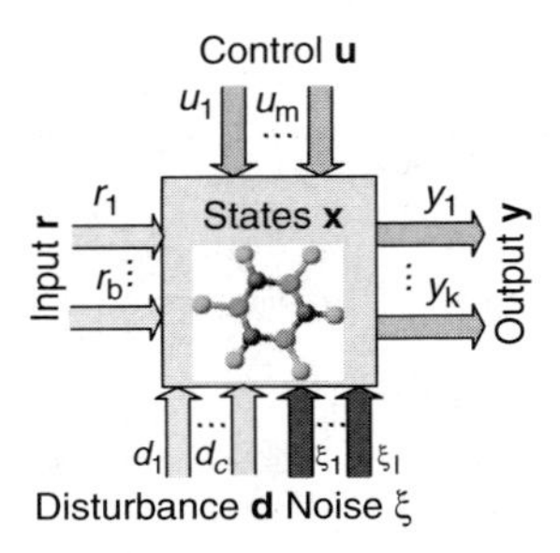

FIGURE 6.27
Molecular device with time-varying variables that characterize dynamic and steady-state device performance in the behavioral **six variables**.

field intensities can be considered as $\mathbf{u}(t)$, while wave function, momentum, velocity, displacement, current, and voltage can be the state and/or output variables. For example, for a controllable solid ^{ME}devices, the voltage (potential) at any terminal is well defined with respect to a common datum node (ground). Figure 6.27 shows a multiterminal ^{M}device with *input, control,* and *output* terminals reporting time-varying variables. The disturbances and noise vectors are also documented.

The phenomena exhibited and effects utilized are defined by the device physics. As illustrated, the proposed ^{ME}device is modeled by using the Schrödinger and Poisson equations. The ^{M}devices for admissible inputs and disturbances are completely characterized (described) by the differential equations and constitutive relations. Neglecting disturbances, unmodeled phenomena, and noise, the transient dynamics and steady-state behavior of ^{M}device are described by the behavioral quadruple $(\mathbf{r}, \mathbf{x}, \mathbf{y}, \mathbf{u})$, where $\mathbf{r} \in \mathbb{R}^b$, $\mathbf{x} \in \mathbb{R}^n$, $\mathbf{y} \in \mathbb{R}^k$, and $\mathbf{u} \in \mathbb{R}^m$. We denote by R, X, Y, and U the universal sets of achievable values for each of these vector variables. In general, the ^{M}device response (behavioral) **six variables** is $(\mathbf{r}, \mathbf{x}, \mathbf{y}, \mathbf{u}, \mathbf{d}, \boldsymbol{\xi}) \in R \times X \times Y \times U \times D \times \Xi$, where $\mathbf{d} \in \mathbb{R}^c$ and $\boldsymbol{\xi} \in \mathbb{R}^l$. To perform the device testing, characterization and evaluation, one uses the measurement set $M = \{(\mathbf{r}, \mathbf{x}, \mathbf{y}, \mathbf{u}, \mathbf{d}, \boldsymbol{\xi}) \in R \times X \times Y \times U \times D \times \Xi, \forall t \in T\}$. The electrochemomechanical state transitions (electron transport, conformation, etc.) are controlled by changing $\mathbf{u}$ to: (1) meet the optimal transient achievable performance, and (2) guarantee the desired steady-state characteristics. The measurement set can be found for the ^{M}gates, $^{\aleph}$hypercells, and ^{M}ICs. For example, for the $^{\aleph}$hypercells one has $M = \{(\mathbf{r}_{\aleph}, \mathbf{x}_{\aleph}, \mathbf{y}_{\aleph}, \mathbf{u}_{\aleph}, \mathbf{d}_{\aleph}, \boldsymbol{\xi}_{\aleph}) \in R_{\aleph} \times X_{\aleph} \times Y_{\aleph} \times U_{\aleph} \times D_{\aleph} \times \Xi_{\aleph}, \forall t \in T\}$.

References

1. M. Galperin and A. Nitzan, NEGF-HF method in molecular junction property calculations, *Ann. NY Acad. Sci.*, vol. 1006, pp. 48–67, 2003.

2. M. Galperin, A. Nitzan and M. A. Ratner, Resonant inelastic tunneling in molecular junctions, *Phys. Rev. B*, vol. 73, 045314, 2006.
3. S. E. Lyshevski, *NEMS and MEMS: Fundamentals of Nano- and Microengineering*, CRC Press, Boca Raton, FL, 2005.
4. S. E. Lyshevski, Three-dimensional molecular electronics and integrated circuits for signal and information processing platforms. In *Handbook on Nano and Molecular Electronics*, Ed. S. E. Lyshevski, CRC Press, Boca Raton, FL, pp. 6-1–6-104, 2007.
5. M. Paulsson, F. Zahid and S. Datta, Resistance of a molecule. In *Handbook of Nanoscience, Engineering and Technology*, Eds. W. Goddard, D. Brenner, S. Lyshevski and G. Iafrate, CRC Press, Boca Raton, FL, pp. 12.1–12.25, 2002.
6. R. Landauer, Spatial variation of current and fields due to localized scattereres in metallic conduction, *IBM Jl*, vol. 1, no. 3, pp. 223–231, 1957. Reprinted in *IBM J. Res. Develop.*, vol. 44, no. 1/2, pp. 251–259, 2000.
7. M. Büttiker, Quantized transmission of a saddle-point constriction, *Phys. Rev. B*, vol. 41, no. 11, pp. 7906–7909, 1990.
8. M. Büttiker and R. Landauer, Traversal time for tunneling, *Phys. Rev. Lett.*, vol. 49, no. 23, pp. 1739–1742, 1982.
9. M. Galperin and A. Nitzan, Traversal time for electron tunneling in water, *J. Chemical Physics*, vol. 114, no. 21, pp. 9205–9208, 2001.
10. M. Büttiker and R. Landauer, Escape-energy distribution for particles in an extremely underdamped potential well, *Phys. Rev. B*, vol. 30, no. 3, pp. 1551–1553, 1984.
11. R. Pariser and R. G. Parr, A semiempirical theory of the electronic spectra and electronic structure of complex unsaturated molecules I, *J. Chem. Phys.*, vol. 21, pp. 466–471, 1953.
12. R. Pariser and R. G. Parr, A semiempirical theory of the electronic spectra and electronic structure of complex unsaturated molecules II, *J. Chem. Phys.*, vol. 21, pp. 767–776, 1953.
13. J. A. Pople, Electron interaction in unsaturated hydrocarbons, *Trans. Faraday Soc.*, vol. 49, pp. 1375–1385, 1953.
14. S. M. Sze and K. K. Ng, *Physics of Semiconductor Devices*, John Wiley & Sons, NJ, 2007.

Index

V

W

Z